Lecture Notes in Computer Science 1718

Edited by G. Goos, J. Hartmanis and J. van Leeuwen

Springer
Berlin
Heidelberg
New York
Barcelona
Hong Kong
London
Milan
Paris
Singapore
Tokyo

Michel Diaz   Philippe Owezarski
Patrick Sénac (Eds.)

Interactive Distributed Multimedia Systems and Telecommunication Services

6th International Workshop, IDMS'99
Toulouse, France, October 12-15, 1999
Proceedings

Springer

Series Editors

Gerhard Goos, Karlsruhe University, Germany
Juris Hartmanis, Cornell University, NY, USA
Jan van Leeuwen, Utrecht University, The Netherlands

Volume Editors

Michel Diaz
Philippe Owezarski
Centre National de la Recherche Scientifique
Laboratoire d'Analyse et d'Architecture des Systèmes
7, avenue du Colonel Roche, F-31077 Toulouse Cedex 04, France
E-mail: {michel.diaz/philippe.owezarski}@laas.fr

Patrick Sénac
ENSICA
F-31077 Toulouse Cedex, France
E-mail: senac@ensica.fr

Cataloging-in-Publication data applied for

Die Deutsche Bibliothek - CIP-Einheitsaufnahme

Interactive distributed multimedia systems and telecommunications : 6th
international workshop ; proceedings / IDMS '99, Toulouse, France, October 12 -
15, 1999. Michael Diaz ... (ed.). - Berlin ; Heidelberg ; New York ; Barcelona
; Hong Kong ; London ; Milan ; Paris ; Singapore ; Tokyo : Springer, 1999
(Lecture notes in computer science ; Vol. 1718)
ISBN 3-540-66595-1

CR Subject Classification (1998): H.5.1, C.2, H.4, H.5

ISSN 0302-9743
ISBN 3-540-66595-1 Springer-Verlag Berlin Heidelberg New York

Typesetting: Camera-ready by author
SPIN: 10704591   06/3142 – 5 4 3 2 1 0   Printed on acid-free paper

Preface

The 1999 International Workshop on Interactive Distributed Multimedia Systems and Telecommunication Services (IDMS) in Toulouse is the sixth in a series that started in 1992. The previous workshops were held in Stuttgart in 1992, Hamburg in 1994, Berlin in 1996, Darmstadt in 1997, and Oslo in 1998.

The area of interest of IDMS ranges from basic system technologies, such as networking and operating system support, to all kinds of teleservices and distributed multimedia applications. Technical solutions for telecommunications and distributed multimedia systems are merging and quality-of-service (QoS) will play a key role in both areas. However, the range from basic system technologies to distributed mutlimedia applications and teleservices is still very broad and we have to understand the implications of multimedia applications and their requirements for middleware and networks. We are challenged to develop new and more fitting solutions for all distributed multimedia systems and telecommunication services to meet the requirements of the future information society.

The IDMS'99 call for papers attracted 84 submissions from Asia, Australia, Europe, North America, and South America. The program committee (PC) members and additional referees worked hard to review all submissions such that each contribution received three reviews. Based on the comments and recommendations in these reviews, the PC carried out an online meeting over the Internet that was structured into two discussion and ballot phases. For this, we used a full conference organization system, called ConfMan, developed in Oslo for the previous IDMS, that combines the World Wide Web and e-mail with a database system, and enforces security, privacy, and integrity control for all data acquired, including the comments and votes of the PC. The final result of the discussions and ballots of the PC online meeting was the final program, and the only additional task for us as program co-chairs was to group the selected papers and structure them into sessions.

At IDMS'99 a high-quality program with 4 tutorials, 3 invited papers, 25 full papers and 3 position papers, coming from 13 countries, included topics as user aspects, QoS, distributed applications, new networks, multimedia documents, platforms for collaborative systems, storage servers, flow and congestion control.

This technical program enabled IDMS'99 to follow the tradition of previously very successful IDMS workshops. We would like to express our deepest gratitude to the organizers of the previous IDMS workshops for their confidence, which allowed us to take on the responsibility of having IDMS'99 in Toulouse, and particularly to the previous organizers, Thomas Plagemann and Vera Goebel, who gave us very valuable information and help. We would like to acknowledge the cooperative efforts of ACM and IEEE, and the financial support from CNRS, Région Midi-Pyrénées, DGA, France Telecom and Microsoft, which allowed us to keep the fees of IDMS'99 affordable and to offer a very interesting technical and social program. Finally, we would like to thank all the people that helped us here in Toulouse, and particularly, Daniel Daurat, Sylvie Gay, Marie-Thérèse Ippolito, Joëlle Penavayre, Marc Boyer, Jean-Pierre Courtiat, Laurent Dairaine, and Pierre de Saqui-Sannes.

July 1999   Michel Diaz
Philippe Owezarski
Patrick Sénac

Organization

Program Co-Chairs

Michel Diaz   LAAS-CNRS, Toulouse, France
Philippe Owezarski   LAAS-CNRS, Toulouse, France
Patrick Sénac   ENSICA, Toulouse, France

Program Committee

H. Affifi   ENST Bretagne, France
P.D. Amer   University of Delaware, USA
E. Biersack   Institut Eurécom, France
G. v. Bochmann   University of Ottawa, Canada
B. Butscher   GMD FOKUS, Germany
A.T. Campbell   Columbia University, USA
T.S. Chua   University of Singapore, Singapore
J.-P. Courtiat   LAAS-CNRS, France
J. Crowcroft   University College London, UK
L. Delgrossi   University Piacenza, Italy
C. Diot   Sprint ATL, USA
R. Dssouli   University of Montréal, Canada
M. Dudet   CNET France Télécom, France
W. Effelsberg   University of Mannheim, Germany
F. Eliassen   University of Oslo, Norway
S. Fdida   LIP6, France
D. Ferrari   University of Piacenza, Italy
V. Goebel   UNIK, Norway
T. Helbig   Philips, Germany
J.-P. Hubaux   EPFL, Switzerland
D. Hutchison   Lancaster University, UK
W. Kalfa   TU Chemnitz, Germany
T.D.C. Little   Boston University, USA
E. Moeller   GMD FOKUS, Germany
K. Nahrstedt   University of Illinois, USA
G. Neufeld   UBC Vancouver, Canada
B. Pehrson   KTH Stockholm, Sweden
T. Plagemann   UNIK, Norway
B. Plattner   ETH Zurich, Switzerland
L.A. Rowe   University of California at Berkeley, USA
H. Scholten   University of Twente, The Netherlands
A. Seneviratne   UTS, Australia
R. Steinmetz   GMD, Germany

J.B. Stéfani   CNET France Télécom, France
L. Wolf   TU Darmstadt, Germany
M. Zitterbart   TU Braunschweig, Germany

Referees

S. Abdelatif	K. Drira	P. Owezarski
H. Affifi	R. Dssouli	S. Owezarski
P.D. Amer	M. Dudet	B. Pehrson
J.C. Arnu	W. Effelsberg	T. Plagemann
P. Azéma	F. Eliassen	B. Plattner
V. Baudin	S. Fdida	B. Pradin
P. Berthou	D. Ferrari	V. Roca
E. Biersack	O. Fourmaux	L. Rojas Cardenas
G. v. Bochmann	F. Garcia	L.A. Rowe
M. Boyer	T. Gayraud	P. Sampaio
B. Butscher	V. Goebel	P. de Saqui-Sannes
A.T. Campbell	T. Helbig	H. Scholten
J.M. Chasserie	J.-P. Hubaux	G. Schürmann
C. Chassot	D. Hutchison	P. Sénac
T.S. Chua	W. Kalfa	A. Seneviratne
L. Costa	T.D.C. Little	M. Smirnov
J.-P. Courtiat	A. Lozes	R. Steinmetz
J. Crowcroft	E. Moeller	J.B. Stéfani
L. Delgrossi	C. Morin	J.P. Tubach
M. Diaz	K. Nahrstedt	T. Villemur
C. Diot	G. Neufeld	L. Wolf
A. Dracinschi	R. Noro	M. Zitterbart

Local Organization

Marc Boyer   LAAS-CNRS, Toulouse, France
Laurent Dairaine   ENSICA, Toulouse, France
Daniel Daurat   LAAS-CNRS, Toulouse, France
Sylvie Gay   ENSICA, Toulouse, France
Marie-Thérèse Ippolito   LAAS-CNRS, Toulouse, France
Joëlle Pennavayre   LAAS-CNRS, Toulouse, France
Pierre de Saqui-Sannes   ENSICA, Toulouse, France

Supporting/Sponsoring Organizations

ACM	CNRS	Microsoft Research
IEEE	Région Midi-Pyrénées	Xerox
DGA	France Télécom	

Table of Contents

Invited Paper

Adaptive Applications and Networks

New Trends in IDMS

Advances in Coding

Invited Paper

Conferencing

Video Servers

Position Papers

The Internet 2 QBONE Project Architecture and Phase 1 Implementation

Phillip Emer

NCState.net & North Carolina Networking Initiative

Abstract. As IP-based real-time services move into the mainstream of national and international communications infrastructure, the need for a differentiated services framework becomes more important. The Internet 2 QBONE group is focused on building such a differentiated services framework atop NSF supported high performance networks - namely, Abilene and the vBNS. The QBONE proposes to offer an expedited forwarding (EF) service from campus edge to campus edge. This EF service includes marking and admission control at the ingress edge and forwarding and rate shaping in the core transit network and at the egress edge. Further, the QBONE framework includes a bandwidth broker (BB) function for negotiating diffserv behaviours across domain (AS) boundaries. Finally, QBONE domains are instrumented with measuring and monitoring equipment for verification and traffic profiling (and research).

In this paper, we describe the phase one QBONE architecture, some exercising applications, and some preliminary results. Some results are based on applications sourced in the North Carolina Networking Initiative's (NCNI) NC GigaPOP. Other results are based on interoperability tests performed in association with NC State University's Centennial Networking Labs (CNL).

Hardware Acceleration inside a Differentiated Services Access Node

Juha Forsten, Mika Loukola, and Jorma Skytta

Helsinki University of Technology, P.O. Box 3000, FIN-02015 HUT, FINLAND,
{juha.forsten, mika.loukola, jorma.skytta}@hut.fi

Abstract. This paper describes the hardware implementation of Simple Integrated Media Access (SIMA) access node. SIMA introduces Quality of Service (QoS) in IP networks by using the IPv4 Type of Service (TOS) field as priority and real-time indication field. Hardware acceleration is used to speed up the SIMA cal-culations.

1 Introduction

SIMA is a Differentiated Services (DS) [1] scheme with detailed access node and core network node functions. [2]

The SIMA access node has to calculate the momentary bit rate (MBR) in order to determine how well the user traffic flow confirms to the service level agreement (SLA). The access node can calculate the priority of the IP packet from the relation between the MBR and the nominal bit rate (NBR). The NBR value is indicated in the SLA. After inserting the priority in the IPv4 TOS field the header checksum must be recalculated. As the link speeds increase the time left for these calculations decrease and thus new methods for high-speed calculations are needed. In this implementation the SIMA specific calculations are performed in the special hardware acceleration card. The parallelism between the CPU and the SIMA hardware acceleration card leaves the processor free to handle other tasks such as the packet capture/send and SLA database management.

2 SIMA Specification

SIMA splits the DS field [3] into a 3-bit priority field and a 1-bit real-time/non-real-time indication bit. The QoS offered to the customer is determined by MBR, NBR, and real-time bit (RT). The value of MBR is the only parameter that keeps changing in an active flow. The other two (NBR and RT) are static parameters that are agreed upon in the SLA. The SIMA access node may also mark the RT value based on the application requirements. When RT service is requested, the SIMA network attempts to minimize the delay and delay variation. This is achieved by two queues in the core network routers as indicated in Figure 1.

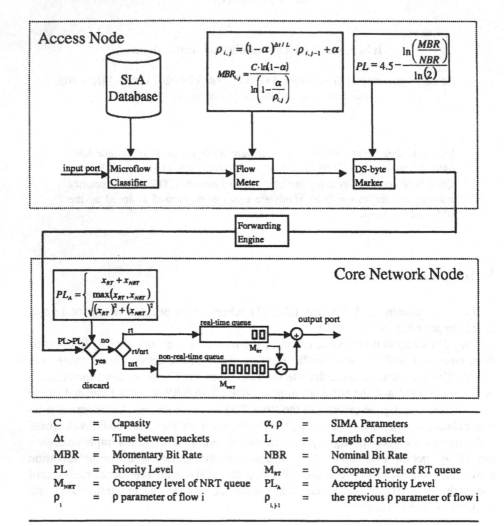

Fig. 1. Forwarding Path Model of SIMA Nodes

The packets are accepted at the core node by comparing Priority Level (PL) against Accepted Priority Level (PLA). All the packets that have been accepted in the scheduling unit are situated either in the real-time or non-real-time queue. The real-time queue has a strict priority over the non-real-time queue.

In the access node the traffic flows are metered. The result MBR of the flow is then checked against the SLA. The packets will get priorities according to the equation presented in Figure 1.

The relation between the MBR and NBR is the key to the applied priority level as shown in Table 1.

Table 1. The Assignment of Priority Level in The SIMA Access Node

Momentary Bit Rate (MBR)	Priority Level
Reserved for non-SIMA traffic	7
< NBR · 0,35	6
NBR · 0,7	5
NBR · 1,41	4
NBR · 2,83	3
NBR · 5,65	2
NBR · 11,31	1
> NBR · 22.62	0

During network congestion the packets marked with higher priority levels will be insulated from the congestion that is experienced at lower priority levels. Real-time connection features decreased delay, but also increased packet loss, because the priorities will decrease during peaks [4].

SIMA network introduces an easy way of charging. With guaranteed service the customer must first predict the parameters of the flow, which is not easy even for an expert. In case of SIMA the SLA consists of just a NBR (which can be different for specific source/destination pairs).

3 Hardware Acceleration

3.1 Architecture of the FPGA Card

A hardware acceleration card (Figure 2) is used in this project to speed up the SIMA calculations. The card is designed in Helsinki University of Technology, Laboratory of Signal Processing and Computer Technology. It contains three Field Programmable Gate Array (FPGA) chips – one for PCI (Peripheral Component Interconnect) controller and two for calculations. The robust architecture of the card makes the interfacing straightforward. Thus implementation of the PCI controller was possible using Altera's FLEX-8000 FPGA device. It is smaller than the other two FPGAs used for calculations (Altera's FLEX-10K devices [5]) but the operating speed is higher allowing the required speed of the PCI bus (33 MHz). There is also an on-board SRAM (128KB) that makes possible to handle data in the card without the use of PC's main memory.

In addition to the FPGAs and the SRAM the acceleration card contains a clock generator device and a FLASH memory chip. The clock generator has two separate outputs for both of the FPGAs. It is fully programmable and capable of clock frequencies from 100kHz to 120MHz. This configuration allows the usage of two independent clocks in the FPGAs. The onboard FLASH contains the programming data for PCI controller FPGA because. When the power is switched on, the SRAM based FPGAs are unprogrammed and there is no way to program it trough PCI bus. Instead, the configuration data for the controller is read from the FLASH memory. Optionally, there is a configuration input for external programmer. That input can be used for example with a Industry Standard Architecture (ISA) bus I/O card enabling programming the PCI controller chip on-the-fly. The architecture of the FPGA card is presented in the Figure 3.

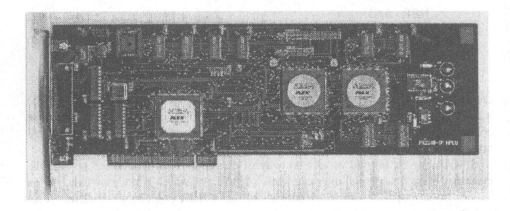

Fig. 2. Hardware Acceleration PCI Card

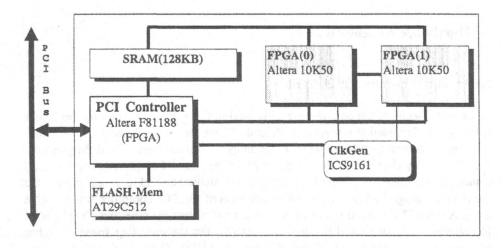

Fig. 3. Architecture of the FPGA card

3.2 Communicating with the Operating System

Communication with the FPGA card is done via PCI bus. The bus offers a 32-bit data path, 33MHz clock speed, and a maximum data transfer rate of 132MB/sec. PCI bus is time multiplexed, so the address and data lines use the same connection. PCI is also part of the plug and play standard allowing auto configuration. [6]

There are two different methods to transfer information between the PC and the FPGA card. The first method uses card's SRAM through memory mapping and can be described as a primary data exchange channel. The second method that uses PCI controller's registers via I/O ports is mainly for control and configuration purposes. When using memory mapping, a part of the card's SRAM is mapped to PC's physical memory address space. This makes it possible to directly access the on-board SRAM. Because the PC's memory window size to the SRAM is smaller than the available memory in the card, a dedicated Page Register is used to select the right memory area before reading or writing to the SRAM. To prevent data corruption, one bit in the Control Register is used to determine if the PC or FPGAs has the privilege to access the SRAM. In addition to these registers there are few other registers that can be used to program the calculation FPGAs, Clock Generator or FLASH memory. Figure 4 shows the card's location in PC's physical address space including I/O ports.

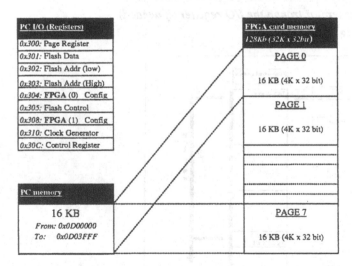

Fig. 4. Memory and I/O layout of the FPGA card

Communicating with the acceleration card requires some low-level operations like direct access to memory and I/O. Linux uses processor's memory protection mode thus omitting user's direct access to the physical memory. To overcome this there are two different ways to access the card. The general approach is to use a kernel level device driver written for that card to handle the low-level access. The other option is to do the access indirectly, mapping a part of a physical memory to a user-process virtual memory with some privileged system functions. The mapped address space can be accessed using these functions. In this project, the latter was chosen because of the simplicity and easy integration with the software. The I/O-ports are directly available when accessed by privileged user like *super user* in UNIX systems. That means that reading and writing the control registers in the FPGA card are trivial: after reserving those I/O-ports from an operating system, they are directly accessible. The communication with FPGA card is illustrated in Figure 5. [7]

Procedure to access the card:

1. *Get the permission to use the I/O addresses from the operating system.*
2. *If the FLASH memory is not used, use ISA I/O card to program the PCI controller in the acceleration card thus enabling the communication with the card.*
3. *Program one or both of the FPGAs through the I/O register(s).*
4. *Use a system call to map the needed physical memory range (the card's memory window) to the user's virtual memory space.*
5. *Set the page register to point to a needed page.*
6. *Enable the access to the card's SRAM for the PC or the FPGAs depending, which one needs it. This is done with the control register.*
7. *Program the clock generator trough the I/O register (if needed).*
8. *Start the FPGAs using control register.*

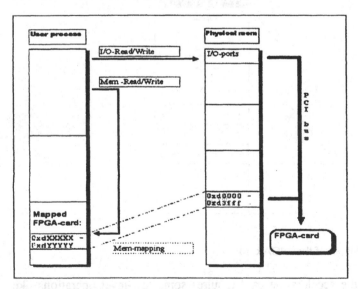

Fig. 5. Accessing the FPGA Card in the Linux Environment

4 Floating Point Notation

In the early prototype integer values were used for SIMA parameters. However the number of bits was limited and insufficient for the needed granularity. This led to very coarse-grained ρ and α values. As a result the allocated priorities did not feature any dynamics and saturated to some static values.

Based on those test cases it was obvious that some special floating point data type was needed. We came up this four different floating point notations presented in Table 2. The e2m4 notation indicates that total six bits are used for the data type, two bits for exponent and four bits for mantissa (see Figure 6). The integer data type was sufficient for MBR values as test customers featured low bit rates. However there is scheme to expand the range of MBR with a similar floating point notation.

Table 2. SIMA Parameters Types

Parameter	Type	Min value	Max value
ρ	e2m4	0.0001	0.9
α	e2m3	0.0001	0.8
$\Delta t/L$	e2m3	0.0001	0.8
C (kB/s)	e4m0	$10^0 = 1$	10^{15}
MBR	int (8-bit)	0	255

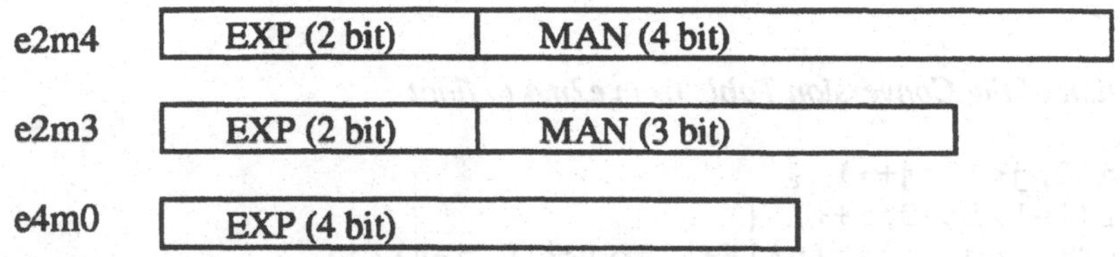

e2m4 | EXP (2 bit) | MAN (4 bit) |

e2m3 | EXP (2 bit) | MAN (3 bit) |

e4m0 | EXP (4 bit) |

Fig. 6. SIMA Floating Point Data Types

The main program running on Linux is written in C language. The data types of SIMA parameters in main program as well as in the SLA are standard C floating point data types. Before sending the parameters to the SRAM on hardware acceleration card the parameters must be converted to the notation used in the FPGA. Similarly as the parameters have been transferred back to the main program there is a need for reverse conversion. Examples of those conversion functions are presented for e2m4 data type. In the reverse conversion the values of the e2m4 data type is the index to a table that returns a standard C float.

Conversion Fucntion from float to e2m4

```
unsigned char float_to_e2m4(float f)
{
    int man,exp,i;

    if(f >= 0.9) {
        man=9;
        exp=3;
    } else {
        if(f <= 0.0001){
            man=1;
            exp=0;
        } else {
            for(i=3;i>=0;i--) {
                f=(float) (f * (float) 10.0);
                if(f>= (float) 1.0) {
                    man=(int) (f + 0.5);
                    exp=i;
                    break;
                }
            }
        }
    }
    return(man+exp*16);
}
```

Initialization of the Conversion Table from e2m4 to float

```
for(j=0;j<=3;j++) {
    for(i=1;i<=9;i++) {
        e2m4_to_float[j*16+i]=pow(10,j-4)*i;
    }
}
```

5 VHDL Code

The starting point for the VHDL code is that the IP header and SIMA parameters have been transferred to the card's SRAM. The VHDL state machine is illustrated in Figure 7. The initial transition from the *init* state will be forced by the control register signal 0x02. At first the FPGA chip will read the input SIMA parameter values from rows 5 and 6 (see Table 6) into its internal signals. After that the FPGA will calculate new ρ, MBR, and priority values in the *calculate* state. The priority value will be determined by the relation of the MBR and NBR values (Table 1). Then the actual IP header will be read from the on-board SRAM (rows 1 to 5) in a loop between the *load* and *update_header_checksum* states. Every time the *update_header_checksum* state is bypassed the new header checksum will be calculated further. Once the last row of the IP header is read the state will transition to *pre_store*. In a loop between *pre_store* and *store* the new updated ρ, updated TOS field value, the updated header checksum will be written to the eighth row. After that the state will transition to *stop*. In the *stop* state the card will notify the main program of the finished task through a control register signal.

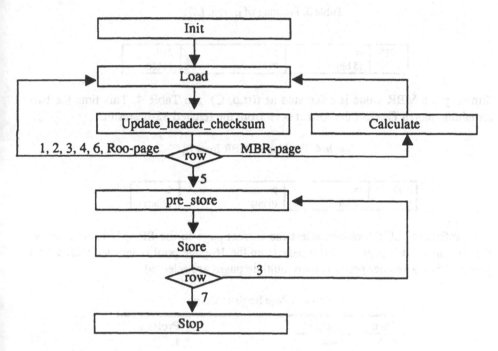

Fig. 7. VHDL State Machine

The checksum algorithm in IPv4 consists of adding one's complements of 16-bit fields together. Each row of the IPv4 header (Table 6, rows 1-5) is divided into two 16-bit fields. The header checksum field is considered zero during the calculation. The adding result is once again complemented. This 16-bit result is written to the header checksum field. The algorithm is very easy to implement with VHDL language since signal and register outputs inside Altera's FPGA chips are available in one's complement in addition to the original register output values.

As shown in Figure 1 the SIMA parameter ρ_i is a function of α, ρ_{i-1}, and $\Delta t/L$. This function (new ρ value) is pre-calculated and placed to the on-board SRAM pages 4-7. The mechanism is known as Look-up-Table (LUT). The VHDL code receives a 32-bit result from the SRAM as it inserts the parameters of the functions as an input address for the SRAM (see Table 3). The two remaining bits of $\Delta t/L$ determine which octet of the 32-bit result contains the correct result is e2m4 from.

Table 3. Fetching of ρ from LUT

"01"	α (5 bit)	ρ (6 bit)	$\Delta t/L$ (3 bit)

Similarly the MBR value is calculated as $f(\alpha, \rho, C)$, see Table 4. This time the two remaining bits of C select the correct octet from the fetched 32-bit result.

Table 4. Fetching of MBR from LUT

"001"	α (5 bit)	ρ (6 bit)	C (2 bit)

The prefixes of LUT look-ups select the correct page in the SRAM. Table 5 shows the content of each page. A single page from the 16-page pool is selected with a 4-bit prefix. When a smaller prefix is used multiple pages are selected.

Table 5. Page Register Usage

Page	Content	Prefix
8-15	Unused	1
4-7	ρ	01
2-3	MBR	001
1	Unused	0001
0	Header + SIMA parameters	0000

Page 0 is used for communication between the main program and the calculation FPGA chips, see Table 6. SIMA parameters are located right after the IP header (rows 6 and 7). As the original IP header is already in the PC memory, it is only necessary to transfer back the modified header fields and the updated ρ value (row 8).

Table 6. SRAM Page 0 Content on the PCI Board

4-bit ver	4-bit hdr len	8-bit TOS	16-bit total length	
16-bit identification			3-bit flags	13-bit fragment offset
8-bit TTL		8-bit protocol	16-bit header checksum	
32-bit source IP address				
32-bit destination IP address				
NBR		RT	α	$\Delta t/L$
ρ		C	Status OK=1 ERROR=0	
Updated ρ		Updated TOS	Updated header checksum	

7 Comparison between Software and Hardware Based Operation

We have measured the performance of SIMA calculation routines both with software-only and FPGA implementation. All the measurements are made using the Intel Pentium processor's assembler instruction RDTSC (read-time stamp counter). It reads the status of the time-stamp counter that keeps an accurate count of every cycle that occurs on the processor. The Intel time-stamp counter is a 64-bit model specific register (MSR) that is incremented at every clock cycle. By taking the value of that counter just before and after the measured routine, it is possible to get accurate timing measurements.

Table 7 shows the measured performance of the SIMA calculations in several different cases. The average IP packet size used in the estimated average throughput calculations is 416 bytes. This packet size has been calculated taking an average over the traffic statistics from different subnets in the University of Buffalo [8].

Table 7. Performance Measurements

Used Method	Time delay of the SIMA calculation	Estimated average throughput
Pure software calculations with 50% processor utilization	54,4 μs	61 Mbps
Pure software calculations with 100% processor utilization	27,2 μs	122 Mbps
FPGA based calculation	14,82 μs	225 Mbps

Figure 8 illustrates how the total FPGA calculation time (14.82 μs) is divided among the different tasks.

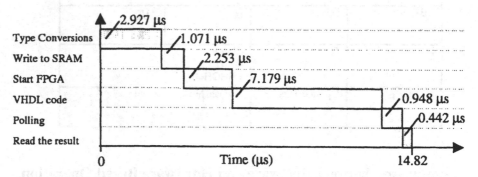

Fig. 8. Consumption of Time for Different Hardware Operations

In order to verify the correct operation of the hardware acceleration card several tests were performed with and without the hardware acceleration. [9] The homepage of Helsinki University of Technology [10] was chosen for this trace. This particular page presents a typical web page containing text and several images. The used SIMA parameters are presented in Table 8.

Table 8. SIMA Parameter Values

Parameter Name	Value
α	0.0001
capacity, C	10000 kB/s
real-time bit RT	0 (=non-real-time)
nominal bit rate, NBR	1 kB/s

Figure 9 illustrates the allocated DP values for each packet of the web page download. The session lasted for 0.22 seconds. During that time a total of 28037 bytes was downloaded. The average bit rate of the flow is 127.44 kB/s and the NBR value was set to 1 kB/s. That is why 84.16 percent (software) of the data was marked with DP values less than four, see Figure 10. The low α value of 0.001 slowed down the decrease of the DP values. If the session had continued longer with the same transmission speed, the rest of the packets would have been marked with the DP value of zero. This kind of parameter setting speeds up the interactive applications with bursty traffic characteristics like web.

When comparing the result obtained with hardware and software approaches the differences were reasonably small even though the hardware approach uses a heavily decreased granularity in the SIMA parameters values. As the result of the SIMA calculations is a 4-bit value varying from 0 to 6 it is possible to perform the calculations with fairly low number of bits.

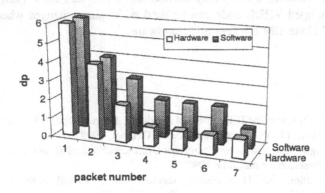

Fig. 9. Assigned DP Values

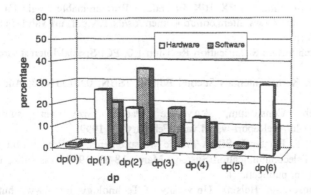

Fig. 10. DP Values Percentages

8 Conclusions

The SIMA equations require a great deal of processor resources especially in high-speed networking environment. The hardware acceleration card has proven to speed up the total throughput considerably. In addition to speed up the hardware approach set the processor free to other duties and thus realized parallelism to the SIMA calculations and packet capture/send. The usage of FPGA chips was discovered to be successful due to the their configurability for specific implementations. The solution can be optimized freely without the restrictions introduced by the fixed processor architecture. [11]

It was found that the decreased granularity in the SIMA parameters values gives fairly good results while featuring a low delay needed for high-speed networking. This proves that the developed VHDL code can be used with minor changes when designing complete stand-alone SIMA access node hardware.

References

1. Blake S., et al., "An Architecture for Differentiated Services", Internet Draft <draft-ietf-diffserv-arch-02.txt>, Torrent Networking Technologies, October, 1998
2. Kilkki K., "Simple Integrated Media Access", Internet Draft <draft-kalevi-simple-media-access-01.txt>, Nokia Research Center, June 1997
3. Nichols K, et al., "Definition of the Differentiated Services Field (DS Field) in the IPv4 and IPv6 Headers", RFC 2474, Cisco Systems Inc., December 1998
4. Loukola M.V, Engdahl T., Forsten J., SIMA Implementation architecture," in Proc. 13th International Conference on Information Networking (ICOIN'99), January 27-29, 1999, Cheju Island, Korea, pp. 7D-3.1–3.6
5. World Wide Web Consortium, "FLEX 10K Embedded Programmable Logic Family Data Sheet, ver. 3.13", http://www.altera.com/document/ds/dsf10k.pdf, url valid: February 17, 1999, Altera Corp.
6. PCI to PCI Bridge Architecture Specification Revision 1.0, PCI Special Interest Group, April 5, 1994
7. Beck M., et al., Linux Kernel Internals Second Edition, ISBN: 0201331438, Addison Wesley, 1998
8. World Wide Web Consortium, "Average Packet size per subnet", http://netop.cc.buffalo.edu/AveReport-w, url valid February 17, 1999
9. Loukola M.V, Engdahl T., Forsten J., SIMA Traffic Measurements," in Proc 7th International Conference on Telecommunication Systems March 18-21, 1999, Nashville, Tennessee, USA (accepted for publication)
10. World Wide Web Consortium, Helsinki University of Technology, http://www.hut.fi/, url valid: February 17, 1999
11. Vuori J.," Optimization of Signal Processing and Network Control in Telecommunications Systems", Doctor's Thesis, Helsinki University of Technology, Finland, 1998

REDO RSVP:
Efficient Signalling for Multimedia
in the Internet

Laurent Mathy, David Hutchison, Stefan Schmid, and Steven Simpson

Lancaster University, UK
{laurent, dh, sschmid, ss}@comp.lancs.ac.uk

Abstract. Alarming reports of performance and scalability problems associated with per-flow reservations have led many to doubt the ultimate utility of RSVP and the Integrated Services Architecture that relies on it. Because we are convinced of the need for some form of resource reservation, to support multimedia communications in the Internet, we have set about trying to improve RSVP. By careful study of the protocol, we have identified areas for improvement, and propose REDO RSVP, a reduced overhead version that includes a fast establishment mechanism (FEM). In this paper we describe the rationale for REDO RSVP and present a detailed analysis of its features and operations. We also analyse REDO RSVP by means of simulations, and show that it offers improvements to the performance of RSVP.

1 Introduction

The Internet was originally designed to offer a best-effort data transfer service. Such a type of service was simple to engineer while particularly well suited to applications whose utility is only loosely bound to the performance in the network. It is now widely recognized that to become a global telecommunication platform with integrated services—a must in the provision of information super-highways—the Internet must evolve to provide proper support to applications requiring stringent qualities of service.

Several proposals have thus been made to support such an evolution of the network. These can be classified in three general categories: explicit QoS support (per flow or with aggregation), service differentiation and IP/ATM integration. Because they address different issues, these proposal will coexist in tomorrow's Internet. Without considering in further detail these solutions and numerous possibilities of integration, the following trend can be observed: the network will be based on a better control of resource usage.

Among resource management mechanisms, those offering the finest grain of traffic control operate on a per-flow basis. These mechanisms, however, suffer from scalability problems as the number of flows with reservations increases. Although this rules out their use within the core of the network, per-flow provisioning can still be used at the edge of the network where the concentration of flows is rather low. In the Internet, the IntServ (Integrated Services) architecture [4] offers a framework for per-flow QoS control which relies on RSVP ("Resource ReSerVation Protocol") [5][12] as the signalling protocol to carry resource requirements between the source and the destination(s) of a flow. Furthermore, several proposals, mainly flow aggregation techniques [10][2] and the DiffServ (Differentiated Services) architecture [3], have been put forward to overcome the state scalability problems in the core of the network. These latter proposals will have to be integrated with the "traditional" IntServ approach to provide some form of end-to-end QoS commitments acceptable to multimedia and real-time applications.

The above mentioned flow aggregation techniques do not necessarily result in any reduction in the number of control messages sent per individual flow in the core of the network and certainly add to the overhead by generating messages per aggregate. Therefore, such a message overhead may still create a computational bottleneck (depending on the number of aggregates supported) in core routers as well as consuming bandwidth. Consequently, the steady-state message overhead in RSVP represents a significant scalability challenge.

Although RSVP was originally designed for resource reservation, several proposals have now been tabled where RSVP is used to carry other types of control information in the network [7][9]. Another example is the possible use of RSVP within the DiffServ architecture [1]. Therefore, we believe that, whether it is for resource reservation or other control/signalling purposes, RSVP will have to operate over routes of various lengths and to satisfy demands exhibiting a broad range of dynamics. Consequently, RSVP's ability to carry control information efficiently across the network in any circumstances will be vital to the effective operation of the Internet.

That is why we propose, in section 2, a modified flow establishment mechanism aimed at improving the resource set-up capabilities of the protocol. In section 3, we also seek ways to improve the message overhead scalability of RSVP in terms of the number of flows it can support. To that effect, we propose a REDuced Overhead (REDO RSVP) technique which is a form of aggregation for the control traffic in order to reduce the "steady state" overhead of RSVP at high loads of traffic. Section 4 presents comparative simulation results between RSVP and the improvements proposed in the paper. Section 5 concludes our discussion.

2   Improving Resource Set-up in RSVP

RSVP uses periodic messages to manage its states [5]:

- a sender sends Path messages (per flow) toward the receiver(s). These messages construct and maintain a "path state" and advertise the characteristics of the flow.
- a receiver sends Resv messages towards the senders. These messages construct and maintain actual reservations.

19

– every intermediate node periodically sends its *own* Path and reservation messages, that is there is no way senders or receivers can force a node to send copies of RSVP messages in the network.

The lapse of time between consecutive Path or Resv messages defines the refresh period of the protocol (in a refresh period, there is one Path and one Resv messages per flow on each link of the path). The default value R for the refresh period is 30 seconds[1]. Because RSVP messages are exchanged unreliably, such a lapse of time between similar RSVP messages seems prohibitively long, since it represents the average amount of time in which the loss of a control message can be corrected at reservation establishment. This results in a very long latency at reservation establishment, as confirmed by the simulation results presented in section 4.

Simply reducing the value of the refresh period is not the right approach, however. Indeed, doing so would increase the control traffic associated with every flow, thus threatening to pose severe scalability problems. Consequently, reducing the refresh period *at establishment time only*[2] (including local repair conditions, see [5]) is considered a better solution.

In modern high speed networks, message losses are mostly due to buffer overflow and thus occur in bursts [6]. We therefore see that proper "inter-spacing" is required between consecutive control messages, to prevent them from encountering the same congestion conditions along their route. This observation rules out the use of a fixed, short establishment period for the sending of consecutive RSVP messages during the establishment phase. Furthermore, in order to avoid unnecessary overhead, we must find a way to discover the end of the establishment phase, that is the moment after which the control messages related to a flow simply refresh the path states and reservations associated with that flow.

The first hurdle to overcome is the lack of an appropriate acknowledgment message in RSVP. A straightforward introduction of such messages in the protocol (as proposed in [11]) would result in major changes to the protocol definition and specification. However, when considering the different RSVP messages, it is clear that the role of an initial Path message is to "prepare" for a subsequent Resv message. The Resv message is then the obvious candidate to acknowledge the Path message. A Resv message indicates a successful reservation *to the sender of the corresponding Path message*. Therefore, any node that has forwarded a Path message, and has received a Resv message from every direct neighbour down the route followed by the corresponding flow, knows that the reservation has been successfully established *downstream*.

We still need to find a way for the receiver of a Path message to discover whether the establishment of a flow is in progress or has been completed. Because upstream nodes will use establishment periods shorter than the refresh period as long as they have not received a proper Resv message, a node can guess the status of a flow from the spacing of the Path messages it receives: if the

---

[1] Each period is chosen randomly in $[R/2, 3R/2]$[5].
[2] Such shortened refresh periods are called *establishment periods* in the rest of the paper.

lapse of time between consecutive Path messages is smaller than the shortest lapse of time allowed in "steady state", then the flow is more than likely being established and a Resv message should be forwarded as soon as possible to complete the establishment procedure (we thus see that the Resv message will be re-transmitted by the last RSVP node that correctly received the previous Resv message). On the other hand, if the time between consecutive Path messages is greater than or equal to the minimum allowed by the "classical" refresh periods (that is $R/2$), then we can suspect that the Path message is simply a refresh and a Resv message should only be sent when the current refresh period expires[3]. Of course, for this technique to be robust in the event of loss of Path messages, the periods used at establishment time must be quite a bit smaller than $R/2$.

We have already ruled out the use of fixed periods at establishment. The other important point is that, if the establishment periods are too short, unnecessary RSVP messages will be sent, which increases the overhead of the protocol. Therefore, the initial establishment period (T_0) should not be smaller than the round-trip-time (RTT) for the RSVP messages, which may have to be estimated.

After sending or forwarding the initial Path message, an RSVP node will wait for a lapse of time equal to the initial establishment period (T_0). If by that time a Resv message has not been received, the node retransmits the Path message (this procedure is applied by all the nodes supporting our technique, so that the copy of the Path message is generate as close as possible to where the loss of the previous RSVP message occurred). In order to be adaptive to a wide range of congestion conditions, the value of the establishment period must be backed-off: we propose to multiply it by a factor $(1 + \Delta)$ at each retransmission of the Path message. As soon as a Resv message acknowledges the establishment of the reservation, the nodes start using the refresh period R for their Path messages. A refresh period equal to R is also used if no Resv messages has been received, but the value of the establishment period has become greater than R. We therefore see that, in any case, the nodes "fall back" to the behaviour prescribed by the "classical" RSVP specification and FEM RSVP is backward compatible with "classical" RSVP.

With T_0 set to 3 seconds and Δ set to 0.3, this timer scheme is equivallent to the staged refresh timers described in [11]. It should be noted that for local repairs, a shorter value of T_0 would be acceptable, since we expect the new portion of the route to be fairly short. Furthermore, such a more aggressive behaviour of the protocol is justified by the fact that local repairs apply to existing flows.

---

[3] The period used by a node to send Resv messages is the refresh period defined in "classical" RSVP. The concept of establishment period timer does not apply to Resv messages.

3 Reducing the Overhead

The concept of soft-state was originally introduced in RSVP to deal easily with a number of conditions [12]. These conditions all fall into one of the following categories:

1. changes in routes,
2. reclamation of obsolete resources,
3. dynamic membership of multicast groups,
4. loss of control messages,
5. temporary node failures.

However, it soon appeared that the soft-state mechanism used in RSVP was too slow to deal with conditions of type 1 or 3, and the mechanism of local repair (see section [5]) was then introduced to improve the protocol's responsiveness to such conditions. Furthermore, in section 2, we presented an improved method to deal with loss of control messages (at establishment time). This leaves the soft-state in charge of the reclamation of obsolete resources and of dealing with some temporary node failures.

3.1 "Steady State" Overhead in "Classical" RSVP

In "classical" RSVP, periodic refresh messages have a keep-alive function which results in an overhead that is linear in terms of the number of established flows. This overhead thus increases both the bandwidth requirement and the CPU usage of the protocol, which results in scalability problems. This "steady state" overhead of RSVP is therefore a prime target when seeking to reduce the overall overhead.

When considering node or link failures, we see that refreshing each flow individually is inefficient. This is because of both the definition of a session in RSVP and the way IP routing works: all the data flows of a given session, visiting the same router at any given time, follow the same downstream path and are therefore *collectively* affected by any route change or any network failure. We could therefore seek ways for any RSVP node to refresh simultaneously several flows, and indeed all the flows, shared with a direct neighbour. This corresponds to an aggregation of control information, and is therefore independent of the number of flows. However, there is one condition for this technique to work properly: teardown messages *must* be delivered reliably. Indeed, if a teardown message on a flow was lost, the associated states or resources would be kept partially alive and would then waste resources indefinitely. If RSVP was modified to provide reliable teardown of flows, the risk of "resource leak" would be avoided and the steady-state message overhead of RSVP could then be dramatically reduced, solving the message overhead scalability problem.

3.2 New Focus for the Soft-State

As we saw in section 2, receipt of Resv messages indicates successful establishment of a reservation downstream of the node that received these messages.

Therefore, if a node implements local repair, the "exceptional" conditions that have still to be taken care of are network failures and the loss of control messages used to reclaim obsolete resources.

Once flows are established[4], network failures can easily be detected by implementing the concept of soft-state *per neighbour*: neighbours periodically exchange *heartbeats* so that the absence of too many consecutive heartbeats is interpreted as a network failure. Note that such a mechanism allows the detection of every type of failure from the signalling protocol point of view: link and router failure, as well as the failure of the RSVP process in a neighbouring node. In parts of the network using point-to-point links between nodes, there is only one neighbour per link, so the mechanism consists of a periodic check of each link. On broadcast links, the heartbeats could be sent to a well known multicast address so that only one heartbeat would be required from each node per refresh period.

When implementing per-neighbour soft-state, a node only sends Resv messages in two cases: in response to Path messages; or after receiving a Resv message changing the reservation on a flow. Similarly, after having received a Resv message, a node only forwards new Path messages or Path messages modifying the path state of a flow. Any other RSVP messages are treated in accordance with the RSVP specification.

The benefit of per-neighbour soft-state as opposed to per-flow soft-state is that it generates control messages at a fixed rate, independent of the number of established reservations, as illustrated in figure 1. This makes it more scalable than its per-flow counterpart while potentially providing much faster reaction times.

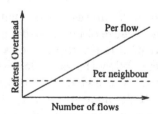

Fig. 1. Soft-State Overhead

We now need to devise a way to exchange teardown messages reliably. However, there is no need for complex end-to-end acknowledgment semantics: after all, a signalling protocol carries information hop-by-hop, and we can now rely on the heartbeats to detect the failure of a node (and therefore to react properly to any possible damage resulting from such failure conditions). Introducing the concept of hearbeats in RSVP is probably enough of a "revolution", so we strive to avoid changing the existing message types as well as defining any new

---

[4] Before reservation is completed, FEM or "classical" RSVP is enforced.

message types, and in particular *specific* acknowledgment messages such as Tear-Acks. On the other hand, nothing prevents the heartbeats from carrying some form of identification (i.e. sequence number field). Then, if each heartbeat sent on a link carries a copy of some or all the teardown messages that were previously sent on this link, reliable exchange of teardown messages between neighbours can be guaranteed by having nodes piggybacking acknowledgment of received heartbeats in their own heartbeats. A node will keep copying a teardown message in its heartbeats as long as a heartbeat containing it has not been acknowledged by all the neighbours on the same link.

Of course, this means that each teardown appears at least twice on each link. However, because flows requiring reservations will usually be long-lived (e.g. flows belonging to multimedia sessions), such an extra overhead at teardown will usually be far smaller than the "steady state" overhead of "classical" RSVP. It should also be noted that because the loss of a teardown message is only corrected, in "classical" RSVP, by the expiration of a lifetimer which will usually be several minutes long (see section [5]), the new teardown scheme proposed here will be much more efficient at resource reclamation than "classical" RSVP and will therefore improve resource usage in the network.

A protocol specification including local repair, FEM and per-neighbour soft-state (including reliable teardown messages) is called REDuced Overhead RSVP (REDO RSVP).

3.3 Compatibility with "Classical" RSVP

REDO mode should only be applied between REDO nodes. If a REDO node does not receive, or stops receiving, heartbeats from one of its neighbours, then "classical"/FEM RSVP must be used to communicate with that particular neighbour. Furthermore, as we will see in the next section, it is sometimes necessary for REDO nodes to revert to "classical" mode for certain flows, even when heartbeats are correctly exchanged.

The following rules are followed by a REDO node applying "classical" mode to some of its flows with one of its neighbours:

- if the REDO node is *upstream* of its neighbour, **upstream classical mode**(UCM) is applied to the flows concerned:
  - per-flow soft-state is applied to reservations;
  - periodical Path messages are sent downstream.
- if the REDO node is *downstream* of its neighbour, **downstream classical mode**(DCM) is applied to the flows concerned:
  - per-flow soft-state is applied to the path state;
  - periodical Resv messages are sent upstream.

At any time, either or both upstream or downstream classical modes can be applied to a flow by a REDO node.

REDO mode can only be applied to a flow between two REDO nodes when:

1. heartbeats are exchanged between these nodes (and the nodes have been synchronized, see below);
2. for the corresponding flow:
   - the node in upstream classical mode has received a Resv message acknowledging its Path messages; we say that node enters **upstream REDO mode** for that flow.
   - The node in downstream classical mode has received a Path message; we say that node enters **downstream REDO mode** for that flow.

The rules about the sending of Path and Resv messages in REDO mode, described in section 3.2, ensure correct operation of the protocol. However, when a node receives an RSVP message that changes the state (path state or reservation) associated with a flow, the node operates that flow in "classical" mode, until the above mentioned rules to enter REDO mode are met again.

However, in order to avoid inconsistent states in the network, the following rule must *always* be observed: when a REDO node starts, or re-starts, sending heartbeats to one of its neighbours, *synchronization* of these nodes must be completed before REDO mode can be applied to any flow between these nodes. In other words, during the synchronization period, all the flows between the nodes being synchronized *must* be operated in "classical" mode. The synchronisation is illustrated in figure 2. The synchronisation is only considered complete when the stabilisation period has expired. The role of the stabilisation is to prevent re-incarnated control messages, which could have been queued (e.g. in device driver buffers) but not delivered before the start of the synchronisation, from wrongly triggering REDO mode in a node. Similarly, the contention ensures that the upstream node (relatively to a flow) enters REDO mode after the downstream one (if this was not the case, the absence of Path messages would cause the reservation to time-out in the downstream node). The length of the stabilisation and contention timers should therefore be greater than the maximum packet lifetime (MPL) in the network. A value of 30 seconds to 2 minutes is proposed. This synchronisation mechanism is applied to each direction of traffic between the nodes.

It should be noted that a REDO node that does not receive any heartbeat from any of its neighbours on a given interface will behave totally like a "classical"/FEM RSVP node on that interface and should therefore refrain from sending heartbeats. We therefore see that *backward compatibility* is guaranteed, with REDO nodes "bridging" the two RSVP worlds. This allows for a progressive deployment of REDO RSVP.

3.4   Exception Handling in REDO RSVP

With REDO RSVP, as soon as a reservation has been established, and as long as no "special" conditions appear in the network, nodes simply exchange "empty" heartbeat messages. We now turn our attention to the kind of "special" conditions the protocol has to deal with and describe how REDO RSVP handles them.

25

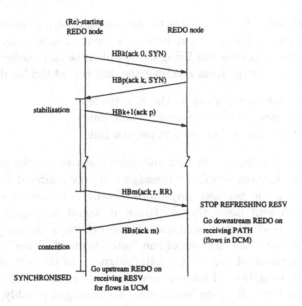

Fig. 2. Heartbeats (HB) during synchronisation between REDO nodes.

Route Change REDO nodes swiftly respond to route changes by using local repair and FEM on the new segment of route. We note that the use of FEM is essential in such a case: because any route change affects several sessions, there are potentially many flows to repair, so the newly visited nodes are likely to see a "burst" of control messages which could result in temporary congestion of the signalling traffic.

When a node which has initiated a local repair has received Resv messages from all its new neighbours downstream, it then knows that a flow has been repaired. This also means that it is now safe to tear down the old reservation on the old route. Because Path teardown messages follow normal IP routing, "classical" RSVP has no way of sending such a message down the old route. On the other hand, REDO RSVP can place such messages in the heartbeats destined for the old "next-hop" node and can therefore promptly reclaim the resources on the old route. Any node receiving a teardown from its neighbour registered as the *previous hop* in its path state copies the teardown in its heartbeats[5] downstream. If a node receives any teardown ("classical" or within a heartbeat) from a neighbour which is not the previous hop indicated by the path state, the teardown should be discarded, to avoid tearing down resources along portions of the old route which are still part of the new route. We therefore see that REDO RSVP can, in the event of route changes, reclaim resources faster than "classical" RSVP.

---

[5] In a unicast session, a node receiving a teardown in a heartbeat can alternatively *first* forward the information in a "classical" teardown message before propagating it within heartbeats.

On the other hand, if the initiator of the local repair has not received any Resv messages from some of the new next-hop nodes, one cannot conclude whether resources on the old route are still being used. This is because several conditions can prevent a Resv message from reaching the initiator of the local repair:

1. there are insufficient resources on the new route;
2. repair Path messages have been lost on new links;
3. repair Resv messages have been lost on new links.

The important point is that in the first and third cases above, the path state has been repaired, so that any teardown message will stay confined to the portion of the old route that is not used any more. However, in the second case, if a teardown was ever sent down the old route, it would propagate all the way to the receivers. To avoid such spurious release of resources, we propose that when sending the last Path message of the "establishment phase" of a flow (see section 2), the initiator of the local repair instructs, in its next heartbeat, its direct neighbours along the old route to revert to "classical" RSVP operation on the corresponding flows. Because hearbeats are exchanged reliably, so are these messages. The net result of this procedure is that, if during a route change, some flows are not repaired within the operation of FEM, REDO RSVP reverts back to "classical" RSVP for these flows, so that its performance is *in the worst case* equal to the performance of "classical" RSVP[6].

Network Failures Expiration of the soft-state timer associated with a neighbour is interpreted as a link or node failure. In such a situation, "classical" RSVP operations are reverted to for the flows handled by that neighbour. That is, upstream classical mode is applied to flows for which that neighbour is a downstream node, while downstream classical mode is applied to flows for which that neighbour is an upstream node.

In order to deal properly with "asymmetrical" link failures, a REDO node whose soft-state timer associated with one of its neighbours has timed out, should refrain from sending heartbeats towards these neighbours. REDO operations can only resume between these nodes after they have been re-synchronized (see section 3.3). Also, the rule of synchronisation ensures that, in case of a node or REDO RSVP process failure, the synchronisation will be triggered from the failure point on reset. However, in the event of a link failure, all REDO processes involved may refrain from sending heartbeats. In such a case, the two following techniques are suggested to discover the recovery from failure:

1. a synchronisation phase is triggered as soon as any "classical" RSVP message (with appropriate version number) is received from a neighbour involved in the failure;
2. periodical "synchronisation probes" are sent to the neighbours involved in the failure.

---

[6] Of course, as soon as the reservation for that flow is completed downstream of a node, that node enters normal REDO mode with its downstream neighbours.

On a broadcast link, where a well known multicast address may be used for the exchange of heartbeats (section 3.2), it is impossible to refrain from sending heartbeats to a specific neighbour. In such a case, synchronisation information concerning a particular neighbour must be present in every heartbeat sent to the multicast address.

The strategy of "going classical" was chosen in the event of a failure because REDO nodes do not know if, when and how the routing protocol is going to work around the fault. As a consequence, in the event of network failures, REDO RSVP performs at least as well as "classical" RSVP. Actually, in the event of network failures, REDO RSVP offers possibilities not supported by "classical" RSVP.

Indeed, REDO nodes directly connected upstream and downstream of a fault could propagate information about the failure towards respectively the senders and receivers, as well as possibly informing the local routing daemon. When used with advanced routing protocols offering alternative routes, this option would allow to by-pass a failure swiftly. In particular, for fault-tolerant systems having stand-by routes, REDO RSVP would enable fast recovery from faults.

Note that, because of the possibly relatively high frequency of the heartbeats, the values of the soft-state timers have to be chosen in such a way as to minimize the risks of erroneous failure detection. Furthermore, the scheme would benefit if nodes gave preferential treatment to heartbeats.

Transient Failures A node or REDO RSVP process failure which is detected neither by the heartbeat mechanism nor the routing protocol is called a transient failure. This can happen when recovery from the failure is achieved within the lifetime of the per-neighbour soft-state (which should be configurable per neighbour) or the response time of the routing protocol.

Upon reset, the REDO RSVP process will initiate a synchronisation phase with each of its neighbours (see section 3.3). This will have the effect of forcing "classical" operation (with FEM) in the neighbours for the flows that were handled by the REDO RSVP process that had failed, which of course results in a swift re-establishment of the reservations for these flows in the recovered node.

4 Simulation Results

4.1 FEM RSVP

We have simulated the external behaviour of both "classical" and FEM RSVP, in order to compare them. Our simulations consisted of repeated reservation establishments between a sender and a receiver, over routes of various length and under various loss conditions.

In these simulations, the loss process on each direction of a link is represented, independently, by a two-state model. One of the states represents congestion (i.e. loss) periods while the other one represents no-loss periods. The loss process spends an exponentially distributed time in each state, with these exponential

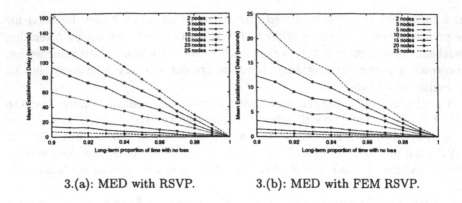

3.(a): MED with RSVP.   3.(b): MED with FEM RSVP.

Fig. 3. Mean Establishment Delay (MED)

distributions set so that the mean congestion period is 200 ms and the loss process spends a long-term proportion of time of 90% and more in the no-loss state. Such a model was chosen because of its ability to mimic loss bursts in a simple way.

Configurations comprising respectively 2, 3, 5, 10, 15, 20 and 25 nodes (including the sender and the receiver) were considered. For every configuration, 1000 flows were established and no delay was introduced in nodes and links to isolate the time overhead introduced by the external (i.e. observable) operation of the protocol. Finally, the default of 30 seconds was used as the average value of the refresh periods in RSVP, while for FEM RSVP T_0 and Δ have the values proposed in section 2. Figures 3.(a) and 3.(b) show the measured mean establishment delay and clearly demonstrate the gain in performance obtained with FEM RSVP. Note the low establishment delay achieved by FEM RSVP over short routes with low per-hop success probability. This result is of particular interest in the case of a change of route, since local repairs produce bursts of control messages which can temporarily congest the signalling channel.

4.2 REDO RSVP

We have measured the waste of resources incurred by both "classical" and REDO RSVP. By "waste of resource", we mean the average time that a resource is reserved in a node while the corresponding flow is not in use by the application. Such a waste occur at both resource establishment and teardown. At establishment, the waste is due to the receiver-oriented nature of the protocol: the reservation has to make its way up towards the source while the reserved resources will only be used when the reservation actually reaches that source. REDO RSVP minimises this type of resource waste by using the FEM extension at flow establishment. At teardown, the waste is essentially due to losses of teardown messages. REDO RSVP speeds up resource reclamation by implementing reliable exchanges of teardown messages within the use of its periodic heartbeats.

The simulations presented in this section were performed under the same conditions as the ones in the previous section, but with an average period H of 2 seconds for the heartbeats. Indeed, to avoid synchronisation of the heartbeats [8], the time between consecutive heartbeats is chosen randomly in $[H/2, 3H/2]$. In these circumstances, the overhead generated by the heartbeats is equivalent to the "steady-state" overhead of 15 "classical" RSVP flows. Figure 4 show the mean resource waste incurred per flow in any RSVP node along the route of that flow. The results show that REDO RSVP reduces resource waste in the nodes by an order of magnitude.

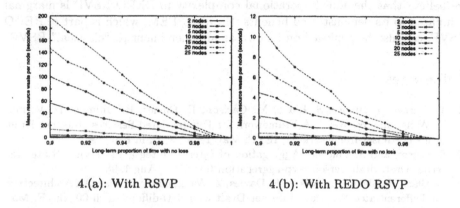

4.(a): With RSVP   4.(b): With REDO RSVP

Fig. 4. Mean per flow Resource Waste per Node.

5 Conclusions

We have presented FEM, a Fast resource Establishment Mechanism that is not only more robust to the conditions in the network than the establishment mechanism currently used in RSVP, but also establishes resources faster and without any unnecessary increase of control traffic. FEM is a very simple mechanism that relies totally on the Path and Resv messages defined in RSVP and thus does not require major amendments to the protocol specification.

The soft-state mechanism in RSVP is a simple way to deal with some conditions in the network. Unfortunately, it does not provide a good response to every error condition (see section 3). It also has the drawback of incurring an important "steady-state" overhead that jeopardizes the scalability of RSVP. We have proposed a way of overcoming these problems by "re-thinking" the use of the soft-state mechanism: the main idea of our REDO RSVP is that if the soft-state is applied per-neighbour instead of per-flow, the steady-state overhead is reduced and is independent of the number of flows in the network.

REDO RSVP responds to each situation in the network in a specific way, including reverting to "classical" RSVP operation in conditions where per-flow soft-state is deemed the most appropriate and simple solution. This ensures that REDO RSVP consistently exhibits performance which is better than, or equal to, that of "classical" RSVP. No change to the messages currently used in "classical" RSVP is required in REDO RSVP, which instead relies upon a new message type. REDO RSVP can thus be seen as a super-set of the mechanisms defined in "classical" RSVP. This guarantees backward compatibility and allows for a progressive deployment of REDO RSVP in the Internet.

Almost the entire complexity of REDO RSVP resides in its synchronisation mechanism. Because the synchronisation is only performed occasionally, we believe that the added operational complexity in REDO RSVP is marginal compared to its demonstrated benefits. However, FEM, which is part of REDO RSVP, can also be deployed on its own as an amendment to "classical" RSVP.

References

1. Y. Bernet, J. Binder, S. Blake, M. Carlson, E. Davies, B. Ohlman, D. Verma, Z. Wang, and W. Weiss. A Framework for Differentiated Services. Internet Draft draft-ietf-diffser-framework-00, IETF, May 1998. Work in Progress.
2. S. Berson and S. Vincent. Aggregation of Internet Integrated Services State. Internet Draft draft-berson-rsvp-aggregation-00, IETF, Aug 1998.
3. D. Black, S. Blake, M. Carlson, E. Davies, Z. Wang, and W. Weiss. An Architecture for Differentiated Services. Internet Draft draft-ietf-diffserv-arch-00, IETF, May 1998. Work in Progress.
4. R. Braden, D. Clark, and S. Shenker. Integrated Services in the Internet Architecture: an Overview. RFC 1633, IETF, Jun 1994.
5. R. Braden, L. Zhang, S. Berson, S. Herzog, and S. Jamin. Resource ReSerVation Protocol (RSVP)—Version 1 Functional Specification. RFC 2205, IETF, Sep 1997.
6. I. Cidon, A. Khamisy, and M. Sidi. Analysis of Packet Loss Processes in High Speed Networks. *IEEE Trans. Info. Theory*, 39(1):98–108, Jan 1993.
7. B. Davie, Y. Rekhter, E. Rosen, A. Viswanathan, V. Srinivasan, and S. Blake. Use of Label Switching with RSVP. Internet Draft draft-ietf-mpls-rsvp-00, IETF, Mar 1998. Work in Progress.
8. S. Floyd and V. Jacobson. The Synchronization of Periodic Routing Messages. *IEEE/ACM Trans. Network.*, 2(2):122–136, Apr 1994.
9. O. Fourmeaux and S. Fdida. Multicast for RSVP Switching. *Telecommunication Systems Journal*, 11(1-2):85–104, Mar 1999.
10. R. Guérin, S. Blake, and S. Herzog. Aggregating RSVP-based QoS Requests. Internet Draft draft-guerin-aggreg-rsvp-00, IETF, Nov 1997. Work in Progress.
11. P. Pan, H. Schulzrinne, and R. Guérin. Staged Refresh Timers for RSVP. Internet Draft draft-pan-rsvp-timer-00, IETF, Nov 1997. Work in Progress.
12. L. Zhang, S. Deering, D. Estrin, and D. Zappala. RSVP: A New Resource ReSerVation Protocol. *IEEE Network*, 7(5):8–18, Sep 1993.

QoS-aware Active Gateway for Multimedia Communication

Klara Nahrstedt and Duangdao Wichadakul

Computer Science Department
University of Illinois at Urbana-Champaign
{klara,wichadak}@cs.uiuc.edu

Abstract. With the rapid growth of variety in networked multimedia applications over the Internet, active gateways between senders and receivers become more and more desirable. However, there are at least two major concerns to use the active network concept for multimedia communication at a gateway: (1) support of bounded end-to-end configuration delays when a gateway is dynamically configured with QoS services during media transmission, to maintain minimum QoS degradation (e.g. jitters) and (2) support of bounded reconfiguration delays for fault tolerance handling when a gateway goes down and a new gateway needs to be reconfigured. We have designed, implemented and experimented with a QoS-aware active gateway architecture which addresses the two above described concerns. Our experiments and results show that an active gateway with flexible QoS services can be configured in an efficient manner and QoS guarantees can be preserved during the configuration/reconfiguration time if no additional security services such as authentication have to be performed at each reconfiguration request. In case of authentication our results show that the QoS will be degraded over a period of time and upgraded once the reconfiguration process is finished.

1. Introduction

The variety of current multimedia networked applications over the Internet demands more and more flexible and reconfigurable services to provide customized environments for different types of networks, operating systems, and users. This particular goal can be achieved with the active network concept which allows for transmission and dynamic loading of programs into any network node, hence allows for new type of flexibility and reconfigurability of services along the communication path.

In this paper we will investigate the *active gateway* node which will be used along the multimedia communication path. Our goal is to configure and reconfigure the gateway, once a multimedia application wants to use it, with QoS-specific services such as frame dropping service, minimum frame rate enforcement service and others. For the configuration capability we will use the active network concept, which allows the gateway to transmit, load, and execute requested QoS services on multimedia traffic. However, there are couple of concerns which need a closer investigation: (1) the multimedia applications require end-to-end Quality of Service (QoS) guarantees such as end-to-end delay, bounded delay jitter, during their data transmission, hence the active configuration/reconfiguration capability of services during the transmission needs to be QoS-aware; (2) in Internet networks a gateway can go down, or can get allocated to higher priority traffic, hence the multimedia communication path might need to be re-routed during the transmission, which means that a reconfiguration of a new gateway with suitable QoS services must be executed without significant QoS degradation. This means that the reconfiguration and path setup must be QoS-aware and performed under strict timing constraints. [1]

We will present an active gateway architecture which will allow us to sustain end-to-end QoS guarantees during transmission when QoS services are being configured at a gateway, or QoS services are being configured at a new gateway in fault tolerance. Our framework has the characteristics such as: (1) It operates in connection-oriented mode and uses the discrete approach of the active network concept, which means that the executable code is sent separately from the user data. The reason for this choice is that we believe this combination can reduce the overhead of all packets processing at an active gateway node even though it takes more time to setup the connection as our results show. (2) It provides all necessary components for handling QoS. It provides additional knowledge at the active gateway such as dependencies of multimedia data units, useful QoS services, and parameters for each individual customized connection request. Examples of these services and parameters are dynamic switching service for handling congestion, and enforcement services for minimum frame-rate, minimum required bandwidth and maximum acceptable delay jitter such as the frame dropping service. (3) It allows for object delegation and dynamic linking not only at the gateway, but also at the end-nodes which increases further the flexibility and customization of services.

Our results show that we can provide flexibility and efficiency in gateways within requested time bounds. However our results also show that if reconfiguration requests are concatenated with security mechanisms such as authentication, larger delays need to be taken into account and degradation of QoS will occur during this period.

The outline of this paper is as follows: In Section 2, we present an extensive related work, done in the area of active networks. Section 3 discusses the overall architecture where our active gateway framework resides, Section 4 describes the details of the active gateway design and implementation, Section 5 presents achieved results, and Section 6 concludes the paper.

---

[1] The timing constraints for configuration/reconfiguration can be determined by the application behavior and by the network behavior. For example, if the multimedia communication relies on the underlying RSVP protocol, where the refresh rate is 5 seconds, then the end-to-end configuration and path setup need to be performed below 5 seconds. Otherwise, the underlying network gets lost.

2. Related Work

Active network concept is a broadly explored concept to bring flexibility into the network. The goal of the active networks is to make network nodes programmable in an efficient and safe way. The active network research pursues several research directions: (1) active network architectures which study flexibility and efficiency of programmable network nodes, (2) safety and security to be included in programmable network nodes, (3) programming languages for active networks to enforce flexibility, efficiency and security, and (4) operating system support for active nodes. The **active network project** at MIT [1, 2] proposes an active network architecture using the ANTS toolkit [3] and follows the *capsule approach* with an appropriate API to ensure a safe execution of the code within the capsule. Capsule approach means that each user packet includes the user data and at least an instruction code or a reference to an executable code. ANTS is implemented in user space using JAVA programming language, and uses proper code distribution mechanism. The required code, not available at the active network node, is dynamically downloaded from the previous network node through which the capsule just traveled. Applications utilizing the active network architecture are address spoofing [4], web caches [5], online auction and active reliable multicast [6]. The **SwitchWare project** [7, 8] at the University of Pennsylvania develops a three layer active network architecture which includes: *active packets* to replace the traditional IP packets, *active extensions* for provision of services at active network nodes, and a *secure active routers infrastructure*. SwitchWare is implemented using PLAN [9] and Caml programming languages which are enhanced with various security capabilities. Applications using the SwitchWare architecture are reliable and unreliable datagram delivery services, RIP-routing, ARP-like address resolution and active bridging [10]. The **active network project** [11, 12] at the Georgia Institute of Technology develops an active network architecture to control congestion in the network nodes. This architecture uses *connection-less capsule approach* for code distribution. It is implemented within the augmented SunOS IP kernel to support active packet processing. The **NetsScript project** [13] at the Columbia University develops an active network architecture which addresses *large scale programmable networks*, *dynamically programmable networked devices*, and a language, called *NetScript*. The programmable network consists of a collection of Virtual Network Engines connected via Virtual Links, which support broadcasting, and SMART operation server for active operations. Applications using the NetScript architecture are remote network monitors, SNMP agents and an ATM signaling protocol. The **Smart Packets project** [14] proposes an active network architecture to apply this concept to network management and monitoring. The Active Network Encapsulation Protocol (ANEP) [15] distributes smart packets to individual network nodes and a corresponding virtual machine at the active nodes provides a safe environment for their execution.

Various extensible operating systems appeared to support active networks. Examples are Exokernel [16], SPIN [17], and Scout [18]. **The Liquid Software** project [19] at the University of Arizona implements an infrastructure of moving dynamic functions through out the network. The major objectives are *fast compilation* of the machine-independent code, *mobile searching* to explore availability of liquid software, and *integration* of the fast compilation and mobile searching into a customizable platform for liquid software support.

3. Overall Architecture

Our objective is to provide an architecture which allows (1) for active configuration and reconfiguration of gateways with respect to QoS services (dynamic loading of QoS services) such as a frame dropping service both during the connection setup of a multimedia communication path and during the media transmission, (2) for reconfiguration of a gateway with respect to equivalent QoS services in case a primary active gateway goes down and the multimedia communication must be redirected through a new gateway. Especially, our design is driven by the goal to achieve seamless configuration and reconfiguration of QoS services.

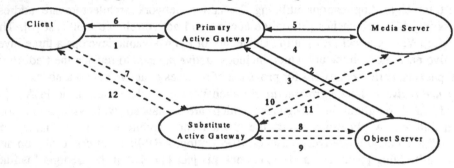

Fig. 1. Overall Architecture

The overall architecture, where we study the design of active gateways, consists of several nodes as shown in Figure 1.

1. **Client Node** receives the streaming video from a media server. This node is the initiator of a new multimedia connection request between the client and the media server. The connection request carries among other information also a request for QoS services needed at gateways along the path.
2. **Media Server Node** generates and/or stores video. This node waits for a connection request and sends continuous media to the client node in streaming mode.
3. **Active Gateway Node** is an intermediary node which processes and forwards packets between the clients and servers. This node has the capabilities of configuration and reconfiguration, and dynamically loads QoS services according to a multimedia connection request or change in network load.

4. **Object Server Node** is a service node which stores executable codes of various QoS services. This node waits for a service request from a configuration or reconfiguration process at a gateway and upon request it delivers the requested QoS service executable code if the code exists and the downloading request can be ,satisfied within timing constraints placed by the configuration and reconfiguration process.

In our architecture, as shown in Figure 1, we also consider a **substitute active gateway**, which will be used in case the primary active gateway needs to be replaced.

The **active configuration protocol** within this architecture works as follows: Step 1: The client sends a new multimedia connection request for video service to the media server along the primary active gateway path, including timing constraints on the duration of the QoS service configuration, and QoS services necessary for gateways along the path. The required QoS services are customized individually. The primary active gateway accepts the connection request, and finds out what QoS services need to be configured. Step 2: The primary active gateway issues a service request to the object server. Step 3: The object server accepts the request, finds the executable code of the requested QoS services and sends it back to the requesting primary active gateway, where the customized QoS services are configured. Step 4: The primary active gateway sends a new multimedia connection request to the media server, where the media server sets up all resources for video streaming. Step 5: Media server sends a control message to the primary active gateway with acceptance response (READY_TO_SEND packet) with respect to the requested video service. Step 6: The primary active gateway forwards the connection response to the client node.

The **active reconfiguration protocol** repeats functionally the same steps as discussed above. In Figure 1 the reconfiguration and redirection of the multimedia flow through the substitute active gateway follow the Steps 7 through 12.

Note that each performed step within the configuration and reconfiguration protocol is under timing control to satisfy the configuration timing constraints. If the accumulated configuration time (Step 1 through current Step) is greater than the requested timing constraint, an error message will be sent to the client to notify about timing violations for configuration and reconfiguration processes which can result in QoS degradation.

4. Active Gateway Design

The active gateway is the core component of our architecture; hence we will concentrate on its design in more detail. Furthermore, we will discuss gateway's important assisting entities which contribute to the active network capability such as the *object server* and *dynamic loader*.

4.1 Active Gateway Architecture

The active gateway architecture consists of several components: state information and structures, a protocol and QoS services downloaded by specific QoS handlers as shown in Figure 2.

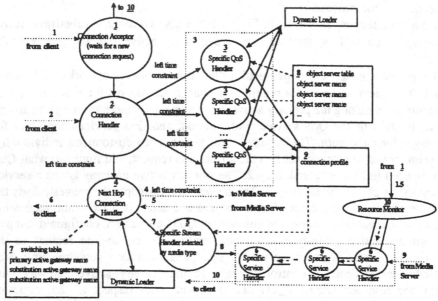

Fig. 2. Active Gateway Architecture

Components: The active gateway consists of several components. First component, where the client's request is delivered, is called the *Connection Acceptor* which waits for a new (connection) request and activates a resource monitor if none exists. The second component, *Connection Handler*, issues the customized QoS service requests via *Specific QoS Handlers*. Each *Specific QoS Handler* requests a QoS service from the *Object Server* using the *Dynamic Loader* if the requested executable service code is not found in the current active node. If all requested but unavailable QoS services can be downloaded from the object server and can be configured in a timely fashion, the *Next Hop Connection Handler* is invoked to send a new connection request to the next hop which can be a media server or next hop active gateway. The downloaded QoS services will be activated as *Specific Service Handlers* by the *Specific Stream Handler*, using the soft-state information in the connection profile, if the next hop connection handler gets the READY_TO_SEND packet from the media server. The active gateway uses information from the *Resource Monitor* to perform self-adaptation.

State Information and Structures: The active gateway keeps soft states for a connection in the connection's *connection profile*. An example of soft state information is the connection's supported frame rate. The connection profiles are created and updated by specific QoS handlers and used by the specific stream handler. Second important structure is the *switching table* which is used by the next hop connection handler to look up the next hop address. Third structure in the active gateway is the *object server table*. This table has the same structure as the switching table but it keeps the object server addresses. The dynamic loader uses this structure when new QoS services need to be downloaded and configured at the active gateway.

Configuration Protocol: The client's connection request arrives at the gateway's connection acceptor, where the acceptor checks the request and sets a resource monitor (Step 1) if none exists. Once the connection acceptor accepts the connection, it forwards the request to the connection handler (Step 2). The request is further forwarded to specific QoS handlers who check if the requested QoS services are available or not. The unavailable QoS services in local active gateway will be downloaded from the object server using the dynamic loader and be configured by the specific QoS handlers at the gateway (Step 3). The soft states of the connection are stored/updated in the connection profile. Once this step is done, a new connection request is sent to the media server or to the next hop gateway (Step 4). In case the media server accepts the connection request, the media server responds with a control packet and the next hop connection handler forwards the packet to the client or the previous hop gateway (Step 5 and 6). Furthermore, the next hop connection handler activates a specific stream handler (Step 7) which then activates specific service handlers (Step 8).[2]

QoS Services: Our current gateway design supports three types of QoS services: (1) frame dropping service, (2) enforcement service of minimal frame rate, and (3) dynamic switching service. The *frame dropping service* is configured at the active gateway in order to control the delay jitter and overall end-to-end delay. If the delay jitter experienced at the gateway violates a required bound, the frames are dropped. Because we are dealing with a multimedia active gateway, we can differentiate among the frames and drop according to the importance of the frames within the media stream. The *enforcement service* for minimal frame rate per connection is based on the comparison between the desired minimal frame rate and the monitored frame rate. If violations of the minimal frame rate occurs and persists, then the connection is suspended. The *dynamic switching service* is the representative of self-adaptable services in the network nodes. This service deals with congestion control and allows for switching between individual specific service handlers. For example, if a congestion status is received from the resource monitor, the default service handler will switch dynamically to another specific service handler such as the frame dropping service.

---

[2] Note that the Steps 9 and 10 in Figure 2 are not executed during the configuration time, but during the transmission of the media. The Steps 9 and 10 mean that data are moved from the media server or the next hop gateway through specific service handlers at the gateway, and are forwarded to the client or the previous hop gateway.

4.2 Object Server

The active gateway needs the object server for provision of mobile services (see Figure 3). The object server consists of three major components: *Object Request Acceptor*, *Connection Handler* and *Specific Handler*. The object request acceptor waits for an object download request where the object is an executable code of a QoS service. If a new object download request is accepted, the connection handler handles the request, using the specific handler (object request handler) to retrieve the executable code from the local storage. Once the object is found it is sent to the active gateway. Note that the operations at the object server are under time constraints as they are part of the configuration and reconfiguration process. The connection handler receives the time constraint for its duration from the active gateway and it monitors its own performance to satisfy the timing bounds.

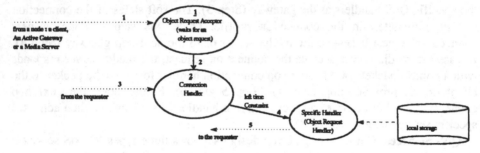

Fig. 3. Object Server Architecture

4.3 Dynamic Loader

Dynamic Loader is one of the crucial components in the overall active gateway architecture which contributes in a major way to the dynamic configuration and reconfiguration. The dynamic loader is based on *the modified provider concept* as follows:

The actual *provider concept* is a concept for implementing flexibility in the Java Cryptography Architecture [20]. The provider concept allows a program code to *dynamically select* a specific provider and a specific algorithm with a generic security mechanism. For example, we can select the specific algorithm MD5 with the generic security mechanism MessageDigest which is implemented by the specific provider SUN. We introduce the *modified provider concept* which supports only the dynamic selection of a specific algorithm with a generic mechanism. The dynamic selection of a specific "QoS service" provider is not necessary in our case because there exists currently only a single provider. However, the modified provider concept can be expanded back to the provider concept. The modified provider concept allows the dynamic loader to select dynamically a specific service handler corresponding to the media type encoding and the soft states in the connection profile at the run time. Note that the modified provider concept is also applied to the dynamic loader components at the client nodes and media server nodes and allows for flexibility at these nodes as well.

The dynamic loader consists of two main components: *Main Loader* and *Object Server Connection Handler*. The main loader is activated when a component (e.g. active gateway's specific handler) requests for a specific service (e.g. download a specific QoS service). The main loader responds for finding the requested service and returning an instance to the requesting entity in a timely fashion. If the requested service does not exist in the local storage, the main loader activates the object server connection handler to contact the object server and get the requested executable code of the QoS service. Once the transmission of the executable code is finished, the main loader passes the new service instance to the requesting component.

5. Experiments

Experimental Platform and Implementation: The gateway software is implemented on SUN Sparc stations (Ultra 1, Ultra 2) using JAVA programming language (JDK1.1.5). The individual SUN Sparc stations are connected via 10 Mbps Ethernet. On top of this platform we implement a single client/server video-on-demand application which transmits a video stream from the media server to the client upon client's request. The video stream passes through the active gateway as discussed in Section 3. The transmitted video is a 2.77M MPEG video with average frame size equals to 2.47K. The frame rate is 5 fps. The content of the video stream is the Simpson cartoon animation stream. The underlying transport protocol stack, over which the connection setup and media transmission are carried, is the TCP/IP protocol stack.

Performance Metrics: In this implementation we study various performance metrics in order to evaluate our objective *"flexible and efficient configuration and reconfiguration of active gateways"*. We measure the following performance metrics:

1. **End-to-end configuration delay** between a client and a media server: This delay represents the elapsed configuration time since the client sends a new (connection configuration) request until the client receives a READY_TO_SEND packet.
2. **Dynamic service delegation and linking time:** This time represents the total time used for dynamically loading a specific QoS service by an active gateway from the object server node and configuring the service at the active gateway.
3. **Configuration time within an active gateway:** This time represents the accumulated configuration time used by each component in the active gateway before issuing/forwarding a connection request to the media server.

Experimental Scenarios: We have run the VOD application in several scenarios and studied the above discussed performance metric parameters:

1. *Scenario*: We setup a connection for VOD transmission from the media server through the primary active gateway to the client. In this scenario we assume that when a new (connection configuration) request is made, no security check is made on the request, it means that we do not perform any authentication on the new request for QoS service configuration.

2. *Scenario*: We setup a connection for VOD transmission from the media server through the primary active gateway to the client, and perform authentication on the new request for QoS service configuration.

3. *Scenario*: We assume that the currently used primary active gateway is going down, which means that we need to reconfigure the substitute active gateway and redirect the multimedia traffic from the media server to the client through the substitute active gateway. In this scenario, we assume that the reconfiguration request is not authenticated.

4. *Scenario*: We assume reconfiguration of the substitute active gateway, and redirection of the multimedia traffic from media server to client through the substitute active gateway. In this scenario we assume that an authentication is performed on the reconfiguration request.

Results: We performed several experiments combining above described scenarios and performance metrics, however due to space limitation we will limit the presented results only to primary proxy.

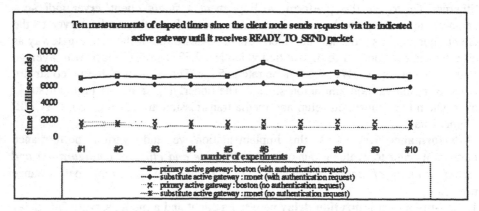

Fig. 4. End-to-end configuration delays between a Client and a Media Server (Combined four Scenarios)

1. *Experiment*: In this experiment we run all four scenarios and we measure the **end-to-end configuration delay** of a successful configuration/setup between the client and the media server. Figure 4 shows the individual results and we can see that the average end-to-end configuration delays without the authentication request are below 2 seconds which is three to four times better performance than the end-to-end configuration delays with authentication. The overheads of the authentication are caused by the JDK 1.1.5 Security package. This clearly indicates that the overhead is too high if our upper time constraint bound on the

end-to-end configuration delay is 5 seconds. If the reconfiguration does not happen within this period, the actual underlying reservation might expire. Another important observation one needs to make is that there is only a slight time difference between end-to-end configuration delay using the primary active gateway and the end-to-end reconfiguration delay using the substitute active gateway. This indicates that switching between gateways is connected with some overhead but in case of the Scenario 1 and 3 the configuration and reconfiguration delays are still in timing bounds below 5 seconds.

The following experiments (Experiment 2-3) run individual scenarios, measure specified performance metrics, show more detailed results and reinforce the results shown in Figure 4.

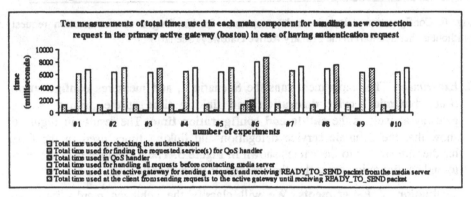

Fig. 5. Configuration delays in each component for handling a new connection request (with authentication) in the Primary Active Gateway (Scenario 2)

2. *Experiment*: In this experiment we run the Scenario 2 and measure the **configuration time, dynamic service delegation and linking time** during configuration at the primary active gateway, and the **end-to-end configuration time**. In Figure 5, we present configuration delays spent on each component within the primary active gateway. According to the graphs, the authentication generates the largest portion of the total times used for handling requests at an active gateway. The dynamic service delegation and linking times create only 5% portion of the total configuration times. All end-to-end configuration delays are over the upper time bound of 5 seconds.

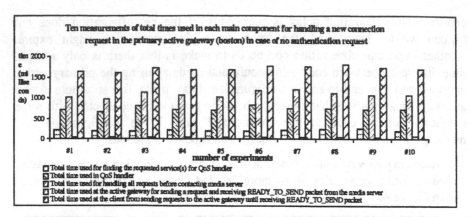

Fig. 6. Configuration delays in each component for handling a new connection request (without authentication) in the Primary Active Gateway (Scenario 1)

3. *Experiment*: This experiment runs the Scenario 1, and measures **configuration time, dynamic service delegation and linking time** at the primary active gateway as well as the **end-to-end configuration time.** The results in Figure 6 show that the dynamic service delegation and linking times used in the QoS handler are similar to the values shown in Figure 5. The difference is in the end-to-end configuration times due the absence of the authentication.

Evaluation of Experiments: We will classify the achieved results into two categories, and discuss each category separately. The first category comprises the results of experiments which do not perform authentication at each configuration and reconfiguration request. The second category includes the results of experiments which perform authentication check at each configuration and reconfiguration request.

1. *Category*: For this set of experiments (Experiment 3), the results indicate that using active network concept to configure and reconfigure active gateways (primary or substitute), using dynamic loading service, is applicable and we can provide *flexible customization* of QoS services at the gateways for multimedia communication. The flexible customization of QoS services can be done within required time bounds below 5 seconds. A larger configuration delay is acceptable if the quality of the connection improves. Furthermore, our goal is to bound the configuration time, so that the configuration can happen during the multimedia transmission if a new QoS service needs to be installed or another substitute active gateway needs to be reconfigured without letting the user know. In this case the duration of the configuration and reconfiguration time influences how long a multimedia transmission service will be disrupted and QoS degradation may occur. The current delays indicate acceptable timing delays within 1-2 seconds and perceptually the user of the VOD application does not notice major QoS degradations.

2. *Category*: The results in Experiments 2 show large delays for authentication, hence large and unacceptable delays for the overall end-to-end configuration times. Because of the unacceptable delays, the authentication upon configuration request is not performed on the current active gateways. As mentioned above, the large overheads are due to the current available JDK 1.1.5 Security Package and we believe that the next generation of JDK security implementation will improve on these overheads and authentication will become part of the active gateway framework to ensure a secure active environment. In between, there are several alternative solutions: (a) establish two connections between the client and media server through primary and substitute active gateways and perform authentication only during the initial connection setup where the large end-to-end configuration delay does not interfere with the multimedia transmission; (b) establish one connection between the client and the media server through the primary active gateway and perform authentication only during the initial connection setup. In case the primary gateway goes down, we inform the user, tear down the connection and start establishment of a new connection with authentication through the substitute gateway; (c) a centralized security management authenticates ahead users' accesses to any gateway within a considered network and gives the users permission to establish connections and configure QoS services any time. The suggested solutions have different trade-offs among number of connections to keep up, reconfiguration times, management complexity and scalability.

6. Conclusions

The major contribution of this work is the provision of a flexible active gateway architecture using the active network concepts. Our active gateway architecture allows flexible configuration and reconfiguration of QoS services for multimedia communication in a timely fashion. Our objective, to bound the timing duration of the configuration and reconfiguration delays so that this active capability can be performed during the multimedia transmission with minimal QoS degradation was achieved. Our results show that the active gateway framework without security can be timely bound below 2 seconds in LAN environment, and acceptable end-to-end configuration delays as well as dynamic service delegation and linking times can be achieved to support minimal or no QoS degradation.

References

1. Tennenhouse D., et al., "From Internet to ActiveNet," MIT, 1996.
2. Tennenhouse DL and W. DJ, "Towards an Active Network Architecture," *Multimedia Computing and Networking (MMCN 96)*, 1996.

3. Wetherall, D., J. Guttag, and D.L. Tennenhouse, "ANTS: A Toolkit for Building and Dynamically Deploying Network Protocols," *IEEE OPENARCH'98*, April 1998.
4. Van, V., "A Defense Against Address Spoofing Using Active Networks," MIT, 1997.
5. Legedza, U., D. J. Wetherall, and J. Guttag, "Improving The Performance of Distributed Applications Using Active Networks," *IEEE INFOCOM'* 98, 1998.
6. Li-wei H. Lehman, Stephen J. Garland, and D.L. Tennenhouse, "Active Reliable Multicast," *IEEE INFOCOM'98*, 1998.
7. Smith, J., et al., "SwitchWare: Towards a 21st Century Network Infrastructure," Department of computer and information science, University of Pennsylvania.
8. Smith, J., et al., " SwitchWare: Accelerating Network Evolution (white paper)," Department of computer and information science, University of Pennsylvania.
9. Michael Hicks, et al., "PLAN: A Programming Language for Active Networks," *submitted to ICFP'98*. 1998.
10. D. Scott Alexander, et al., "Active Bridging," *Proceedings of the ACM SIGCOMM'97*. September 1997. Cannes, France.
11. Bhattacharjee, S., K.L. Calvert, and E.W. Zegura, "Implementation of an Active Networking Architecture," 1996, College of Computing, Georgia Institute of Technology: Atlanta, GA.
12. Bhattacharjee, B., K.L. Calvert, and E.W.Zegura, "On Active Networking and Congestion," 1996, College of Computing, Georgia Institute of Technology: Atlanta, GA.
13. Yemini, Y. and S. Silva, "Towards Programmable Networks (white paper)," *to appear in. IFIP/IEEE International Workshop on Distributed Systems: Operations and Management*, October 1996, L'Aquila, Italy.
14. Beverly Schwartz, et al., "Smart Packets for Active Networks," *in OpenArch*, March 1999.
15. D. Scott Alexander, et al, "Active Network Encapsulation Protocol (ANEP)," July 1997.
16. Dawson R. Engler, M. Frans Kaashoek, and J.O.T. Jr., "Exokernel: An operating system architecture for application-level resource management," *in Proceedings of the Fifteenth Symposium on Operating Systems Principles*, December 1995.
17. Brian Bershad, et al., "Extensibility, Safety and Performance in the SPIN Operating System," *in Proceedings of the 15th ACM Symposium on Operating System Principles (SOSP-15)*. Copper Mountain, CO.
18. Allen B. Montz, et al., "Scout : A Communication-Oriented Operating System," June 17, 1994, Department of Computer Science, University of Arizona.
19. Hartman, J., et al., "Liquid Software: A New Paradigm for Networked Systems," June 1996, Department of Computer Science, University of Arizona.
20. Robert Macgregor, et al., *Java Network Security*, Prentice Hall, 1998.

A Study of the Impact of Network Loss and Burst Size on Video Streaming Quality and Acceptability

David Hands and Miles Wilkins

BT Laboratories, Martlesham Heath, Ipswich, Suffolk, UK.

david.hands@bt-sys.bt.co.uk, miles@msn.bt.co.uk

Abstract. As multimedia IP services become more popular, commercial service providers are under pressure to maximise their utilisation of network resources whilst still providing a satisfactory service to the end users. The present work examines user opinions of quality and acceptability for video streaming material under different network performance conditions. Subjects provided quality and acceptability opinions for typical multimedia material presented under different network conditions. The subjective test showed that increasing loss levels results in poorer user opinions of both quality and acceptability. Critically, for a given percentage packet loss, the test found that increasing the packet burst size produced improved opinions of quality and acceptability. These results suggest that user-perceived Quality of Service can be improved under network congestion conditions by configuring the packet loss characteristics. Larger bursts of packet loss occurring infrequently may be preferable to more frequent smaller-sized bursts.

1 Introduction

Quality of Service (QoS) is defined by the International Telecommunications Union (ITU) in Recommendation E.800 as the "collective effect of service performances that determine the degree of satisfaction by a user of the service" [1]. Within this definition, QoS is closely related to the users' perception and expectations. These aspects of QoS are mostly restricted to the identification of parameters that can be directly observed and measured at the point at which the service is accessed – in other words the users' perception of the service.

E.800 also defines "Network Performance" (NP) as the parameters that are meaningful to the network provider. These NP parameters are expressed in terms that can be related directly to users' QoS expectations. This is subtly different to the network engineering meaning of QoS, which has traditionally been concerned with technical parameters such as error rates and end-to-end delay. The service provider can measure the provided network QoS (i.e. the operator's perception of QoS), map this to E.800 NP parameters and then compare this with the measured customers' perception of QoS.

The user's perception of QoS will depend on a number of factors, including price, the end terminal configuration, video resolution, audio quality, processing power, media synchronisation, delay, etc. The influence of these factors on users' perception of QoS will be affected by the hardware, coding schemes, protocols and networks used. Thus, the network performance is one factor that affects the users' overall perception of QoS. For multimedia applications, previous work has identified the most important network parameters as throughput, transit delay, delay variation and error rate [2]. The required network service quality does not necessarily coincide with the conditions at which the effects of delay, error rate and low throughput are perceptible to the user. Previous results [3], [4], [5] have shown that users can accept both loss and delay in the network, up to a certain point. However, beyond this critical network performance threshold users will not use the service. It has also been shown that a significant component of the total end-to-end delay experienced by the user is caused by traffic shaping in the network [6].

IP networks, such as the Internet and corporate intranets, currently operate a 'best-effort' service. This service is typified by first-come, first-serve scheduling of data packets at each hop in the network. As demands for the network resources increase, the network can become congested, leading to increased delays and eventually packet loss. The best effort service has generally worked satisfactorily to date for electronic mail, World Wide Web (WWW) access and file transfer type applications. Such best effort networks are capable of supporting streaming video. Video streaming requires packets to be transported by the network from the server to the client. As long as the network is not too heavily loaded the stream will obtain the network resources (bandwidth, etc) that it requires. However, once a certain network load is reached the stream will experience delays to packets and packet loss. Best effort networks assume that users will be co-operative – they share the available network resource fairly. This works if the applications can reduce their network traffic in the face of congestion. Current video streaming applications used in intranets do not do this – they will continue to send traffic into a loaded network.

Data loss, whether caused by congestion in network nodes (IP routers or ATM switches), or interference (e.g. electrical disturbance), is an unfortunate characteristic of most data networks. Network loss is generally not a problem when downloading data to be stored at the client, as the user can wait for all the data to be sent and any missing data can be resent. However, in real-time video streaming, often requiring the transfer of large amounts of data packets across a network, any loss of data directly affects the reproduction quality (i.e. seamlessness) of the video stream at the receiver. If traffic is buffered at the receiver it may be possible for the re-transmission of missing data to be requested. However, this is only feasible if the buffer play-out time is sufficiently long for this re-transmission to occur.

The impact of network data loss on the quality of real-time streamed video services is dependent on three factors:

- Amount of loss. As the percentage of data packets lost increases, the quality of service tends to decrease.
- Burst-size. Data loss can occur as a number of consecutive packets (for a particular stream) being dropped.
- Delay and delay variation. Ideally packets will be received with the same inter-packet time intervals as when they were sent. However, the end-to-end network delay experienced by each packet will vary according to network conditions. Thus some packets may not arrive when expected. Receivers are required to buffer received traffic in order to smooth out these variations in arrival time.

In most video streaming applications, the buffer is sufficiently large to accommodate the majority of variations in network delays. Therefore, most losses likely to affect the users' perception of video streaming performance are due to the amount of loss and the burst size. Research using both human subjects and computer simulations has shown that the amount of loss can have a dramatic effect on quality perception and the readiness of users to accept a service [3], [4], [5], [7]. However, in this research the burst size was not controlled (i.e. burst size could vary within each level of loss). The relationship between loss and burst size and the effect on users' opinions of service quality and acceptability is not known.

In addition to the network loss characteristics, the perceived quality of service can be influenced by the manner in which a particular application streams the audio/video media. For example, audio and video information may be sent together in an IP packet or as two separate streams. The effect of network QoS on different streaming methods and the resulting affect on user perceptions of service quality will influence the design of the next generation of real-time streaming applications.

This paper examines the impact of different levels of network performance on users' perception of video streaming service quality and acceptability. Data obtained from five different industrial and academic intranets were analysed to determine the performance of each network. Using the results of the network performance analysis, a subjective test was run in which subjects provided quality and acceptability judgements for test clips presented under a number of network conditions. The test focuses on the effects of network loss and burst size on judgements of quality and acceptability. The implications of the test results are discussed with particular reference to future network design to support acceptable user perceptions of video streaming services.

2   Examination of Network Performance

Before subjective testing could begin a range of realistic values for packet loss and burst lengths was required. The applications under study are designed to be used in corporate intranets rather than the public Internet. The requirement was to determine the loss characteristics that these applications might encounter in an intranet. A number of tools are available for measuring network performance. However, these generally do not generate 'realistic' traffic. It is unlikely that arbitrary streams of packets will experience the same loss/delay as real applications. For this reason a test tool that generates traffic with similar characteristics to the applications under study was created.

Clients and servers were connected to a stand-alone Ethernet LAN. The traffic generated by a test video sequence was captured by a third machine running the Unix tool *tcpdump*. The network traffic characteristics (packet size, packet inter-arrival time and bandwidth) were determined by analysis of the collected traffic. The captured network traffic was analysed to determine each applications approximate traffic characteristics, as shown in Table 1.

The MGEN/DREC toolset [8] was used to generate and collect traffic streams. A script was written for each application which caused the MGEN tool to generate traffic with characteristics (packet size, packet inter-arrival time and bandwidth) similar to the actual application. The traffic generated by the MGEN scripts was analysed to verify that it was indeed similar to that produced by the real applications. The MGEN scripts were used to perform testing at 5 partners' intranets. At each site a number of 5 minute tests were run consecutively, totalling several hours continuous testing. The DREC tool was used to capture the received traffic. The MGEN tool generates a timestamp and sequence number in each packet. Analysis of the traffic logs revealed the delay, delay variation and packet loss caused by the intranet under test. The logs from all the intranets under study were analysed to determine the range of packet loss characteristics that had been experienced.

Application A (450kbit/s)	Audio & Video	5763 byte packet every 100ms
Application B (1.2Mbit/s)	Audio	2 consecutive 1000byte packets every 100ms
	Video	12 consecutive 1100byte packets followed by 100ms gap

Table 1. Network traffic characteristics for both applications.

3   Subjective Evaluation of Network Performance

The purpose of the subjective quality experiment was to investigate the effects of loss and burst size on human opinions of video quality and its acceptability. A total of 24 subjects took part in the experiment. All subjects had experienced video over the corporate intranet. This experience was considered important so that the subjects' frame of reference would be computer mediated video rather than broadcast television.

3.1   Test Equipment

A small test network (see Figure 1) was built to perform the experiments. The key component is a PC with two Ethernet cards and software that bridges traffic between the two network cards. This software (called Napoleon) was developed at BT Labs and allows the behaviour of the bridging function to be manipulated. The bridge can be configured to introduce controlled amounts of packet loss and delay. The number of consecutive packets lost can also be defined. One Ethernet port was directly connected to a client PC configured with the application viewer programs. The other Ethernet port was connected to an Ethernet hub. A further two PCs (300MHz PII) were attached to the hub. These acted as the content manager and server; video sequences were stored on the server and the content manager controlled the organisation of files held on the server. Thus, by configuring the Napoleon bridge, the loss characteristics of the network between video server and client could be manipulated. Using the results from the intranet characterisation described above the network performance of a 'typical' corporate network could be emulated.

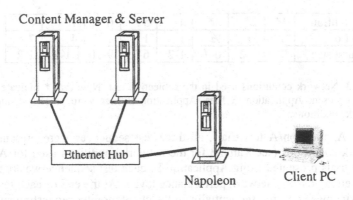

Fig. 1. Test environment used in the subjective test.

The server was installed with two video streaming applications, termed Application A and Application B. Video files could be accessed and viewed from the client using appropriate viewers for both applications. Audio was presented using Altec Lansing ACS41 speakers.

3.2 Method

Seven different audio-video sequences were recorded from off-air broadcasts as MPEG-1 files at 1.3 Mbit/s. Four clips were used as practice sequences and three clips as test sequences. The practice sequences were presented using the range of network characteristics present in the test sequences. The selection of test material was based on typical business-oriented video streaming content. Each of the recorded sequences was edited down to 10s duration and stored on the server. The details of each sequence were entered into the content manager as both Application A and Application B files, thereby enabling the client to access the files from the server. The content of each test sequence is provided in Table 2. The video viewer window on the client was set to 160 x 120 pixels for both applications.

Sequence	Video Content	Audio Content
News1	Studio, female newsreader, head and shoulders, plain backdrop	Presenter reports a recent crime story.
News2	Outside broadcast, man enters building followed by several photographers	Commentary reporting some legal proceedings
Finance	Male presenter, coloured backdrop, scrolling stock prices	Presenter reports business story

Table 2. Description of audio-video content for the test sequence used in the subjective test.

Expert examination of the performance of Application A and Application B files prior to the test found that Application B was more tolerant of higher loss levels. Indeed, the performance of Application A appeared unusable above 1% loss levels. Table 3 shows the network conditions used in the present study.

Condition	1*	2*	3*	4*	5	6	7	8
% Loss	0	½	½	1	1	1	4	7
Burst size	0	1-2	6-7	1-2	6-7	9-10	1-2	1-2

Table 3. Network conditions used in the subjective test. Note that * indicates the conditions used to present Application A files. Application B files were presented under all possible network conditions.

For Application A test trials, all three test sequences were presented under four network conditions (see Table 3 for the network conditions used for Application A). In test trials presented using Application B, each test sequence was presented at each of the eight possible network performance levels. At the end of each test presentation, four windows appeared sequentially on the client monitor requesting subjects to make three quality ratings (overall quality, video quality, and audio quality) and an acceptability rating. Table 4 shows the questions used in the test. Quality assessments were made using the 5-grade discrete category scale (Excellent, Good, Fair, Poor, Bad) [9], [10], [11]. Acceptability judgements were made using a binary scale ('Acceptable' and 'Not Acceptable'). Thus, for each trial, subjects made four responses (overall quality rating; video quality rating; audio quality rating; acceptability). A LabView program controlled the presentation of quality and acceptability questions and was used to collect and store the response data. At the end of the experiment, subjects were asked to complete a short questionnaire (see Appendix). The aim of the questionnaire was to obtain feedback on the criteria subjects used to arrive at opinions of quality and acceptability.

Question type	Question text
Overall quality	"In your opinion the OVERALL quality for the last clip was"
Video quality	"In your opinion the VIDEO quality for the last clip was"
Audio quality	"In your opinion the AUDIO quality for the last clip was"
Acceptability	"Do you consider the performance of the last clip to be:"

Table 4. Questions used in subjective test.

4 Results

For each test sequence the subjective test data showed a similar response pattern for each network condition. Therefore, to simplify interpretation of the test results, the results presented here are averaged across all three test sequences for each network condition. This section will examine subjects' quality responses first, followed by a description of the acceptability data.

4.1 Quality Ratings

Application B

Each subject's responses to the three quality questions (overall, video and audio quality) for each test trial were collected and Mean Opinion Scores (MOS) calculated. The results for each quality question are shown in Figure 2. Inspection of Figure 2 shows that both the amount of loss and the loss burst size were important to subjects' quality opinions and this was true for all three quality questions. Increasing network loss generally led to a reduction in quality ratings. Not surprisingly, zero loss resulted in the best quality ratings. Quality ratings tended to become progressively poorer as the amount of loss in the network was increased from 0.5% through to 7%. The poorest quality ratings were found with 7% loss.

Examination of the effect of network performance on the separate quality questions shows that subjective responses to the video quality question were poorer than responses to the audio quality question. Video quality was rated worse than audio quality in every network condition, and the difference was especially pronounced at higher loss levels. Overall quality opinions appear to be derived from some form of averaging between video quality and audio quality.

The effect of burst size is especially interesting (see Figure 3 below). At both the 0.5% and 1% loss levels, quality ratings were worse for the 1-2 burst size condition compared to the 6-7 packet burst size. Larger burst sizes associated with 1% loss (namely burst sizes of 6-7 and 9-10) produced improved quality ratings compared to a burst size of 1-2 (1% loss). A burst size of 6-7 resulted in slightly better quality ratings compared to the largest burst size (9-10). Further, larger burst sizes at the 1% loss level were rated as roughly equivalent to 1-2 burst size with 0.5% loss.

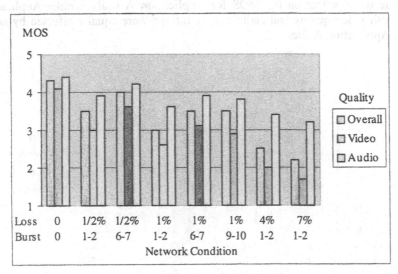

Fig. 2. MOS for each network condition for the Application B test trials.

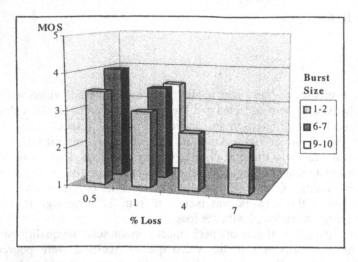

Fig. 3. Overall quality ratings for each network condition.

Application A

MOS derived from the Application A trials are shown in Figure 4. The Application A MOS data follow a very different trend from the Application B data. At zero loss level, the quality ratings are similar to those obtained with zero loss using Application B. However, as soon as any loss is introduced into the network, quality ratings for Application A trials were severely affected. The impact of increasing the level of packet loss on quality ratings for Application A trials was minimal, perhaps due to the low quality baseline provoked by even very small levels of loss. Similarly, burst size had little or no effect on the MOS for Application A trials. Unlike Application B trials, both video quality and audio quality ratings were equally affected by network loss to Application A files.

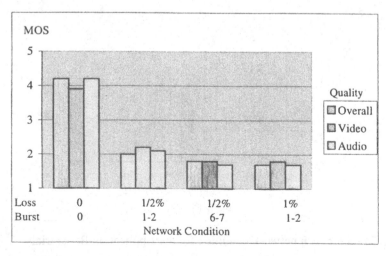

Fig. 4. MOS for each network condition for the Application A test trials.

53

4.2 Acceptability Responses

Acceptability opinions for each test condition were based on a binary choice: acceptable or not acceptable. The acceptability data was collected from subjects' responses for each test trial. Figure 5 displays the percentage of subjects reporting 'acceptable' opinions for each network condition for both Application A and Application B trials. Figure 5 clearly shows that Application A is especially sensitive to loss. Even small levels of loss result in the video streaming performance to be considered unacceptable (e.g. with 0.5% loss, only 16.7% of subjects found the performance acceptable).Test sequences presented using Application B were found to be considerably more robust to network loss. Figure 5 shows that loss levels up to and including 1% were found to be acceptable by over 70% of subjects.

The acceptability data support the results of the quality ratings, with loss and burst size affecting acceptability of the video streams. As the loss level increased, acceptability decreased, although only the 7% loss level resulted in acceptability falling below 50% of subjects. Larger burst sizes were found to produce improved acceptability for the video streaming, with burst sizes of 6-7 and 9-10 being more effective in maintaining acceptability than a burst size of 1-2. However, increasing burst size from 6-7 to 9-10 produces a decrease in acceptability, suggesting that an optimum burst size exists.

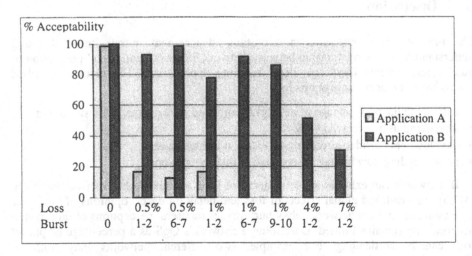

Fig. 5. Percentage of subjects reporting acceptable opinions of video streaming performance for Application A and Application B trials.

4.3 Questionnaire

The responses to the questionnaire (see Appendix) showed that all subjects were aware of both audio and video errors being present in the test sequences. Subjects reported four types of errors: video pauses; blockiness in the video; loss of audio; and audio-video asynchrony. Of these errors, loss of audio information was considered to be the most serious by 14 subjects, 6 subjects identified video errors as the most serious, with 4 subjects reporting audio and video errors to be equally serious. The majority of subjects did not find any type of errors to be disturbing, although 3 subjects noted that rapid pause-play loss behaviour was slightly disturbing (reported by one subject as a 'kind of strobe effect') and 3 subjects stated audio errors as disturbing.

Acceptability opinions were most affected by the loss of meaning, and the majority of subjects reported that provided the audio information remained intact they considered the performance to be acceptable – as long as the video was not too bad. The responses to the acceptability question are rather vague – what does 'as long as the video was not too bad' mean? It may be concluded that audio was of primary importance in the formation of acceptability opinions, but video was an important secondary consideration. A point raised by two subjects was that the material used in this test was audio dominant (i.e. the message could be understood by listening to the audio alone). It would certainly be of some interest to extend the scope of this study to visually dominant material (e.g. sports events). According to the subjects used in this test, the acceptability of video streaming material would be improved by making the computer monitor more like a television (e.g. common responses to this question were to use a larger video window or surround sound).

5   Discussion

The present work represents an important advancement in our knowledge and understanding of the relationship between network behaviour and users' perception of quality and acceptability for video streaming applications. The results described above have practical implications for:

- enhancing the performance of networks employed in real-time video streaming
- improving user perceived QoS
- development of video streaming application technologies
- understanding how human users arrive at quality opinions

IP networks can exhibit various patterns of loss, and although important work [3], [4], [5] has identified global effects of loss and delay on users' opinions of QoS, our understanding of how network behaviour impacts on users' perceptions of quality and acceptability remains limited. Describing a network's QoS as a percentage of packet loss can be misleading. For example, two different networks may both be characterised as exhibiting 5% loss. This paper has shown that the effect on the user perceived QoS will be very different if one network regularly drops a small number of packets and the other network only suffers occasional losses of large numbers of packets. Therefore, in understanding the effects of network behaviour on users' opinions of network performance, it is necessary to obtain network data on the frequency, burst size and temporal characteristics of packet loss. The present work clearly shows that burst size has considerable influence on subjects' opinions of quality and acceptability for video streaming data. Burst sizes of 1-2 were found to be more detrimental to quality and acceptability compared to larger burst sizes. This result suggests that networks exhibiting frequent, small packet losses may provide poorer user perceived QoS compared to a network showing infrequent large packet losses. The implication of this result is that network protocols would be more efficient and provide improved user perceived QoS if they were designed to conform to a mode of loss behaviour in which burst sizes were large but occurred infrequently. Within EURESCOM Project P807 the configuration of network QoS techniques to achieve this aim is under study.

The performance of the two video streaming applications used in the subjective test produced very different user responses for both quality and acceptability. Video streaming using Application A was particularly sensitive to loss whereas the performance of Application B was reasonably robust to all but the highest loss levels. Application A streamed video and audio packets together, and although the application has an in-built priority for protecting audio packets, loss in the network was shown to produce a dramatic drop in quality ratings for both audio and video. As a result, overall quality ratings for Application A trials incorporating some degree of loss typically failed to reach 'fair' on the 5-grade quality scale and few subjects considered the performance to be acceptable. Application B, on the other hand, streamed constantly sized video packets together with occasional audio packets containing a large amount of audio data. The consequence of this method of streaming is that video packets have a much higher risk of being lost compared to audio packets. The subjective test data show that overall quality ratings tended to be arrived at by some form of averaging together of the audio and video qualities. Thus, providing high quality audio had the effect of improving both quality and acceptability ratings. Protecting the audio packets appears to be especially important, given that the questionnaire responses identified audio quality as particularly important for understanding the message contained in the test material, and that audio was reported by subjects as being more critical in forming both quality and acceptability judgements. The streaming technique employed by Application B is clearly advantageous in maintaining user-perceived QoS.

6   Conclusions

This paper examined the effects of network loss and packet burst size on subjective opinions of quality and acceptability. A subjective test, using network parameters derived from real network data, found that both quality and acceptability judgements were affected by the amount of loss exhibited by a network and by packet burst size. Quality and acceptability judgements were reduced (i.e. became worse) as the amount of network loss was increased. Test results comparing different burst sizes for a constant level of network loss showed that small 1-2 burst sizes resulted in poorer user opinions compared to larger burst sizes (packet bursts of 6-7 and 9-10). Increasing the burst size beyond 6-7 did not lead to any further improvement in quality or acceptability judgements. This latter result suggests that an optimum burst size may exist, and that for the present test data the optimum burst size was 6-7. The findings of this test indicate that designing a network to lose larger numbers of packets simultaneously, provided packet loss occurs relatively infrequently, can enhance users' perceptions of QoS.

There is still a great deal of research required before the impact of network behaviour on users' opinions of QoS can be fully characterised. For example, the present work employed audio dominant material. It would be useful to understand the effects of loss and burst size on video dominant material. This is particularly important for streaming technologies such as Application B, when video packets have a far greater probability of being lost compared to audio packets. The present work has identified burst size as an important determinant of users' perceptions of quality and acceptability. However, only three levels of burst size were used (1-2, 6-7 and 9-10). To fully understand burst size effects, and define the optimum burst size for different types of material, a more extensive experiment is required using a broader range of material and burst sizes. Finally, the present work employed short, 10 second test sequences. The use of longer test sequences (e.g. 5 minutes) would enable the examination of network loss and burst size under more realistic test conditions.

Acknowledgements

This work was undertaken as part of EURESCOM [12] project P807 ('JUPITER2') The authors thank the project partners for their assistance in providing network performance data and in helpful discussions on many aspects of the work. We are also grateful for the technical support provided by Uma Kulkarni, Margarida Correia and David Skingsley of the Futures Testbed team at BT Labs.

References

1. ITU-T Rec. E.800: Terms and definition related to quality of service and network performance including dependability. Geneva (1994)
2. Fluckiger, F.: Understanding networked multimedia. London: Prentice Hall (1995)
3. Watson, A., Sasse, M.A.: Evaluating audio and video quality in low-cost multimedia conferencing systems. Interacting with Computers, 3 (1996) 255-275
4. Apteker, R,T., Fisher, J.A., Kisimov, V.S., Neishlos, H.: Video acceptability and frame rate. IEEE Multimedia, Fall, (1995) 32-40
5. Hughes, C.J, Ghanbari, M., Pearson, D., Seferidis, V., Xiong, X.: Modelling and subjective assessment of cell discard in ATM video. IEEE Transactions on Image Processing, 2 (1993) 212-222
6. Ingvaldsen, T., Klovning, E., Wilkins, M.: A study of delay factors in CSCW applications and their importance. Proc. of 5th workshop Interactive Distributed Multimedia Systems and Telecommunications Services, LNCS 1483, Springer (1998)
7. Gringeri, S., Khasnabish, B., Lewis, A., Shuaib, K., Egorov, R., Basch, B.: Transmission of MPEG-2 video streams over ATM. IEEE Multimedia, January-March, (1998) 58-71
8. http://manimac.itd.nrl.navy.mil/MGEN/
9. ITU-R Rec. BT 500-7: Methodology for the subjective assessment of the quality of television pictures. Geneva (1997)
10. ITU-T Rec. P.800: Methods for subjective determination of transmission quality. Geneva (1996)
11. ITU-T Rec.P.920: Interactive test methods for audiovisual communications. Geneva (1996)

12.European Institute for Research and Strategic Studies in Telecommunications:
http://www.eurescom.de/

Appendix

Questionnaire

1) Did you notice any errors in the clips you have seen? If yes, please indicate the type of errors present.

2) Of the errors you reported in answering question 1, which do you consider to be most serious and why (e.g. audio errors, video errors)?

3) Did any errors cause you discomfort?

4) What factors were most important in determining your opinion of acceptability?

5) What additional properties, if incorporated into the current application, would improve your opinion of acceptability?

Transport of MPEG–2 Video in a Routed IP Network *

Transport Stream Errors and Their Effects on Video Quality

Liang Norton Cai[1], Daniel Chiu[2], Mark McCutcheon[1], Mabo Robert Ito[1], and Gerald W. Neufeld[2]

[1] University of British Columbia, Electrical and Computer Engineering Department
[2] University of British Columbia, Computer Science Department mjmccut@cs.ubc.ca

Abstract. The effects of packet network error and loss on MPEG–2 video streams are very different from those in either wireless bitstreams or cell–based networks (ATM). We report a study transporting high–quality MPEG–2 video over an IP network. Our principal objective is to investigate the effects of network impairments at the IP layer, in particular packet loss in congested routers, on MPEG–2 video reconstruction. We have used an IP testbed network rather than simulation studies in order to permit evaluation of actual transmitted video quality by several means. We have developed a data–logging video client to permit off–line error analysis. Analysis requires video stream resynchronization and objective quality measurement. We have studied MPEG–2 video stream transport errors, error–affected video quality and error sources. We have also investigated the relationship of packet loss to slice loss, picture loss, and frame error. We conclude that slice loss is the dominant factor in video quality degradation.

1 Introduction

Multimedia services involving the delivery of digital video and audio streams over networks represent the future of conventional home entertainment, encompassing cable and satellite television programming as well as video–on–demand, video games, and other interactive services. This evolution is enabled by the rapid deployment by telcos (ADSL) and cable providers (cable modems) of high–bandwidth connections to the home. The ubiquity of the Internet and the continuous increase in computing power of the desktop computer together with the availability of relatively inexpensive MPEG–2 decoder plug–in cards have made MPEG–2 based, high quality video communications an interesting possibility.

Many recent studies have investigated the transport of MPEG–2 video over ATM. However, with the explosive growth of the Internet, intranets, and other IP–related technologies, it is our belief that a large majority of video applications

* This work was made possible by grants from Hewlett–Packard Canada and the Canadian Institute for Telecommunications Research

in the near future will be implemented on personal computers using traditional best–effort IP–based networking technologies. "Best–effort" implies that the link bandwidth, packet loss ratio, end–to–end packet delay and jitter can vary wildly depending on the network traffic conditions. While not critical to most data–related applications, these present major hurdles to real–time digital video and audio (**DAV**) delivery. On the other hand, realtime DAV are non–stationary stochastic sources which are often considered to require guaranteed Quality of Service (**QoS**).

In order to gain a better understanding of the characteristics of MPEG–2 video traffic in a best–effort IP network, and how reconstructed MPEG–2 video quality is affected, an experimental Video–on–Demand (**VoD**) system has been employed. The system is comprised of a video file server, a PC–based router, and several client machines which are connected via switched Fast Ethernet (Figure 1). Different load conditions for the IP–based network are generated, using highly bursty background traffic sources. Video quality at the client is measured objectively using Peak–Signal–to–Noise Ratio (**PSNR**) calculations and also subjectively using the Mean Opinion Score (**MOS**) (see Section 4.5). We examine the correlation between IP packet loss and video quality.

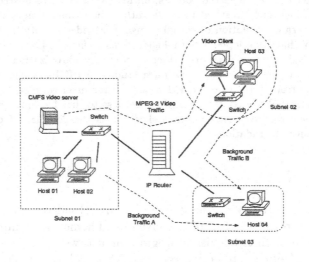

Fig. 1. VOD Test Network

The remainder of this paper is organized as follows. In Section 2 we place our work relative to other research in networked MPEG transport. Section 3 provides a brief description of the effect of transport errors on MPEG–encoded digital video. Section 4 describes our experimental VOD system and measurements. Section 5 presents the results of the transport of MPEG–2 video over IP experiments. Finally, our conclusions are presented in Section 6.

2 Previous Work

Despite published work on the effects of channel errors on the MPEG bitstream, and how these errors affect perceived video quality at the client, there have been relatively few studies of *network* effects on MPEG quality. Current fiber optic or copper media packet networks differ greatly in their error and loss behaviour from "classical" distribution systems such as wireless satellite downlink. Most importantly, packet networks operate at very high signal-to-noise ratio and experience exceedingly low bit error rates. Conversely, the greatest cause of cell or packet loss in data networks is congestion at network nodes, not a factor in conventional video signal distribution.

Some of the work relating network Quality of Service to video quality perceived by an end-system viewer has assumed that individual *video frame* loss is the dominant impairment. Ghinea and Thomas [1] emulate this effect by evaluating viewer perception and understanding of multimedia (MPEG-1 video and audio) at several different frame rates (5, 15 and 25 frames per second). However, as we show in Section 5.3, slice loss is actually more important than frame loss in degrading video quality and viewability. Measures designed to protect against frame loss without improving on network conditions leading to slice loss are unlikely to be useful in practice.

Most of the work regarding network effects on MPEG video or voice and video streams has focussed on ATM networks. Zamora et al. [2] [3] [4] have used the Columbia University VoD testbed and both ATM LAN and WAN setups to study the effects of traffic-induced impairment in ATM switches on MPEG-2 subjective video quality. This work uses MPEG-2 Transport Stream (**TS**) packets encapsulated in AAL5 PDUs carried in ATM cells [5] [6], transported over Permanent Virtual Circuits (**PVC**) defined within a Permanent Virtual Path (**PVP**) having guaranteed service parameters. The effects of high levels of ATM network impairment are emulated using a Hewlett-Packard Network Impairment Emulator module, which generates bit errors, cell loss and PDU loss according to several scenarios. They analyze the effects of varying PDU size in the presence of different errors. Two types of video client are used, a set-top box with fixed PDU size of 2 MPEG-2 TS packets and a Windows NT client with variable-length PDU capability (up to 20 TS packets per PDU). The studies employ 2 types of program material, VCR-quality (2.9Mbps) and HDTV-quality (5.4Mbps) in both CBR and VBR modes. Interfering cross-traffic comes from a 5Mbps bidirectional videoconference data stream, another CBR video server, or a bursty IP-ATM source which is either on (attempting to utilize full path rate) or off.

This is a complicated system and gives rise to complex results. In oversimplified summary, Zamora et al. find that:

- PHY bit error rates of less than 10^{-5} are required for "good" MPEG picture quality
- PDU loss rates of 10^{-5} generally give good MPEG picture quality; PDU loss rates of 10^{-3} produce "slightly annoying" degradation (i.e. ITU MOS rating

3), while at PDU loss rates around 10^{-4} the video degradation depends sensitively on the buffering characteristics of the client.

– Greater video degradation is observed for large (14 TS packets) PDUs than small (4 TS packets) PDUs in the presence of moderate PDU loss (10^{-3}); however, the choice of PDU size and the nature of the outcome is heavily influenced by client buffering capabilities and encapsulation choice, so the result not easily extrapolated to different situations.

This is an interesting body of work, though its conclusions are not generally relevant to MPEG over IP networks. Many results are conditioned by the encapsulation used - MPEG–2 TS in ATM cells, in several block sizes. The MPEG–2 Transport Stream is not content–aware; TS packets are fixed size and not assigned according to slice boundaries or those of other picture elements. Thus PDU loss will degrade more slices than is the case for the MPEG–RTP–MT–UDP–IP encapsulation we have used, and as we show, slice loss is the dominant factor influencing perceived video degradation. The test system was used to generate various levels of single- and multi–bit errors and delay–variance profiles, but as discussed above, we believe that these factors are not of much relevance to VoD over IP networks. The effect of PDU size during degraded conditions in this system depends mostly on the mode of encapsulation and nature of particular client network buffers, so does not easily generalize to other systems. Finally, use of a PVP with QoS guarantee contrasts strongly with the situation in today's IP networks, where no service guarantees of any kind are offered.

The work most similar to ours of which we are aware is that of Boyce and Gaglianello [7]. These authors encapsulate MPEG–1 streams into RTP packets over UDP/IP transport/network layers, using the IETF RFC2250 [11] format as we do, always aligned on slice boundaries, though there are differences in slice size settings. This results in some of their packets, exceeding the Ethernet MTU and being fragmented, while our packets are all limited to less than the MTU size. Two stream rates were used, 384 kbps and 1 Mbps, and streams were transmitted from each of 3 sites on the public Internet to the Bell Labs in Holmdel, N.J. Packet loss statistics were accumulated as functions of date and time, packet size, and video frame error rate.

Boyce and Gaglianello draw several conclusions. They find that average packet loss rates across the public Internet were in the range of 3.0% to 13.5%, with extremes ranging from 0% to 100%. Due to spatial and temporal dependencies in the MPEG stream, 3% packet loss translates to as much as 30% frame error rate. However, these authors made no subjective evaluation of video quality, and used only a very simplified notion of objective quality; any error in a given frame, whether the loss of the frame or some slight block error arising as propagated error from an earlier lost frame, scored as an errored frame. This is clearly quite different from a PSNR calculation. Their conditional packet loss probability curves show the most common lost burst length to be one packet, but with significant probability of longer bursts. They also showed a significantly higher packet loss probability when the packet size exceeds the network MTU.

In areas of overlap, we are in general agreement with the work of Boyce and Gaglianello. However, because they used relatively low bit–rate video across the public Internet, while we used higher–rate video (6 Mbps) forwarded by an IP router with controllable levels of generated cross traffic, we believe that our results are better able to illuminate details of video quality effects due to packet loss. In the case of packet loss vs. frame error rate, we obtain a much stronger relationship than do these authors despite using the same calculation method, finding 1% packet loss rates correspond to 30%-40% frame error levels.

3   MPEG–2 Video Coding and Transport Errors

The ISO/IEC MPEG–2 standard outlines compression technologies and bit–stream syntax for audio and video. ISO/IEC 13818–2 [8] defines a generic video coding method that is capable of supporting a wide range of applications, bit rates, video resolutions, picture qualities, and services. MPEG–2 defines three main picture types: **I** (intra), **P** (predictive), and **B** (bidirectionally–predicted) pictures. I–pictures are coded independently, entirely without reference to other pictures. They provide the access points to the coded bit stream where the decoding can begin. P–pictures are coded with respect to a previous I- or P–picture. B–pictures use both previous and future I- or P–pictures. The video sequences are segmented into groups of pictures (**GOP**).

Within a transported MPEG–2 bitstream, packet loss induced errors may propagate both intra–frame and inter–frame. An error can corrupt resynchronizing boundaries and propagate through multiple layers, resulting in a very visible erroneous block in the corrupted frame. Such errors may persist through multiple frames if the corrupted frame happens to be an I or P frame. Depending on where the error occurs in the MPEG–2 bitstream, the resulting video quality degradation may be quite variable. However, when packet loss is severe, the probability of picture header loss is high. A picture header loss results in an undecodeable picture. If the lost picture is a reference picture, longer–lasting video quality degradation will occur due to loss propagation.

4   Experimental System and Measurements

4.1   Video Client–Server

The University of British Columbia's Continuous–Media File Server (**CMFS**) [9] is used as the video server. The CMFS supports both CBR and VBR streams as well as synchronization for multiple concurrent media streams (i.e., lip–synchronization of audio and video or scalable video). Our server machine is a 200 Mhz Pentium Pro PC running FreeBSD 3.0, with a 2GB SCSI–II Fast/Ultra hard drive. We used either a software or a hardware MPEG–2 encoder to compress the video sequences as MPEG–2 video elementary streams (**ES**) and stored them on the video server.

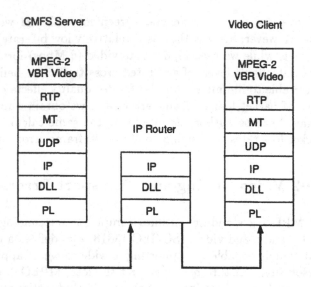

Fig. 2. Protocol Architecture of the VOD System

The protocol architecture of the experimental VOD system is given in Fig. 2. The Real–Time Transport Protocol (**RTP**), specified in RFC1889 [10], is used to encapsulate the video data. We chose to use RTP because it is based on Application Level Framing (**ALF**) principles which dictate using the properties of the payload in designing a data transmission system. Because the payload is MPEG video, we design the packetization scheme based on 'slices' because they are the smallest independently decodable data units for MPEG video. RFC–2250 [11] describes the RTP payload format for MPEG–1/MPEG–2 video.

The CMFS uses its own Media Transport (**MT**) protocol, a simple unreliable stream protocol utilizing UDP for transmission of data as a series of byte–sequenced datagrams. This allows the receiving end to detect missing data. While RTP uses packet sequence numbering to detect packet loss, MT detects the quantity of data rather than just the number of packets lost. Lost packets are not retransmitted since this is usually not possible in realtime systems. The maximum UDP datagram size is set so that it can be sent in one Ethernet frame (i.e. size < 1500 bytes) to avoid fragmentation.

On the client side, we have developed a data–logging video client for off–line error analysis, which runs on a 200 Mhz Pentium MMX PC. Upon arrival of each RTP packet, the client takes a timestamp and stores it along with the RTP packet to a logfile on disk. No realtime decoding of the MPEG–2 video stream is done. An MPEG–2 software decoder is used for non–realtime decoding of each frame in order to perform objective video quality analysis. For subjective quality analysis, we use a REALmagic Hollywood2 MPEG–2 hardware decoder to view the transmitted video streams.

4.2 Network

Our network backbone consists of a 100Base-T Ethernet switch and a PC–based IP router (200 Mhz Pentium Pro PC running FreeBSD 2.2.5). IP forwarding in the unmodified FreeBSD kernel is strictly "best effort", utilizing FIFO output queues on each network interface. When aggregate traffic in excess of an output link's capacity results in queue saturation, new packets are simply discarded (tail–drop). Cho's *Alternate Queuing* [12] modifications provide the capability to implement more sophisticated queuing disciplines.

The switch supports VLANs, allowing its use with multiple independent IP subnets for routing tests. We use three VLANs connected by the IP router. The video server and a host, which is used as a background traffic generator source, are placed on one subnet. The video client and another computer, which is used as the background traffic sink, are placed on the second subnet. Another host on a third subnet is used to generate background traffic through the router.

4.3 MPEG–2 Video Streams

For subjective video quality evaluation we have selected studio quality Main–Profile @ Main–Level (MP@ML) CCIR 601 CBR video streams which have 704x480x30Hz sampling dimensions, progressive frames, 4:2:0 chroma format, and a 6Mb/s constant bit rate. These streams have been encoded with a single slice spanning the frame width (30 slices per frame).

To ensure the generality of our study, it is important that test streams cover the characteristics of most real world MPEG–2 encoded video. The clips we selected present large differences in spatial frequency and motion. Cheer, Ballet, and Susi represent High, Medium, and Low activity and spatial detail, respectively. For clarity, we refer to the clips as video–H, video–M and video–L, respectively. MPEG–2 video can be seen as a stream composed of short sub–periods with varying levels of motion and spatial detail. The overall subjective video quality may be characterized by the single worst sub–period. For this reason, we use test streams that are 10 seconds (300 frames) in length. Table 1 provides detailed statistics of the test streams.

Table 1. Video Stream Statistics

	Frames	Slices	Bytes			Packets		
			video–H	video–M	video–L	video–H	video–M	video–L
I	21	630	866782	664211	864736	843	660	880
P	80	2400	2458287	2983630	2854488	2374	2701	2472
B	199	5970	2930873	2598643	2532044	3415	3114	3037
Total	300	9000	6255942	6246484	6251268	6605	6475	6317

4.4 Stream Re-synchronization and Error Concealment

In a highly congested IP network, multiple frames may be lost due to sequentially dropped packets, but this is more likely to occur when the lost packets contain MPEG–2 bitstream headers, resulting in undecodable or unrecoverable frames. In either case, comparison of video frames between the original and the transmitted streams becomes difficult because temporal alignment is lost. Temporal alignment between the two streams is also required for PSNR measure, so re–synchronization is necessary. Another important factor in providing accurate measurements is how lost frames are handled. This may involve leaving a blank or patching with error concealment techniques. A blank could be a black, grey or white frame; various error concealment techniques are available.

We have chosen to substitute a black frame for a lost frame. In our experience, error concealment techniques affect measurement results so significantly that the real characteristics of measured video streams are masked. Our analysis detects frame loss in the transmitted video stream and substitutes a black frame for the lost frame, thus resynchronizing the original and the transmitted video sequences.

4.5 Measurements

We measure MPEG GOP header loss, Picture header loss, and slice loss to determine the probability of loss under different network loads. With better insight into loss probability, video quality degradation can be minimized, for example, by duplicating critical data structures in the MPEG coded data stream (i.e. Picture headers). Such measurements will also help determine the optimal packet size to use. A balance exists where packets are not so large that a single lost packet results in extensive MPEG data loss and not so small that too much overhead is incurred.

At the application level, the client log files created by the data–logging video client are parsed to calculate the client video buffer occupancy and the MPEG frame interarrival times. The MPEG–2 Elementary Stream (**ES**) data is extracted from the RTP packets contained in these log files. MPEG–2 ES data contained in any RTP packet that arrived later than the time at which it was supposed to have been taken from the buffer to be decoded is discarded. The MPEG–2 ES data is then decoded using a software decoder in order to perform objective video quality measurements.

We use Peak Signal-to-Noise Ration (**PSNR**) [13] as an objective measure of the degree of difference between the original video stream and the corrupted stream. PSNR correlates poorly with human visual perception, or subjective quality. Nevertheless, it provides accurate frame to frame pixel value comparison. We use the Mean Opinion Score (**MOS**) as a "true" evaluation of the video quality through an informal viewer rating on a 5–point scale (see Table 2) [1].

---

[1] Scale defined per Recommendation ITU-R BT.500.7, "Methodology for the Subjective Assessment of the Quality of Television Pictures", Oct. 1995

We always rate the original video stream as level 5 in MOS–style evaluations, then visually compare the corrupted stream against it. The test streams are encoded and decoded using the same codec. In this way, we eliminate effects of the MPEG–2 codec (such as the introduction of noise to the frames by the lossy compression process) from the tests, since we are only interested in transport loss issues.

Table 2. Subjective Quality Scale (MOS)

Scale	Impairment
5	Imperceptible
4	Perceptible, but not annoying
3	Slightly Annoying
2	Annoying
1	Very Annoying

5 Experimental Results

The PSNR and MOS results from each video stream under test are expressed as a function of IP network packet loss ratio. The router shown in Figure 1 is stressed by moderately to heavily loading with cross–traffic so as to obtain various packet loss rates. The characteristics of the stream and the statistical correlation between PSNR, MOS and packet loss are investigated.

5.1 Quantitative MPEG–2 Video Quality to IP Packet Loss Mapping

We are primarily interested in effects of IP packet loss on reconstructed MPEG–2 video, in other words, MPEG–2 video transport errors. Our review of existing work [14] indicates that video block artifacts, repeated movement and flickering due to the loss of slices or frames are the most visible types of quality degradation.

Figure 3(a–c) shows PSNR vs. packet loss plots for video–H, video–M and video–L. Figure 3(d–f) are the corresponding MOS vs. packet loss plots. Since a human viewer will be the recipient of the video service, the MOS subjective video quality measurement is considered to be the **true** video quality. We define video with an MOS between 5 and 3 as the Low Degradation Zone (LDZ), between 3 and 2 as the Annoying Degradation Zone (ADZ), and between 2 and 0 as the Very-annoying Degradation Zone (VDZ). We use Low Degradation Point (LDP) and Annoying Degradation Point (ADP) to specify two quality boundaries. With video quality above the LDP, the viewer will occasionally observe corrupted slices. With video quality between LDP and ADP, besides seeing a large number

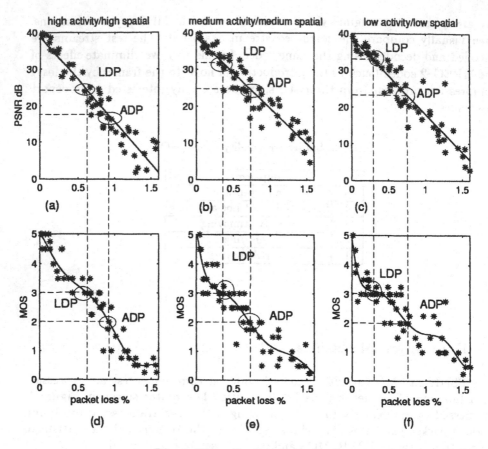

Fig. 3. Video Quality vs. Packet Loss

of corrupted slices, the viewer will also observe occasional flickering due to the loss of one or more frames. With video quality below ADP, the viewer will experience multiple damaged frames accompanied by strong flickering.

Table 3 is a mapping of Quality Boundary to PSNR and Packet loss ratio. For example, the VDZ zone of video-M normally results from greater than 0.7% packet loss.

The Low Degradation Zone (MOS 5 to 3) is of the most interest because it is in the range of acceptable video quality for viewers. By observing the LDZ of Figure 3(d–f), one finds that the slope in Figure 3(d) is not as steep as the one in Figure 3(e) and (f). A steeper slope in the MOS/packet loss plane indicates a higher rate of video quality degradation. We note that in the acceptable video quality range (MOS between 5 and 3), the lower–activity/lower–spatial detail MPEG–2 video stream is more susceptible to IP packet loss impairment. Quantitatively, video–H reaches the LDP at a packet loss rate of 0.60% while video–M reaches LDP at a loss rate of 0.40% and video–L at 0.28%. PSNR values for the LDP are 24.9 dB for video–H, 31.0 dB for video–M and 33.1 dB for video–L.

The foregoing indicates that it is satisfactory to examine only low–activity/low–spatial detail MPEG–2 streams for the worst–case scenario video quality evaluation of video streams transported over a lossy IP network.

5.2 Correlation of PSNR and Packet Loss

Many studies have indicated that PSNR as an objective measurement of video quality does not correlate well with the true subjective video quality. This conclusion is quantitatively verified by our experiments. As shown in Figure 3, an MOS value of 3 corresponds to 24.9 dB PSNR in video–H but 33.1 dB PSNR in video–L, a 26% difference in value.

On the other hand, as depicted by Figure 3(a–c), the PSNR of the three video streams show linear correlation with packet losses. The regression coefficients for rate and constant are obtained by using a simple linear regression estimation. For video–H, video–M and video–L, the coefficients are (-0.040, 1.6), (-0.046, 1.8), and (-0.43, 1.7), respectively. The similarity in the coefficients means that PSNR to packet loss mapping of the three video streams are almost identical. Thus from Figure 3, at a packet loss ratio of 0.3%, the PSNR values are 31.8, 32.9 and 32.7, a spread which is not statistically significant.

5.3 Dominant Contributor to Video Quality Degradation

An MPEG–2 video bit stream is a prioritized stream of coded data units. Some headers, for example, the P–frame Picture header, are more important than the other headers, such as the I–frame Slice header or the B–frame Picture header. Packet loss protection techniques normally add redundancy back into the original bit stream, increasing the demand on bandwidth. Without knowing how highly prioritized MPEG–2 video bit streams are affected by packet loss, blindly engineered error protection techniques are an inefficient means to deal with video quality degradation. We must investigate what elements in the MPEG–2 bit stream should be protected from IP packet loss.

Figure 4(a,b) are plots of PSNR and MOS vs. slice loss, Figure 4(c,d) are plots of PSNR and MOS vs. picture loss, respectively. By comparing the slice loss related results against the picture loss related results, it is clear that the variance of the video quality (PSNR and MOS) due to picture loss is much greater than that due to slice loss. The video quality degradation is closely related to slice

Table 3. Quality Boundary Mapping

Quality Boundary	PSNR (dB)			Packet Loss (%)		
	video–H	video–M	video–L	video–H	video–M	video–L
LDZ (MOS 5–3)	40.0–24.9	40.0–31.0	40.0–33.1	0.0–0.6	0.0–0.4	0.0–0.3
ADZ (MOS 3–2)	24.9–17.5	31.0–24.8	33.1–23.3	0.6–0.9	0.4–0.7	0.3–0.8
VDZ (MOS 2–0)	< 17.5	< 24.8	< 23.2	> 0.9	> 0.7	> 0.8

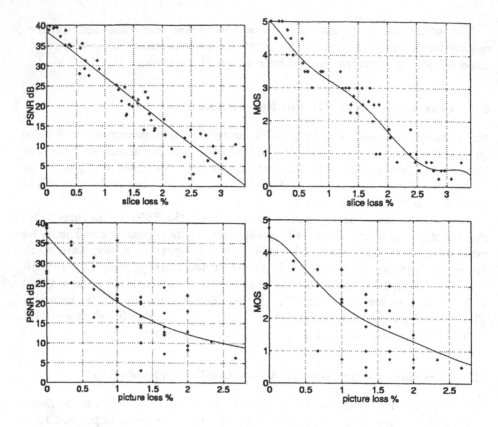

Fig. 4. a) Slice Loss vs. PSNR b) Slice Loss vs. MOS c) Picture Loss vs. PSNR d) Picture Loss vs. MOS

loss ratio but not to picture loss ratio. Alternatively, picture loss is not well correlated with either PSNR or MOS.

In an MPEG–2 video, a lost slice normally presents a corrupted horizontal bar, a block artifact, or a localized jerky motion, while a picture loss can be seen as video hesitating, repeating, or flickering. In our experiments, when network congestion increases, the video quality degrades through the increased occurrence of corrupted bars or block artifacts. By the time viewers perceive flickering or frame repeating, the video quality is already so badly damaged by corrupted slices as to be not worth watching.

Through quantitative measurement and subjective assessment, our study indicates that when MPEG–2 video is transported in IP packets, in the range of fair video quality, the loss of **slices** is the dominant factor contributing to the degradation of the video quality rather than the loss of pictures. Our study also found that in the packet loss range of less than 1.5%, single packet losses account for the majority of the losses. This suggests that it is not effective to protect picture level data and parameters without first reducing slice losses.

5.4 Characteristics of MPEG–2 Data Loss vs. Packet Loss

The MPEG–2 ES data is packetized according to the encapsulation scheme specified in IETF RFC–2250. Because the slice in MPEG–2 is the error re–synchronizing point and the IP packet size can be as large as 64 KB, RFC–2250 selects the slice as the minimum data unit and creates fragmentation rules to ensure that the beginning of the next slice after one with a missing packet can be found without requiring that the receiver scan the packet contents.

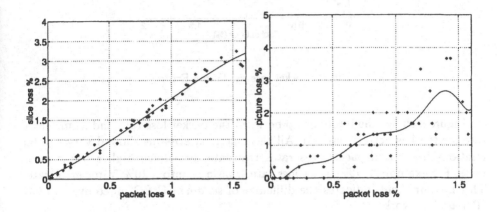

Fig. 5. a) Packet loss vs. Slice loss, b) Packet loss vs. Picture loss

Figure 5(a) is a plot of slice loss vs. packet loss. Due to the RTP packet payload format, slice loss is linearly related to packet loss. Loss of an IP packet means loss of at least one slice. It also means that data loss granularity is much higher than single bit or even a short burst of bits.

Figure 5(b) is a plot of the picture loss vs. packet loss. The variation in picture loss count is quite large. If compared with Figure 3(d), the LDZ, ADZ, and VDZ correspond to the picture loss level of 0–0.2%, 0.2–2% and 2–4%. Picture loss rates higher than 2% produce very annoying flickering. I- and P–frames account for 33.2% of the total frames in the test stream. On average, 2% frame loss will result in about 2 I- or P–frame losses from our 300 frame test program.

The frame error rate, as calculated in [7], is a measure of the number of frames that contain errors. Errors are due to packet loss in the current frame or in a frame that the current frame is predicted from. Packet loss errors spread within a single picture up to the next resynchronization point (i.e. slice). This is referred to as spatial propagation and may damage any type of picture. When loss occurs in a reference picture (I–picture or P–picture), the error will remain until the next I–picture is received. This causes the error to propagate across several non–intra-coded pictures until the end of the Group–of-Pictures, which typically consists of about 12–15 pictures. This is known as temporal propagation and is due to inter–frame prediction.

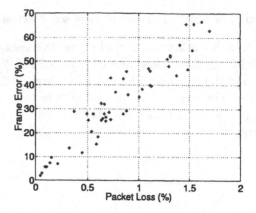

Fig. 6. Packet loss vs. Frame Error

Figure 6 shows the relationship between the packet loss rate and frame error rate for the low activity stream. All three activity streams showed very similar relationships. Small packet loss rates translate into much higher frame error rates. For example, a 1% packet loss rate translates into a 40% frame error rate. This measurement indicates the difficulty of sending MPEG video over a lossy IP-based network.

6 Conclusions

A quantitative mapping between MPEG-2 video quality and IP packet loss has been derived. Using such a mapping, network service providers may be able to predict perceived video quality from IP packet loss ratios.

We have identified block artifacts, repeated movements and flickering as the most visible quality degradations in lossy IP network environments. Thus we believe that low-activity/low-spatial detail MPEG-2 video clips are suitable for worst-case scenario quality evaluation when studying the effects of packet loss.

We have shown that PSNR as an objective measurement of video quality is poorly correlated with the subjective assessment. However, PSNR as a numerical comparison method was found to correlate linearly with packet loss in the high, medium, and low activity test streams.

When MPEG data is transported in IP packets, slice loss rather than picture loss is the dominant factor contributing to video artifacts at a given packet loss rate. By the time a viewer perceives flickering or frame repeating, the video quality is already badly damaged by corrupted slices and unwatchable.

Cells used in ATM networks are small compared to IP packets. Depending upon encapsulation chosen, when an ATM cell is lost, only one or a few macroblocks of a slice may be lost; an IP packet loss will result in one or more slices being lost. Small packet loss rates translate into much higher frame error rates. Therefore, the traditional Forward Error Correction (**FEC**) schemes designed for single bit or short bursts are no longer effective. New schemes are needed to provide data protection and recovery.

References

[1] G. Ghinea and J. P. Thomas, QoS Impact on User Perception and Understanding of Multimedia Video Clips, in Proceedings of ACM Multimedia '98.

[2] J. Zamora, D. Anastassiou and S-F Chang, Objective and Subjective Quality of Service Performance of Video–on–Demand in ATM–WAN, submitted for publication to Signal Processing: Image Communication, July 1997.

[3] J. Zamora, D. Anastassiou, S-F Chang and K. Shibata, Subjective Quality of Service Performance of Video–on–Demand under Extreme ATM Impairment Conditions, Proceedings AVSPN-97, Sept. 1997.

[4] J. Zamora, D. Anastassiou, S-F Chang and L. Ulbricht, Objective and Subjective Quality of Service Performance of Video–on–Demand in ATM–WAN, submitted for publication to Elsevier Science, Jan. 1999.

[5] ATM Forum Service Aspects and Applications, Audio/Visual Multimedia Services: Video on Demand v1.0, af-saa-0049.000, Jan. 1996.

[6] ATM Forum Service Aspects and Applications, Audio/Visual Multimedia Services: Video on Demand v1.1, af-saa-0049.001, Mar. 1997.

[7] J. M. Boyce and R. D. Gaglianello, Packet Loss Effects on MPEG Video Sent Over the Public Internet, in Proceedings of ACM Multimedia '98.

[8] ISO/IEC International Standard 13818; Generic coding of moving pictures and associated audio information, November 1994.

[9] G. Neufeld, D. Makaroff, and N. Hutchinson, Design of a Variable Bit Rate Continuous Media File Server for an ATM Network, IST/SPIE Multimedia Computing and Networking, pp. 370–380, San Jose, January 1996.

[10] H. Schulzrinne, S. Casner, R. Frederick, and V. Jacobson, RTP: A Transport Protocol for Real–Time Applications, RFC 1889, January 1996.

[11] D. Hoffman, G. Fernando, and V. Goyal, RTP payload format for MPEG1/MPEG2 video, RFC 2250, January 1998.

[12] K. Cho, A Framework for Alternate Queueing: Towards Traffic Management by PC–UNIX Based Routers, In Proceedings of USENIX 1998 Annual Technical Conference, New Orleans LA, June 1998.

[13] K. R. Rao and J. J. Hwang, Techniques and Standards For Image, Video and Audio Coding, Prentice Hall PTR, New Jersey, 1996.

[14] L. N. Cai, Taxonomy of Errors Study on Real–Time Transport of MPEG–2 Video Over Internet, TEVIA Project Report, UBC Electrical and Computer Engineering Department, January 1998.

Specification and Realization of the QoS Required by a Distributed Interactive Simulation Application in a New Generation Internet

Christophe Chassot[1], André Loze[1], Fabien Garcia[1]
Laurent Dairaine[1,2] *, Luis Rojas Cardenas[1,2]

[1] LAAS/CNRS, 7 Avenue du Colonel Roche
31077 Toulouse cedex 04, France
{chassot, alozes, fgarcia, diaz @laas.fr}

[2] ENSICA, 1 Place Emile Blouin
31056 Toulouse cedex, France
{dairaine, cardenas@ensica.fr}

Abstract : One of the current research activities in the networking area consists in trying to federate the Internet new services, mechanisms and protocols, particularly the IETF Integrated Service (Int-Serv) Working Group ones, and the ATM ones, in order to allow new multimedia applications to be guaranteed with an end to end Quality of Service (QoS). Study which is presented in this paper has been realized within a national project whose general goal was to specify and then implement the QoS required by a distributed interactive simulation in an heterogeneous WAN environment consisting in an ATM interconnection of three distant local platforms implementing the IntServ propositions. The first part of the results exposed in the paper is related to the QoS a distributed interactive simulation (DIS) application has to require so as to be correctly distributed in the considered network environment. The second part of the paper is dedicated to the end to end communication architecture defined within the project in order to provide a QoS matching DIS applications requirements.

1. Introduction

1.1. Work context

These latest years, technological improvements in both computer science and telecommunications areas led to the development of new distributed multimedia applications involving processing and (differed or in real time) transmission of all media kinds. Communication requirements of such applications make it necessary the use of new networks providing both high speed transmissions and guaranteed qualities of service (QoS) : throughput, transit delay, etc. Among these applications, *Distributed Interactive Simulation* (DIS) applications have specific features (interactivity, hard time constraints, …) that make difficult both their traffic characterization and the QoS specification to require from the network.

In order to distribute these new applications on a high scale, the mostly used and analyzed wide area network (WAN) solutions are the Internet on one side and ATM based networks on the other side. In spite of (or due to) its popularity, the current Internet does not allow an adequate treatment of multimedia applications features and requirements ; these latest years, several works have been initiated so as to solve the problem : particularly, the Integrated Service (*IntServ*) working group [1] of the IETF (Internet Engineering Task Force) has recently proposed some extensions of the current Internet service model [2][3], and also published several papers related to mechanisms and protocols [4][5] allowing to implement the newly defined services. On the other side, several local ATM platforms have been developed in different places, particularly in France in three research laboratories : the LAAS/CNRS (Toulouse), the INRIA (Sophia Antipolis) and the LIP6 (Paris) ; in 1997, the interconnection of these platforms led to the birth of the MIRIHADE platform (today SAFIR) on which several experiments have been realized (performances measurements, multimedia applications testing, etc.). One of the main research activities in the networking area consists in trying to federate Internet and ATM worlds in order to provide applications with an end to end guaranteed QoS, independently of the physical networks used. The study exposed in this paper has been realized within the "DIS/ATM" project[1] involving a French industrial (Dassault Electronique, now Thomson/CSF-Detexis) and three laboratories and Institute (LAAS/CNRS, LIP6 and INRIA). The general goal of the project was to specify and then implement the QoS required by a DIS application in a WAN environment consisting in an ATM interconnection of three distant local platforms implementing the *IntServ* propositions. Within the project, the LAAS-CNRS has been mainly involved in the following three parts :

1. Characterization of the QoS a DIS application has to require from the network so as to be distributed in a WAN environment ;
2. Design of an end to end communication architecture :
   – providing a QoS matching DIS applications requirements ;
   – distributed over an Internet version 6 (IPv6) network environment, able to provide the IETF *Intserv* service models (*Guaranteed Service* and *Controlled Load*);
3. Integration of the targeted DIS application within this architecture in a PC/Linux machine/system.

1.2. Paper content

Results exposed in this paper are related to the two first of the previous points. The following of the paper is structured as follows : the first part of the paper (section 2) consists in an analysis of the end to end QoS a DIS application has to require in a WAN environment, in order to be *coherently*[2] distributed. The second part of the paper (section 3) proposes a classification of the DIS protocol data units (PDUs) that matches application requirements, but also allows the design of an end to end communication architecture based on an architectural principle mainly applied these latest years in the multimedia architecture context.

[1] The DIS/ATM project (convention n° 97/73-291) is co-financed by the French MENRT (Ministère de l'Education Nationale, de la Recherche et de la Technologie) and the DGA (Direction Générale de l'Armement).
[2] This term will be defined in the appropriate section (section 2).

The third part of the paper presents an overview of the defined end to end architecture (section 4). Finally, conclusions and future works are presented in section 5.

1.3. Related work

Several work themes have been recently studied within the DIS context. Only the major ones are introduced in this section.

An important part of the work realized these latest years is related to the "scalability" problem associated with the deployment of DIS applications between several hundred of sites [6] [7]. Particularly, the LSMA Working Group has provided documentation on how the IETF multicast protocols, transport protocols and multicast routing protocols were expected to support DIS applications. Besides the *dead reckoning*, several techniques have been proposed in order to reduce the bandwidth required ; they can be divided into two parts :

- aggregation techniques, that consists in either grouping information related to several entities into a single DIS PDU [8] or reassembling several PDUs into a single UDP message [9] ;
- filtering techniques, that classifies DIS PDUs into several fidelity levels, depending on the coupling existing between sending and receiving entities [10]. Following the same goal in the distributed games area, [11] proposes mechanisms allowing the deployment of the *MiMaze* applications on the Internet.

Another theme is related to the definition of an architecture (HLA/RTI[3]) whose goal are (particularly) [12] to facilitate simulators interoperability and to extend the application area to the civilian one. A last theme concerns DIS traffics measurements ; several studies have been published [13] [14], but none of them allows a forma characterization model.

2. DIS applications and networking QoS

The goal of this section is to define the QoS a DIS application has to require from a WAN so as to be *coherently* distributed through the network. Following this goal, we first show the limits of the QoS specification formulated in the DIS standard in a WAN environment. Then, another QoS expression is proposed allowing to solve the problems associated to the previous limits.

2.1. Coherence notion

Traffic generated by a DIS application may be divided into three parts :
- traffic corresponding to the broadcasting of the simulated entities state (*Entity State* PDU, etc.) and events (*Fire* PDU, *Detonation* PDU, *Collision* PDU, etc.) ;

---

[3] High Level Architecture / Real Time Infrastructure

- traffic corresponding to the communication between participants or emissions to captors : radio communications (*Signal* PDU) and electromagnetic interactions (*EM-emission* PDU, etc.) ;
- traffic corresponding to the distributed simulation management : simulation initialization, participants synchronization, etc. (*Start* PDU, *Resume* PDU, etc.).

With regard to the DIS standard [15][16][17], a DIS application is correctly distributed if any simulation site has both a *temporal* and a *spatial coherent view* of the distributed simulation ; in other words, this means that :

- any event (fire, detonation, etc.) occurred at time t_0 on a given site has to be known by the other sites at least T time units after the date t_0 (temporal coherence) ;
- the state (position, orientation, etc.) of any entity simulated on a given site has to be "almost" identical to the one extrapolated on any other site (by application of the same *dead reckoning* algorithms), the "almost" term meaning that the difference between those two states has to be less than a maximal value defined as the *threshold* value in the DIS standard (spatial coherence).

In order to illustrate this second point, let's consider the example of an entity A representing a plane simulated on a site S_1. At a given time, another site S_2 has a spatially coherent view of A if the difference between the knowledge S_2 has from the state of A and the real state of A (i.e. the one simulated on site S_1) does not exceed a given *threshold* value, defined in the DIS standard as the maximal acceptable error done on the position (Th_{Pos}) and the orientation (Th_{Or}) on each entity. The underlying Fig. 1 illustrates on a same plan the position and the orientation of the entity A (supposed here to be a plane), respectively on site S_1 (plane on the left) and on site S_2 (plane on the right). While the position error E_p and the orientation error E_o do not exceed their respective threshold (Th_{Pos} and Th_{Or}), S_1 and S_2 have a spatially coherent view of A. Out of one of those thresholds, it is necessary for S_2 to refresh the state of A by means of a state information (typically an *Entity State* PDU) sent by S_1.

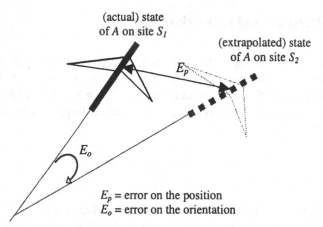

Fig. 1. Position and orientation of an entity

2.2. DIS standard QoS specification

As they are expressed within the DIS standard, DIS application requirements are the following ones :

- the application has to be provided with a multicast transport level service ;
- the QoS associated with the end to end DIS PDUs transport is defined by means of three parameters whose value depends on the *coupling* of the corresponding entities with the other entities :
  - reliability, defined as the maximal PDU loss rate ;
  - maximal end to end transit delay ;
  - maximal jitter for radio type PDUs.

Let's first recall the *coupling* notion. Assume A be an entity simulated on a given site. According to the DIS standard, one of the two following situations may be identified : (1) A is *loosely coupled* with all other entities or (2) A is *tightly coupled* with at least another (local or distant) entity. As an example (given in the standard) :

- two distant tanks provide an example of loosely coupled entities ;
- several tanks in formation at a fast speed provide an example of tightly coupled entities.

From this situation, the QoS associated with the transfer of the *Entity State* PDUs is defined as it follows :

- a transit delay less than or equal to 100 ms and a loss rate less than or equal to 2 % is required for the transfer of PDUs associated with tightly coupled entities ;
- a transit delay less than or equal to 300 ms and a loss rate less than or equal to 5% is required for the transfer of PDUs associated with loosely coupled entities.

2.3. DIS standard QoS specification limits

Two major limits may be identified from the previous QoS specification :

(L1) First, it does not allow the design of an end to end architecture based on the following principle : « one end to end channel per user flow, providing a specific QoS associated with the features and the constraints of the transported flow ».

(L2) The second limit is that the spatial coherence of the simulation can't be respected at any time. Let's detail this second affirmation.

Assume two sites S_s and S_r belonging to the same DIS exercise ; S_s and S_r are supposed to be perfectly synchronized. Assume now an entity A simulated on site S_s, whose state is maintained on site S_r by application of a *dead reckoning* (DR) algorithm. In order to simplify future explanations, A state will be reduced in the following to the position of A according to only one dimension.

The underlying Fig. 2 illustrates on a same plan the evolution in time of the extrapolation error[4] E made on A position, respectively on site S_e (upper part of the figure) and on site S_r (lower part of the figure) ; both sites are supposed to have a same reference time and to apply a same DR algorithm. Let's recall that only S_e is able to calculate the real state of A.

Black (respectively gray) plumps on Fig.2 correspond to the sending Ts dates (respectively receiving Tr dates) of the *entity state* (ES) PDUs refreshing A. We suppose that the time origin corresponds to the sending date of the first ES PDU (Ts_0). Thus, Tr_0, Tr_1, Tr_2 and Tr_3 correspond to the receiving dates of the ES PDUs sent at Ts_0, Ts_1, Ts_2 and Ts_3.

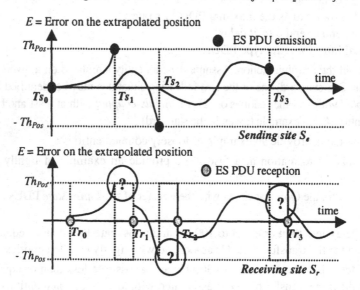

Fig. 2. Extrapolation error evolution

Let's analyze Ts_1, Ts_2 and Ts_3 dates.

- Starting from Ts_0, the error E between the extrapolated and the actual position of A is growing up (in absolute value) and raises its maximal acceptable value (Th_{Pos}, corresponding to the position *threshold* value) at Ts_1 : at this time, an ES PDU is generated by S_e (DR application), indicating the actual position of A ; the error E becomes null on site S_e.

- Starting from Ts_2, the error E between the extrapolated and the actual position of A varies between $(-Th_{Pos})$ and $(+Th_{Pos})$ without exceeding one of those two values ; after a *Heart Beat Timer* (HBT) spent time, i.e. at $Ts_3 = Ts_2 + 5$ seconds (default value for the HBT in the DIS standard), an ES PDU is automatically generated by S_e, indicating as in the previous case the actual position of A ; error E becomes null on site S_e.

Let's now analyze the evolution of error E on site S_r.

---

[4] The extrapolation error E is obviously defined as the geometrical difference between the actual and the "dead reckoned" state of the considered entity.

– Starting from Tr_0, the error E between the extrapolated and the actual position of A is identical to the one on S_s ; particularly, E raises its maximal acceptable value Th_{Pos} at time Ts_1, corresponding to the ES sending date on S_s. However, A position is not updated before Tr_1, the time interval separating Ts_1 from Tr_1 corresponding to the ES PDU transit delay through the network. It then appears an indetermination on E value between Ts_1 and Tr_1 : during this time period, E may exceed the Th_{Pos} value, and then potentially generate a spatial coherence violation.

– Analogously, a same indetermination may occur between Ts_2 and Tr_2, and between Ts_3 and Tr_3, illustrated on Fig. 2 by the gray zones.

It results from this analysis that the current standard QoS definition makes it possible a spatial coherence guaranty, but possibly in each $[Ts_i, Tr_i]$ time interval : during those time periods, the error E may transitorily exceed in absolute value the maximal acceptable value (DIS standard *threshold*). Such a case is illustrated on Fig. 3.

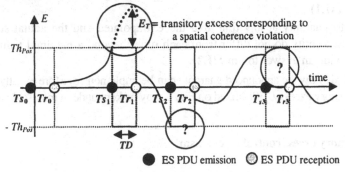

Fig. 3. Transitory excess illustration

Two questions may then be raised : (1) is this phenomenon a problem for a DIS exercise ? (2) in this case, is it possible to solve the problem ; in other words, is it possible to keep the transitory excess E_T illustrated on Fig. 3 under a given value ?

As far as the first question is concerned, the transitory excess becomes problematic as soon as the transit delay (*TD* on Fig. 3) is non negligible with regard to the mean time interval separating two ES PDU consecutive receptions. In this case, it then results that the receiving sites may have an incoherent view of the simulation during non negligible time periods, and even « almost all time » if the refreshing period is negligible compared with the ES transit delay ; of course, this risk does not occur in a LAN environment, but has to be taken into account when the application is distributed in WAN environment such as the one considered in our study. We now formulate a QoS proposition aimed at allowing the respect of a maximal transitory error value.

2.4. QoS proposition

In order to solve the previous two limits (**L1** and **L2**), the QoS specification that we propose is based on the following points :

- as in the DIS standard, the application has to be provided with a multicast transport level service ;
- QoS associated with the end to end DIS PDU transport is defined by means of the same three parameters than the ones defined in the DIS standard :
  - reliability, defined as the maximal PDU loss rate (total reliability meaning a 0 % loss rate ; null reliability meaning that a 100 % loss rate is acceptable) ;
  - maximal end to end transit delay (possibly infinite) ;
  - maximal jitter for radio PDU.

But and differently from the standard specification, we propose :

- to drop the *coupling* notion which is used in the standard to define QoS parameter values ;
- to evaluate QoS parameters value :
  - only from the information contained in each PDU ; in that, we provide an answer to limit **(L1)** ;
  - so as to guaranty <u>at any time</u> that the extrapolated and the actual state of any simulated entity do not differ from each other beyond a maximal value ; in that, we provide an answer to limit **(L2)**.

Let's now study how the announced guaranty can be implemented. More exactly, let's study how the transitory excess E_T illustrated on Fig. 3 may be kept under a maximal value.

2.5. Transitory excess control mechanism

Consider the underlying Fig. 4. Obviously, the transitory excess E_T may be kept under a maximal E_{Tmax} value as soon as the ES PDUs transit delay does not exceed the TD_{max} value illustrated on the figure.

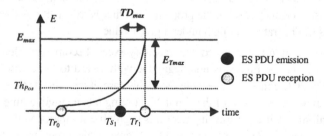

Fig. 4. Transitory excess control

Intuitively, it is possible to conceive that the E_T value depends on the dynamic (i.e. the acceleration) of the corresponding entity : the higher this dynamic, the higher E_T (potentially). Consequently, one can also conceive that the E_{Tmax} value also depends on the dynamic of the entity. Let' now formalize this approach of the problem.

On each receiving site, the influence of the transit delay (TD) on the extrapolated state of a distant entity is related to the difference (in absolute value) $e_p(t)$ between the actual position $P_a(t)$ and the extrapolated position $P_{DR}(t)$ of the considered entity (relation 1).

$$\begin{cases} \mathbf{P}_a(t) = \int_{Ts_i}^{t} du \int_{Ts_i}^{u} \mathbf{A}_a(\tau) \cdot d\tau + (t - Ts_i)\mathbf{V}_i + \mathbf{P}_i \\ e_\mathbf{p}(t) = \left\| \mathbf{P}_a(t) - \mathbf{P}_{DR}(t) \right\| = \left\| \int_{Ts_i}^{t} du \int_{Ts_i}^{u} [\mathbf{A}_a(\tau) - \mathbf{A}_i] \cdot d\tau \right\| \end{cases} \quad (1)$$

where :
- $\mathbf{A}_a(t)$ gives the actual acceleration of the considered entity at time t ;
- $\mathbf{P}_i$, $\mathbf{V}_i$, and $\mathbf{A}_i$ respectively give the actual position, speed and acceleration of the entity at time Ts_i.

Let's note that $\mathbf{P}_i$, $\mathbf{V}_i$, and $\mathbf{A}_i$ are three specific fields of the ES PDU sent at time Ts_i.

If the entity has a bounded acceleration ($\| \mathbf{A}_a(t) \| \leq A_{max}$), then $e_\mathbf{p}(t)$ can be bounded during each $[Ts_{i+1}, Ts_{i+1}+TD]$ time interval. $e_\mathbf{p}(t)$ being less than or equal to Th_{Pos} while $t \leq Ts_{i+1}$, it is then possible to major $e_\mathbf{p}(Ts_{i+1}+TD)$ by a sum of three terms (relation 2) :

$$\begin{aligned} e_\mathbf{p}(Ts_{i+1} + TD) \leq & \left\| \int_{Ts_i}^{Ts_{i+1}} du \int_{Ts_i}^{u} [\mathbf{A}_a(\tau) - \mathbf{A}_i] \cdot d\tau \right\| \\ & + \left\| \int_{Ts_{i+1}}^{Ts_{i+1}+TD} du \int_{Ts_i}^{Ts_{i+1}} [\mathbf{A}_a(\tau) - \mathbf{A}_i] \cdot d\tau \right\| \quad (2) \\ & + \left\| \int_{Ts_{i+1}}^{Ts_{i+1}+TD} du \int_{Ts_{i+1}}^{u} [\mathbf{A}_a(\tau) - \mathbf{A}_i] \cdot d\tau \right\| \end{aligned}$$

Following the *dead reckoning* mechanism on the sender site, the first term is obviously majored by Th_{Pos}.

The third term is majored by :
- $(\frac{1}{2}) A_{max} \cdot TD^2$ for a linear extrapolation ($\mathbf{P}_{DR}(t) = \mathbf{P}_i + \mathbf{V}_i \cdot (t - Ts_i)$)

- $A_{max} \cdot TD^2$ for a quadratic extrapolation ($\mathbf{P}_{DR}(t) = \mathbf{P}_i + \mathbf{V}_i \cdot (t - Ts_i) + \frac{1}{2} \mathbf{A}_i \cdot (t - Ts_i)^2$)

It is then to major the second term, corresponding to the product between TD and a constant representing the difference between the entity actual speed and its extrapolated speed at time Ts_{i+1}, sending date of the ES (relation 3) :

$$\begin{aligned} TD \cdot \int_{Ts_i}^{Ts_{i+1}} [\mathbf{A}_a(\tau) - \mathbf{A}_i] \cdot d\tau \\ = [(\mathbf{V}_{i+1} - \mathbf{V}_i) - \mathbf{A}_i \cdot (Ts_{i+1} - Ts_i)] \cdot TD \quad (3) \\ = (\mathbf{V}_a(Ts_{i+1}) - \mathbf{V}_{DR}(Ts_{i+1})) \cdot TD \end{aligned}$$

This difference raises its maximal absolute value when the difference between the entity actual speed and its extrapolated speed is also maximal during the whole $[Ts_i, Ts_{i+1}]$ time interval. Under those conditions, the second term of relation (2) is majored by $\sqrt{2A_{max} \cdot Th_{Pos}} \cdot TD$ for a linear extrapolation and by $2\sqrt{A_{max} \cdot Th_{Pos}} \cdot TD$ for a quadratic expression.

As a conclusion, it is then possible to major $e_\mathbf{p}(t)$ by the following expression (relation 4) :

$$e_p(t) \leq \begin{cases} \left(\sqrt{Th_{Pos}} + TD \cdot \sqrt{\dfrac{1}{2} A_{max}} \right)^2 & \text{for a linear extrapolaton} \\[2em] \left(\sqrt{Th_{Pos}} + TD \cdot \sqrt{A_{max}} \right)^2 & \text{for a quadratic extrapolaton} \end{cases} \qquad (4)$$

In order to keep the transitory error E_T under a maximal E_{Tma} value, it is then necessary and sufficient to guaranty for each ES PDU a maximal end to end transit delay whose value is given by the following expressions (relation 5) :

$$TD_{max} \leq \begin{cases} \dfrac{\sqrt{Th_{Pos} + E_{Tmax}} - \sqrt{Th_{Pos}}}{\sqrt{0,5 \cdot A_{max}}} & \text{for a linear extrapolaton} \\[2em] \dfrac{\sqrt{Th_{Pos} + E_{Tmax}} - \sqrt{Th_{Pos}}}{\sqrt{A_{max}}} & \text{for a quadratic extrapolaton} \end{cases} \qquad (5)$$

In other words, it is possible to ensure a spatial coherence guarantying the entity position inside a uncertainty circle whose radius is $Th_{Pos} + E_{Tmax}$, provided the ES PDUs end to end transit delay is less or equal than TD_{max}.

The previous discussion may be easily extended to the case of an extrapolation on the orientation. Results are similar, but provide more complex expressions due to the rotation matrix. Let's finally note that :

- for a given $Th_{Pos} + E_{Tmax}$, if Th_{Pos} decreases, then TD_{max} increases : temporal constraints are then easier to satisfy (from the network point of view), but ES PDUs are sent more frequently (the throughput grows up) ;
- with a quadratic extrapolation, the TD_{max} value is smaller than with a linear one ; consequently, temporal constraints are then more difficult to satisfy (from the network point of view) but ES PDUs are sent less frequently.

In both cases, relation (5) allows to deduce the transit delay expected from the underlying network, from features and constraints (acceleration and thresholds) expressed at the application level. Furthermore, this is accomplished without any modification of the standardized DR algorithm : the sending of an ES PDU occurs when the extrapolation error value (for the position or the orientation) goes beyond a fixed threshold. A modification of this sending criterion (by adding a threshold mechanism for the speed or the acceleration) may result in a reduce of either the generated ES PDU number or the required temporal constraints.

3. DIS PDU classification

Starting from this QoS proposal, a DIS PDU classification is now formulated. Two goals are pursued within this classification :

- the respect <u>at any time</u> of the simulation coherence ;

- the design of an end to end architecture based on a well known architectural principle for the support of multi-flow applications (such as multimedia applications) : "establishment of an end to end channel providing a specific QoS for each application level flow".

Four classes have been defined, the fourth one being subdivided into sub-classes :

- **The *discrete* PDU class** (*Collision*, *Fire* and *Detonation* PDU), that requires as QoS parameters :
  - a total reliability ; such a QoS results from the *discrete* feature of the concerned PDUs that makes it impossible their extrapolation ; as an example, if a *Fire* PDU is lost, the associated information could not be deduced from the next received PDU (at the opposite of *Entity State* PDUs) ;
  - a maximal end to end delay associated with the desired interactivity (we kept the 100 millisecond (ms) recommended by the DIS standard for PDUs whose associated entity is in strong interaction with at least another one).
- **The simulation management PDU class** (*Start*, *Stop*, ... PDU), that requires as QoS parameters :
  - a total reliability ; indeed, as they are defined in the DIS standard, PDU exchange rules do no precise the DIS software behavior when a simulation management PDU loss occurs : implicitly, the underlying network is supposed to be perfectly reliable ;
  - an unbounded end to end transit delay (no specific temporal constraint).
- **The signal PDU class** (only one PDU : the Signal PDU), that requires as QoS parameters :
  - a "strong" reliability depending from the audio coding algorithm ;
  - a 250 ms maximal transit delay ms, corresponding to the value recommended for a real time distributed audio conferencing system ;
  - a 50 ms jitter as it is specified in the DIS standard.
- **The *continuous* PDU class** (essentially *Entity State* PDUs) that requires as QoS parameters :
  - a total reliability (implicit hypothesis in the previous TD_{max} calculation) ;
  - a maximal end to end transit delay, deduced from the previous relation (5) ; as a consequence, several sub-classes are defined that we summarized Fig. 5.

A_{max} domain	Guaranteed TD_{max}
$[0, A_1]$	TD_1
$[A_1, A_2]$	TD_2
...	...
$[A_{n-1}, A_n]$	TD_n

Fig. 5. Entity State PDU classification

At each maximal acceleration $[A_{i-1}, A_i]$ interval is associated a maximal transit delay allowing a spatial coherence respect for entities whose maximal acceleration is comprised between A_{i-1} and A_i.

4. The end to end resource reservation architecture

Starting from the previous results, our second goal was to design and implement an end to end architecture allowing DIS applications to be distributed with respect to the QoS defined for each of the defined PDU class. In order to reach this goal, two preliminary needs had been identified :

– a network providing both a multicast transfer and QoS guaranties ; within the project, the chosen network has been a new generation Internet through (1) the IETF *Intserv* service models (*Guaranteed Service*, *Controlled Load*), (2) a traffic scheduling similar to the Class Based Queuing (CBQ) one and (3) the RSVP protocol ;

– a characterization of the DIS traffic generated by each site, so as to provide the network with elements required to implement the resource reservation phase.

We now present the main hypothesis and principles related (1) to the characterization of a DIS traffic, and (2) to the end to end architecture itself.

4.1. Traffic characterization principles

The traffic characterization methodology is based on the basic principle that a global DIS traffic may be considered as the aggregation of individual traffics generated by identifiable application level elements (AE) (Fig. 6).

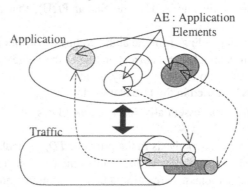

Fig. 6. Global traffic vs. individuals ones

More precisely, three hypothesis are underlying to the defined methodology.

H1 : AEs must be identifiable ; in other words, application might be decomposable into several autonomous parts, each generating its own traffic ; in the DIS application context, AEs identified on a given site have been chosen as the entities (locally simulated) whose *morphology*[5] is the same ; this choice has been validated experimentally.

---

[5] The term *morphology* is defined in the DIS standard as a "set of header fields describing both the physical appearance and behavior of one simulated entity". More precisely, a morphol-

H2 : So as to be statistically reproducible, traffic generated by each AE must depend on application intrinsic parameters ; if this hypothesis seems to be verified in a DIS context (through the basic idea : each entity or entity group generates an individual and well characterized traffic), interactivity between the simulated entities may result in a traffic evolution. This aspect has not been tackled in our study ;

H3 : it exists a correlation between AE individual traffics and the global traffic generated by their composition. At the current time, we only assume this hypothesis valid, without no more demonstration.

From those hypothesis, the applied methodology consists in the following points.

- We first developed a few software's allowing to capture and isolate the DIS traffic corresponding to each identifiable morphology.

- We then generated a reproducible DIS traffic by means of a public software named *DIS Control Station (DIS-CS)* [18].

- By means of simple software's allowing to calculate the RSVP *Tspec* parameters for a given traffic, the *Entity State* traffic generated by each identified DIS morphology has finally been characterized in order to constitute a Traffic Information Base (TIB).

Knowing the morphology and the number of the locally simulated entities, it is then possible, by means of the previous TIB, to evaluate the initial traffic (in trems of ES PDUs) generated by a given site.

4.2. End to end architecture design principles

Starting from the previous results (QoS and TIB constitution), the end to end architecture defined within the project (Fig. 7) relies on the following principles.

- Applying a principle several times used in the design of end to end multimedia architecture [19][20][21][22], a DIS+ level multicast channel (aimed at providing a specific QoS) is established for each identified PDU class.

- A reservation process is then executed in order to make it effective the QoS guaranty on each opened channel. This *static*[6] process is made up of two steps :
  - for each sender : the specification of the to be generated traffic profile ;
  - for each receiver : the resource reservation.

---

ogy is composed of eight fields : *Entity, Kind, Domain, Country, Category, Subcategory, Specific* and *Extra* ; we only keep the first four in our experiments.

[6] *Static* meaning « without re-negotiation » after the beginning of the simulation.

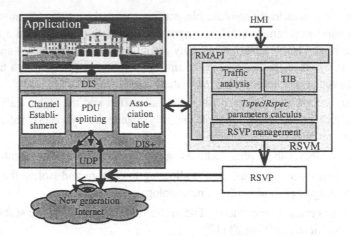

Fig. 7. The end to end architecture

Two different problems are then to be tackled :

- **channels management**, meaning :
  - channels establishment ;
  - DIS PDU splitting and de-splitting in and from the adequate channel.
- **QoS control of each channel**, part dedicated to a Resource reSerVation Manager : the RSVM module, whose main functions are the following ones.

The first function of the RSVM is to allow he DIS application to specify the traffic it is going to generate in terms of owner oriented parameters rather than network oriented ones (typically : RSVP *Tspec* parameters). In order to perform this task, the application is provided with an RSVM level API[7] whose specification is given in section (4.3). The second function is to translate application parameters in terms of *Tspec* parameters ; to perform this task, the RSVM uses a *Traffic Information Base* (TIB), containing the <morphology, *Tspec*> associations as they have been deduced from the DIS traffic characterization. The TIB is basically achieved in a calibration phase occurring before any DIS exercise ; however, it can also be upgraded during a DIS exercise by the way of probes measuring the exchange traffic. The third function is to manage RSVP so as to establish one RSVP session with the adequate QoS (i.e. corresponding to the one required for the transported PDU class) per opened channel.

4.3. Specification of the RMAPI

As we saw in the previous paragraph, the QoS computation is only possible if the application is able to specify its characteristics for a given exercise. In order to perform this task, the application is provided with an RSVM-level API (the RMAPI) which is now presented.

---

[7] API : Application Programming Interface

Five primitives have been defined for the RMAPI (Fig. 8).

Phase	Primitive	Parameters
New entity description	RSVM_set_QoS	source_desc, service_type
Session opening	RSVM_open	App_desc
Sources declaration	RSVM_record_source	App_desc, source_desc, source_num
Resource reservation process	RSVM_start_reservation	App_desc
Session closing	RSVM_close	App_desc

Fig. 8. RMAPI primitives

RSVM_open() allows an application to set an RSVM session. This primitive returns a RSVM session identifier. This identifier will be used in the others primitives to specify an open session.

RSVM_close() is used by an application to close an RSVM session and relax all the related reserved resources.

RSVM_set_QoS() is used by an application to upgrade the TIB. It allows an application to add an entry into the information base, specifying the service type required for a given source (*source_desc*), equivalent to a DIS entity in our case study.

RSVM_record_source() is used by an application to declare the characteristics of a DIS exercise. The RSVM is informed of the existence in the local host of a number of sources (*source_num*) of traffic source referenced in the database by the *source_desc* descriptor. When this primitive is invoked, the RSVM must check the existence of the traffic profile associated to the source descriptor and then upgrade the local traffic list.

RSVM_start_reservation() is used to begin the resource reservation process[8] once the local traffic has been declared. This primitive triggers the setting of the declared traffic into different channels and compute for each channel the necessary QoS and the traffic characteristics (*Tspec*). Then, RSVP is used to send a PATH message that allows the reservation phase beginning. The receivers collect the different PATH messages by the way of RSVP and each RSVM is able to compute the necessary service to be reserved in each channel.

5. Conclusions and future works

This paper has presented part of the work performed within the DIS-ATM project whose general goal was to specify and then implement the QoS required by a DIS application in a

---

[8] In a dynamic version of the RSVM, the **RSVM_start_reservation()** primitive could be invoked many times during an exercise, according to entities death or creation.

WAN environment involving a new generation Internet (*Intserv* service models, associated traffic control mechanisms, RSVP) and ATM.

The major part of the paper deals with DIS application requirements and QoS specification (section 2 and 3) ; results exposed in those two sections may be summarized in three points :

- the first one concerns the DIS standard limits as far as the QoS specification is concerned ; in its current state, this specification does not guaranty the simulation coherence in a WAN environment ; moreover, it does not allow the design of an end to end architecture based the architectural principle "one channel with a specific QoS for each application level data flow" ;

- the second result deals with a QoS specification proposal allowing to solve the previous two limits ;

- the third result consists in a DIS PDUs classification, each class requiring a specific QoS whose parameters value have been given.

On the basis of these results, the end to end architecture defined to satisfy the previous QoS has been exposed in section 4. In the first part of the section, a methodology allowing a *Tspec* characterization of DIS traffics has been introduced. In the second part of the section, the end to end architecture has been exposed, particularly, the Resource reSerVation Manager (RSVM). In the final part of the section, a description of the RSVM-level API has been given.

Different kinds of perspectives are currently studied ; the main two are the following ones :

- The first perspective is to tackle the major architecture drawback : its static aspect. At the current time, once resources have been associated to the defined end to end channels, no modification occurs in the network even if DIS traffic varies (due to entity death or new entity creation). It is then our purpose to specify a resource reservation re-negotiation phase ;

- The second perspective is to extend the proposed architecture in order to take into account a more general application / network environment context, including other real time applications and other Internet service models (i.e. the differentiated services).

These two perspectives are to be examined within a new national project whose main goals are to extend and generalize results obtained within the DIS/ATM project.

6. References

1 IETF : Integrated Service Working Group: http://www.ietf.org/html.charters/intserv-charter.html.
2 Shenker, S., Partridge, C., Guerin, R.: Specification of Guaranteed Quality of Service, RFC 2212, IETF Network WG (1997).
3 Wroclawski, J.: Specification of the Controlled Load Network Element Service, RFC 2211, IETF Network WG (1997).

4   Braden, R., Zhang, L., Berson, S., Herzog, S., Jamin, S.: Resource reSerVation Proto-
    col : Functional Specification, RFC 2205, IETF Network WG (1997).
5   Wroclawski, J.: The use of RSVP with Integrated Services, RFC 2210, IETF Network
    WG (1997).
6   Pillen, J.M., Fellow, P., Wood, D.C.: Networking Technology and DIS, Proceedings of
    the IEEE, vol 83, no. 8 (1995).
7   Pullen, J.M., Myjak, M., Bouwens, C.: Limitations of the Internet Protocol Suite for
    distributed Simulation in the Large Multicast Environment, RFC 2502 (1999).
8   Schricket, S.A., Franceschini, R.W., Petty, M.D.: Implementation Experiences with the
    DIS Aggregate Protocol, Spring Simulation Interoperability Workshop, U.S.A (1998).
9   Russo, K.L.: Effectiveness of Various New Bandwidth Reduction Techniques in Mod-
    SAF, 13th DIS Workshop, (1995).
10 Aggarwal, S., Kelly B.L.: Hierarchical Structuring for Distributed Interactive Simula-
    tion, 13th DIS Workshop (1995).
11 Gauthier, L., Diot, C.: Design and Evaluation of MIMaze, a Multi-player Game on the
    Internet, IEEE multimedia Systems Conference, Austin, USA (1998).
12 Ginder, R. et al.: DIS to HLA Integration, a Comparative Evaluation, 15th DIS Work-
    shop (1996).
13 Pratt, S.: Implementation of the IsGroupOf PDU for Network Bandwidth Reduction,
    15th DIS Workshop (1996).
14 Purdy, S.G., Wuerfel, R.D.: A comparison of HLA and DIS Real-Time Performance,
    Spring Simulation Interoperability Workshop, U.S.A (1998).
15 IEEE: Standard for Distributed Interactive Simulation, Application Protocols, standard
    IEEE n°1278.1 (1995).
16 IEEE: Standard for Distributed Interactive Simulation, Communication Architecture
    Services and Profiles, standard IEEE n°1278.2 (1995).
17 IEEE: IEEE Recommended Practice for Distributed Interactive Simulation, Exercise
    Management and Feedback, standard IEEE n°1278.3 (1996).
18 CAS: http://www.cas-inc.com/
19 Chassot, C., Diaz, M., Lozes, A.: From the partial order concept to partial order multi-
    media connection, Journal for High Speed Networks, Vol. 5, n°2 (1996).
20 Campbell, A., Coulson, G., Hutchinson, D.: A quality of service architecture, ACM
    Computer Communication Review (1994).
21 Nahrstedt, K., Smith, J.: Design, Implementation and experiences of the OMEGA end-
    point architecture, IEEE Journal on Selected Areas in Communications, Vol.14 (1996).
22 Gopalakrishna, G., Parulkar, G.: A framework for QoS guarantees for multimedia ap-
    plications within end system, GI Jahrestagung, Zurich, Switzerland (1995).

A Multicasting Scheme Using Multiple MCSs in ATM Networks

Tae Young Byun[1], and Ki Jun Han[2]

[1] Department of Computer Engineering, Kyungpook National University, Taegu, Korea
tybyun@comeng.ce.kyungpook.ac.kr
[2] Department of Computer Engineering, Kyungpook National University, Taegu, Korea
kjhan@bh.kyungpook.ac.kr

Abstract. In this paper, we proposed a scheme to support multiple MCSs over a single large cluster in ATM networks, and evaluated its performance by simulation. When an ATM host requests joining into a specific multicast group, the MARS designates a proper MCS among the multiple MCSs for the group member to minimize the average path delay between the sender and the group members. This scheme constructs a 2-phase partial multicast tree based upon the shortest path algorithm. We reduced the average path delay in multicast tree using this scheme with various cluster topologies and MCS distribution scenarios, also distributed load among multiple MCSs

1 Introduction

ATM has been widely acknowledged as the base technology as the next generation of global communication, offering a high speed switching scheme for transporting many different types of telecommunication traffic.

Multicast networking support is becoming an increasingly important technology area for distributed or group-based application. But there are many challenges to interwork between legacy network and ATM network such as data format conversion, address resolution, multicast capability so on.

Especially it's not easy to support layer 3 multicast services over connection-oriented ATM networks. Most existing shared and broadcasting media technologies, such as Ethernet, are inherently capable of intra-network multicasting. The network interfaces of this kind of networks can be effortlessly to support the multicast address abstraction. By contrast, mapping the connectionless multicast service onto the Non-Broadcast Multiple Access(NBMA) media, such as ATM, always requires address resolution support for establishing the connection prior to the data transmission. That is, a mechanism is needed to resolve the mapping between high-level protocol addresses and ATM addresses.

The Multicast Address Resolution Server(MARS) [1], which is an extension to the ATM ARP server[2], is proposed to work with both VC-mesh and MCS to manage intra-domain group membership, and to subsequently handle ATM address resolutions. There have been several implementations and extensions based on the MARS model[3, 4, 5, 8].

In [2], a proxy server scheme, called the Multicast Server(MCS), has several advantages such as reduced consumption of the VC resource and easy management of VC compared to VC-mesh, but has disadvantages of a long latency and low data throughput. Also this model suffers from a bottleneck of traffic at a single MCS. So multiple MCS model is suggested to support load sharing in [6, 7].

Multiple MCS scheme proposed in this paper is mainly motivated by minimizing average latency time and load sharing. The core mechanism of the scheme, referred to as the 2-Phases Shortest Path based Multicast Tree(2PSPMT), is primarily designed to assign a proper MCS among the multiple MCSs to a joining ATM host into a specific multicast group. Conceptually, 2PSPMT minimizes average path delay between the sender and the group members via multiple MCSs and shares traffic load among multiple MCSs. It retains the MARS's function as an intra-domain registration of group membership, and serves as a centralized group member distributor over multiple MCSs.

This paper also describes a proof-of-concept simulation of the 2PSPMT scheme with respect to a typical large scale cluster in ATM network configuration. The performance comparisons between random MCS assignment and 2PSPMT based MCS assignment under various cluster topologies and MCS distribution scenarios are presented in terms of the average path delay of multicast tree and load sharing.

2 Motivations

Cluster is logical scope that MARS is responsible for resolving a layer 3 multicast address over ATM networks.

Multiple MCSs model is superior to single MCS as the size of cluster increases for the following reasons.

First, single MCS may be a bottleneck point due to concentration of multicast traffic to it. This increases average path delay between the sender and the receiver, also reduces data throughput over the path. So it's required to scatter multicast traffic across several MCSs to prevent bottleneck at a single MCS. Second, when a single MCS fails due to disorder of system, it's difficult to provide multicasting via the MCS continuously. Under multiple MCS environment, this can be avoided by migrating the role of failed MCS into others instantly.

There are several schemes using multiple MCS based on MARS model [6, 7]. Talpade and Ammar present a scheme to satisfy the load sharing and fault tolerance by involving an enhanced version of the MARS. Byun et al[7, 8], also proposes a load sharing scheme monitoring available link capacity of each MCS. In this scheme, when an MCS suffers from bottleneck, the MARS dynamically partitions a large multicast group at the MCS into several subgroups and assigns subgroups across relatively less loaded MCSs. All these schemes focus only load balancing. So they have several defects for the following reasons.

First, when these schemes don't reflect cluster topology on partitioning large group into subgroups, the average path delay in multicast tree may be very large. This also causes decrease in the end-to-end data throughput.

Second, variation of average path delays between a sender and subgroups via multiple MCSs may be very large.

These disadvantages result from assigning group members to multiple MCSs without considering network topology.

In this paper, we present a multiple MCSs model which contains the sender as a temporary MCS. Also we suggest a multicasting scheme which consists of an MCS selection algorithm based on Dijkstra's shortest path calculation and a multicast construction algorithm. Our scheme considers location information of joining the ATM host and several MCSs to minimize the path delay the between sender and the joined ATM host.

3 Multiple MCS Model

Here, we introduce our multiple MCSs model which is classified into two models : full multiple MCS model and hybrid MCS model as shown in Fig. 1.

In the fully multiple MCSs model, all of registered MCSs serve as a multicasting server in cluster. The sender first transmits multicast data to only registered multiple MCSs, then each MCS forwards it to its own subgroup. In the hybrid multiple MCS model, the sender acts as an MCS temporarily. A specific subgroup is served by the sender during the multicasting period, so we call the sender as ad-hoc MCS because it acts as MCS during the transmission. In Fig. 2, multicasting from the sender to the subgroup G1.3 is similar to VC-mesh model while multicasting from the sender to each subgroup G1.1, G1.2 is similar to the MCS model.

In our scheme, hybrid multiple MCSs is optional because it will produce less average path delay than fully multiple MCSs model but requires more calculation in assigning some group members to ad-hoc MCS(sender) on demand.

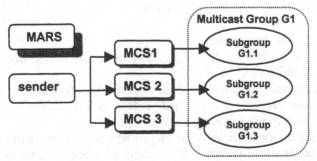

Fig. 1. Fully multiple MCSs Model. A multicast group G1 is split to several subgroups {G1.1, G1.2, G1.3} and each subgroup is supported by a designated MCS. Sender manages a one-to-many VC between sender and multiple MCSs and each MCS maintain a one-to-many VC between the MCS and group members in a subgroup

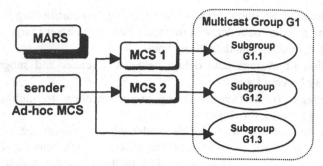

Fig. 2. Hybrid multiple MCSs model. This model is similar to fully multiple MCSs model except that ad-hoc MCS or sender multicasts to subgroup G1.3 directly by setting up one-to-many VC as VC-Mesh model.

4 2-Phases Shortest Path Based Multicast Tree Construction

We first define acronyms for describing our scheme.

- Graph $G = (V, E)$, V : set of nodes, E : set of edges
- s : sender, s $\in$ V
- D : E $\rightarrow$ Z$^+$ weighting function for the link delay.
- SMCS=$\{m_1, ..., m_k\}$, SMCS $\subseteq$ V, set of multiple MCSs
- GM(i), GM(i) $\subseteq$ V – s , set of group members belong to group i
- FOREST$_{1st}$(i) : multicast forest for group i after 1st phase
- PMT$_{1st}$(i, m) : partial multicast tree with root m for group i after 1st phase
- PMT$_{2nd}$(i, s) : partial multicast tree with root s for group i after 2nd phase
- FMT$_{2nd}$(i, s) : final multicast tree with root s for group i after 2nd phase, FMT(i) $\subseteq$ G
- SPL(v_i, v_j) : shortest path list from node v_i to node v_j
- leaf(M) : set of leaf nodes of tree with root M

Also we made some assumptions.
- The locations of multiple MCSs are pre-assigned and static. So being designated, the locations of MCSs don't be changed.
- MARS has additional functionality which acquires the information about the nodes and links across cluster topology, and updates it periodically. MARS always maintains the most recent knowledge of topology.
- A node on graph represents an ATM switch which has many ports connected to ATM hosts or another switches. All of group members, MCSs, and sender are omitted by intention because these are connected the ATM switches directly and the link delay between the host and the switch is very small.
- We consider only link delay as delay factor, and assign a variable delay to the link in proportion to the length of link.

4.1 Locality Degree Calculation

LD(Locality Degree) represents a criterion how far away each group member is located from multiple MCSs.

The smaller LD is , the closer group member is by an MCS. To reduce the distance, or path delay from host A to multiple MCSs, we select an MCS which is closest to host A for serving multicast from or to A.

Fig. 3 shows an example of cluster, which consists of different types of nodes such as MCS, group member, and sender.

In this figure, thick solid lines represent partial multicast trees based on the shortest path between group members and several MCSs, a thick dotted line indicates a partial multicast tree based on the shortest path between the sender and an MCS.

Table 1 illustrates an instance of LD calculation when applying the shortest path algorithm to the cluster shown in Fig. 3. MCS 1, MCS 2, and MCS 3 are responsible for subgroup of {1,2,6,7}, {4. 5}, and {3}, respectively. We italicized relevant entries in Table 1.

Table 1 also shows the shortest path delay, or LD between the sender and each MCS. To help understanding of LD calculation, we assume that a link delay is 1 unit in Fig. 3.

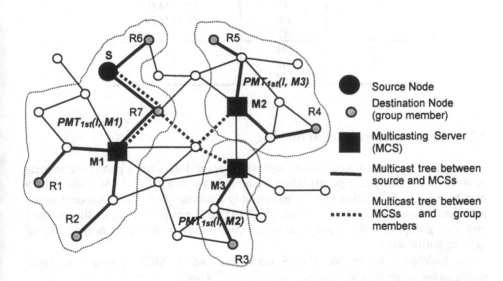

Fig. 3. Example of Multicast Tree Construction. Final multicast tree consists of a forest and a partial tree between sender and multiple MCSs. Forest is composed of several partial trees between multiple MCSs and subgroups.

Table 1. Example of LD calculation in Fig. 3. When two or more MCSs havs same LD value per a group member, a MCS is choosed randomly among them.

From node	To node	Locality Degree
M1	*R1*	*2*
	R2	*2*
	R3	3
	R4	4 (via M2 or M3)
	R5	4 (via M2 or not)
	R6	*3*
	R7	*1*
M2	R1	4 (via M1)
	R2	4 (via M1 or not)
	R3	3 (via M3)
	R4	*2*
	R5	*2*
	R6	3
	R7	2
M3	R1	4 (via M1)
	R2	3
	R3	*2*
	R4	2
	R5	3 (via M2)
	R6	4 (via M2)
	R7	2
S	M1	2
	M2	3
	M3	4

4.2 1ˢᵗ Phase : Multicast Forest Building

In this phase, we form a forest which consists of several trees to be part of a complete multicast tree. This phase occurs when we already know a set of ATM hosts to be a member of specific multicast group and initially construct partial trees between group members and multiple MCSs. When an ATM host newly joins into the existing multicast group, this phase also occurs in order to attach the ATM host to one of existing partial trees.

On building a forest, the MARS determines which MCS is proper to serve multicasting for each group member R_i or new ATM host.

To designate a proper MCS, MARS calculates LD to choose an MCS m_k which reaches the host along with the shortest path $SPL(R_i, m_k)$.

After the procedure at all group members, we get a forest which consists of p trees (p ≤ the number of MCSs).

The forest is updated partly when a new host is included or when one or more existing group members are deleted. Because there is not usually severe change in group membership, the frequency of updating are relatively low. The procedure of forest construction at 1ˢᵗ phase is depicted below.

FOREST$_{1st}$(i) building algorithm

```
INPUT :
  G(V, E); SMCS={m₁,...,mₖ}; GM(i)={v₁,...,v_g}, GM(i) ⊆ V - SMCS
OUTPUT :
```

$$FOREST_{1st}(i) = \bigcup_{j=1}^{k} PMT_{1st}(i,m_j)$$

```
PROCEDURE :
for(i = 1; i <= k; k++)
  MCSGM(mᵢ) = ∅;
for(i = 1; i ≤ g; i++) {
    CSPD=∞; //Current Shortest Path Delay//
    CSPM=0; //MCS which has shortest path to node v, v∈GM(i)//
    for(j = 1; j ≤ k; j++) {
        SPD = CalcLD(vᵢ, mⱼ);//Calculate Locality Degree of vᵢ//
        if(SPD < CSPD) {
          CSPD = SPD;
          CSPM = j;
        }
    } // end of inner-for loop
    AddMCSGM(vᵢ,mⱼ); //Add node vᵢ into MCSGM(i, mⱼ)//
    SaveSPL(vᵢ,mⱼ);  //Save shortest path list from vᵢ to mⱼ//
} // end of outer-for loop
// Construct forest for group i which is consist of all
PMT₁ₛₜ(i,mⱼ), 1 ≤ j ≤ k //
BuildFOREST(i);
```

4.3 2nd Phase : Final Multicast Tree Building

This phase occurs when an ATM host transmits data to a multicast group. Being requested a multicast address resolution by the sender, the MARS constructs a partial multicast tree from the sender to all of MCSs using the shortest path algorithm. After the MARS completes construction of the 2nd tree, there will be a complete multicast tree between the sender and all of group members via multiple MCSs. We depict the construction of a complete multicast tree below.

FMT_{2nd}(i, s) building algorithm

```
INPUT :
  G(V,E); SMCS={m₁,...,mₖ},SMCS ⊆ V; s, s ⊆ V - GM(i) - SMCS;
  FOREST₁ₛₜ(i)
OUTPUT :
  FMT₂ₙ_d(i,s)
PROCEDURE :
```

```
for(j=1; j ≤ k; j++) {
    GetSP(s,mⱼ);        //Calculate shortest path from s to mⱼ//
    SaveSPL(s,mⱼ); } //Save Shortest Path List from s to mⱼ//
if(MCSMode==HybridMCS){ //Check if hybrid multiple MCS mode//
    DetectRP(s,i); //Detect Replicated Path from s to GM(i)//
    AdjustFOREST(i);//Update FOREST(i) considering replicated paths//
}
Build2ndPMT(i,s);    //Construct PMT₂ₙd(i, s)//
//To make FMT₂ₙd(i,s), combine previous FOREST₁ₛₜ(i) and
PMT₂ₙd(i,s)//
BuildFMT(i,s);
```

5 Evaluation Methodology

Thus far, the discussion has centered on description of 2PSPMT scheme. Here, we turn to our evaluation methodology. In this section we describe our network model and our heuristic simulator.

5.1 Cluster Model

The clusters were generated to resemble networks in a manner similar to that of Doar[9], which models some aspects of real networks.

$$P_e(u,v) = \frac{ke}{|g|} \beta \exp\frac{-d(u,v)}{\alpha L} \tag{1}$$

Where
$d(u, v)$: distance between two nodes u, v
L : maximum possible distance between two nodes
α, β : parameters, $0 < \alpha, \beta \leq 1$
$|g|$: the number of nodes in graph G
e : the mean degree of a node
k : a scale factor which keeps the mean degree of each node constant

The connectivity of link is determined by P_e. The values α and β determine the pattern of degree and connectivity of generated cluster, respectively. The delay of a link is set to be linearly proportional to this probability.

5.2 Heuristic MCS Distribution Model

With our guess that various distribution patterns of MCSs affect average path delay, we set up four heuristics on MCS distribution. The criterion of distribution patterns is based on the average path delay from a node to the others in graph as (2).

$$AD(i) = \frac{\sum_{\forall j \in V - \{i\}} D(SPL(i,j))}{|V| - 1}, i \subseteq V \tag{2}$$

Then, we sort all AD(i)s in the ascending order. The sorted list called NAD satisfies (3).

$$NAD[i] \leq NAD[j], 1 \leq i < j \leq MAX_{NodeID} \tag{3}$$

5.2.1 Uniformly Random Distribution

This randomly selected index i in NAD list, chooses a node related to NAD[i] as an MCS. If k MCSs are needed, we do random selection k times over again as (4).

$$SMCS = \{RAND_1(MAX_{NodeID})\} \bigcup \ldots \bigcup \{RAND_k(MAX_{NodeID})\}$$

$$= \bigcup_{i=1}^{k} RAND_i(MAX_{NodeID}), 1 \leq k \leq MAX_{NodeID} \tag{4}$$

5.2.2 Least-Average-Delay-First(LADF)

In LADF, we choose nodes that have small average path delay preferentially as (5). Heuristically, the nodes around center in graph have higher probability to be selected as MCS.

$$SMCS = \{NAD[1], \ldots, NAD[k]\}, 1 \leq k \leq MAX_{NodeID} \tag{5}$$

5.2.3 Greatest-Average-Delay-First(GADF)

In GADF, the node that have great average path delay take precedence in MCS selection. In case of GADF, the leaf nodes in graph tend to be MCS.

$$SMCS = \{NAD[MAX_{NodeID} - k + 1], \ldots, NAD[MAX_{NodeID}]\}$$

$$, 1 \leq k \leq MAX_{NodeID} \tag{6}$$

5.2.4 Stepped-Average-Delay(SAD)

We select nodes which are positioned at regular intervals as MCS. SAD models even distribution of MCSs across cluster relatively while MCSs is likely to concentrate around a place in LADF and GADF.

$$SMCS = \{NAD[1], NAD[\lceil 1 + n/k \rceil], NAD[\lceil 1 + 2*n/k \rceil], \ldots, NAD[\lceil 1 + (k-1)*n/k \rceil]\}$$

$$= \bigcup_{i=0}^{k-1} \{NAD[\lceil 1 + i*n/k \rceil]\} \tag{7}$$

5.3 Criterion for Performance Evaluation

In our simulation, we try to evaluate two factors which may affect performance with multiple MCSs.

First, average path delay of complete multicast tree between the sender and all of group members can be calculated by (8).

$$AMTPD(i,s) = \frac{\sum_{j=1}^{k} \sum_{\forall e \in leaf(m_j)} \{D(SPL(s,m_j)) + D(SPL(m_j,e))\}}{|leaf(s)|} \qquad (8)$$

Second, degree of load sharing is understood by measuring average deviation of group members per MCS. Low average deviation means good load sharing among the MCSs while high value means poor load sharing among them. Average deviation can be calculated by (10).

$$AVGLOAD(i,k) = \frac{\sum_{j=1}^{k} LOAD(MCS_j)}{k} \qquad (9)$$

$$AVGDEV(i,k) = \frac{\sum_{j=1}^{k} (AVGLOAD(i,k) - LOAD(MCS_j))}{k} \qquad (10)$$

6 Simulation Results

The 2PSPMT is validated using a simulation tool called SIMMT(SIMulation tool for Multicast Tree construction). This tool is based on CTCL(Cluster Topology Class Library), which is a C++ library consisting of all necessary classes for an event driven simulation. SIMMT offers all basic modules relating MCS distribution patterns, source selection models, group member distribution patterns. Also, it records detailed results of simulation for numerical analysis. Furthermore, SIMMT contains the tool GSRT(Graphical Simulation Result Tool), which allows the presentation of simulation results in a graphical manner.

The graphic user interface of SIMMT is depicted in Fig. 4. It allows to set up parameters for execution. Also we depicted detailed simulation parameters in Table 2 and eight simulation scenarios in Table 3.

Now, we present simulation results for both random MCS assignment and 2PSPMT-based MCS assignment in the different MCS distribution environments on 20 test networks.

Table 2. Simulation parameters

Parameter	Model or Value
The number of nodes	300
The mean edge degree	5
The ratio of multiple MCSs to n (%)	3, 5, 7, 9
MCS distribution models	Random, LADF, GADF, SAD
The ratio of group members to n (%)	10, 15, 20, 25, 30, 35, 40
Distribution model of group members	Random

Table 3. Simulation scenarios

Scenario	MCS distribution model and MCS assifnment model
1	Random MCS distribution and Random MCS assignment
2	Random MCS distribution and 2PSPMT-based MCS assignment
3	LADF-based MCS distribution and Random MCS assignment
4	LADF-based MCS distribution and 2PSPMT-based MCS assignment
5	GADF-based MCS distribution and Random MCS assignment
6	GADF-based MCS distribution and 2PSPMT-based MCS assignment
7	SAD-based MCS distribution and Random MCS assignment
8	SAD-based MCS distribution and 2PSPMT-based MCS assignment

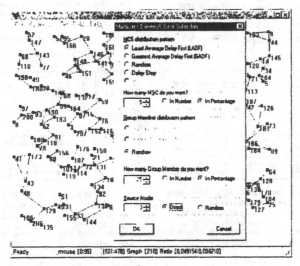

Fig. 4. Screen dump of the graphic user interface.

Simulation results are depicted in Fig. 5 and Fig. 6. From these figures, the necessity for 2PSPMT-based MCS assignment is obvious. Generally, LADF-based schemes have the least delay while GADF-based schemes have the greatest delay. This is due to characteristics of MCS distribution models.

Although average deviation of 2PSPMT-based MCS assignment is greater than that in random MCS assignment, 2PSPMT-based scheme reduces average path delay from the sender to group members by about 8 ~ 18 %.

We can observe good load sharing in all random MCS assignments regardless of the MCS distribution model. This is because the random MCS assignment allows each MCS to include group members with almost the same probabilities due to the characteristics of uniform random distribution. But random MCS assignment can't minimize the average path delay because it does not consider locality of group members and MCSs.

From Fig. 5 and Fig. 6, we can see that SAD offers reasonable performance in terms of average path delay and load sharing.

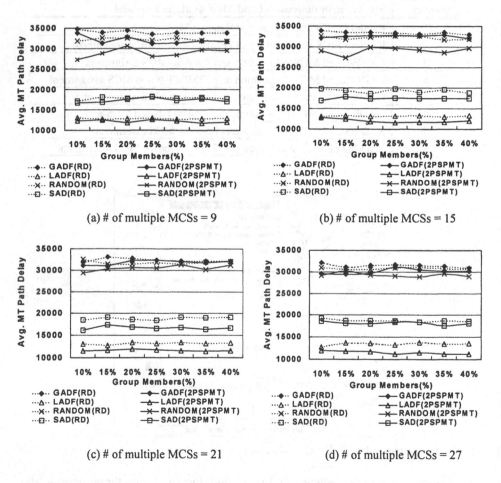

Fig. 5. Comparison of average path delays between multicast tree using 2PSPMT-based assignment and multicast tree using random assignment.

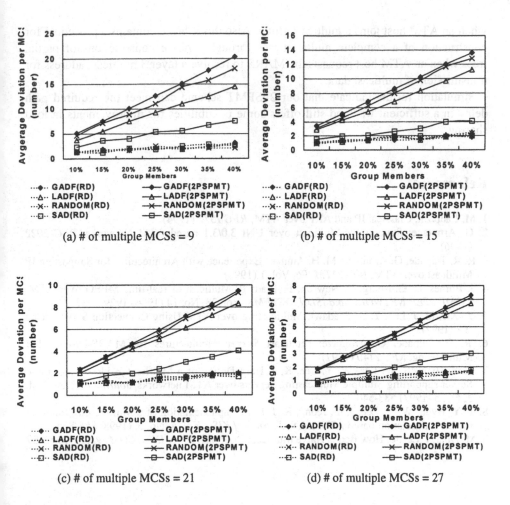

Fig. 6. Comparison of load sharing between multicast tree using 2PSPMT-based assignment and multicast tree using random assignment.

7 Concluding Remarks

An efficient layer 3 multicasting over ATM networks is an important issue for interworking with legacy networks. This requires a new approach to multicasting using multiple MCSs based on the MARS model. The existing single MCS model has several defects in terms of path delay, data throughput, and system robustness. This motivates us to propose a new algorithm for dynamic MCS assignment and multicast tree construction in multiple MCSs environment.

Our multicasting scheme, called 2PSPMT, considers locality of each group member and multiple MCSs, and includes procedure for dynamically proper MCS assignment when an ATM host joins a multicast group. Also this scheme contains a procedure for construction of a complete multicast tree through 2-phase constructions of partial trees when an ATM host requests the MARS to resolve a layer 3 multicast address for transmission of multicast data.

Simulation results indicate that our 2PSPMT scheme can meet the required path delay in a sufficient way and still offers some possibilities for improvements in load sharing.

References

1. M. Laubach: Classical IP and ARP over ATM, *RFC 1577* (1994)
2. G. Armitage: Support for Multicast over UNI 3.0/3.1 based ATM Networks, *RFC 2022* (1996)
3. R. R. Talpade, G. Armitage, M. H. Ammar: Experience with Architectures for Supporting IP Multicast over ATM, *IEEE ATM '96*, Vol. 1 (1996)
4. J. Barnes, D. Ginsburg, D. Newson, D. Pratt: IP Multicast of real-time MPEG over ATM. *COMPUTER NETWORK and ISDN SYSTEMS*, Vol. 28. No. 14 (1996) 1929-1937
5. Y. Xie, M. J. Lee, T. N. Saadawi: Multicasting over ATM Using Connection Server, *ICCI '97* (1997)
6. R. R. Talpade, G. Armitage: Multicast Server Architectures for MARS-based ATM Multicasting, *RFC 2140* (1997)
7. T. Y. Byun, S. W. Lee, W. Y. Lee, K. J. Han: Design of a Multicast Address Resolution Server supporting Multiple Multicasting Servers over ATM networks, *Journal of KISS*, Vol. 3, No. 5 (1997) 533-540
8. T. Y. Byun, W. Y. Lee, J. Y. Jeong, K. J. Han: Implementation of a Multicasting Platform for LAN Emulation-based ATM LAN, *Journal of KISS*, Vol. 4. No. 6. (1998) 855-864
9. M. Doar, I. Leslie: How Bad is Naïve Multicast Routing?, *Proc. INFOCOM '93*(1993) 82-89

A Platform for the Study of Reliable Multicasting Extensions to CORBA Event Service

João Orvalho[1], Luis Figueiredo[2], Tiago Andrade[2] and Fernando Boavida[2]

CISUC – Center for Informatics and Systems of the University of Coimbra
Communications and Telematic Services Group
- Pólo II, 3030 COIMBRA - PORTUGAL
Tel.: +351-39-790000, Fax: +351-39-701266, E-mail: {orvalho, boavida}@dei.uc.pt

Abstract. Conference applications play a key role in workgroup tools, and provide an excellent environment for the development and evaluation of proposals concerning multicasting and reliable communications. A nuclear aspect of conference applications is their ability to use multipoint communication in heterogeneous environments. In the scope this work, a platform that uses several extensions to the CORBA Event Service was developed. This platform is based on ITU T.120 recommendations, and supports reliable multicasting communication for both data and control information. The control information is managed by a service implemented in Java™ , according to the CORBA Event Service model. The present paper introduces and discusses the main features of the proposed platform. Also, provides a general description of the conference framework architecture that was built on the platform, with some emphasis on reliable multicasting communication and on system synchronism. In the final part of the paper, the specification and the implementation of the proposed extensions to the CORBA Event Service are presented.

1 Conference Framework Architecture

The development of the extensions to the CORBA Event Service [1] and the evaluation of their usefulness were performed using a conference prototype developed for this purpose. This conference framework will be described in this section.

1.1. General Description of the Architecture

The conference system is based on the tightly coupled conference paradigm, and has the aim to achieve a high scalability degree. To reach such objectives we have conceived two nuclear services, one for conference control - *Conference Controller* (CC) - and another one for multipoint communication between collaborative applications (Figure 1). The CC service - as responsible for the administration of the conference - implements a minimal set of the basic functionalities of ITU T.124

---

[1] College of Education, Polytechnic Institute of Coimbra
[2] Informatics Engineering Department of the University of Coimbra

recommendation (GCC- Generic Conference Control) [2] [3] [4], and uses a reliable service for multicasting communication. For the applications data multipoint communication, we have chosen to use *JSDT* (Java Shared Data Toolkit) [5]. JSDT is a framework based on the ITU T.122 recommendation, *Multipoint Communication Service* [6, 7], implemented in Java [8] with the option to use reliable multicasting communication protocols (ex: LRMP – Light-Weight Reliable Multicast Protocol [9, 10]).

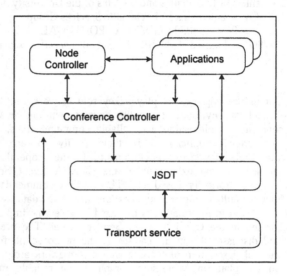

Fig. 1. Component stack of the developed conference system prototype

The conference system relies on a hierarchy of nodes, clients/customers and servers, organised in a tree topology, supported by JSDT links (Figure 2). The servers are the suppliers of the conference services, with a top server and some proxy servers. The collaborative applications reside in the customer nodes, linked to one or more servers, or in a server.

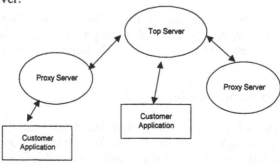

Fig. 2. Typical conference topology

The top server manages the conference resources – sessions, channels and tokens, among others – keeping, as such, a conference information base. The resource requests are made directly to the hierarchic superior, working up in the chain until arriving at the top server. To avoid high response times, proxies have the capacity to give resources, as they keep a replication of the top server resource database. To guarantee the information updating, the top server and the proxies share a reliable multicasting communication channel. This functionality is provided by the CC module, based on the standard OMG specification CORBA *Event Service* [1] (Figure 3).

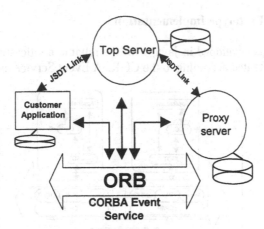

Fig. 3. Information updating using the underlying CORBA Event Service

The CC module provides high resource availability by the use of group abstractions that comprise the top server and the proxy servers. This functionality is supported by replication processes that are based on reliable multicast communication mechanisms.

As the conference scalability is directly related to resource request response times, the servers must speed up the resource response process by placing the answers in a reliable multicasting communication channel with capacity for message filtering.

Due to its nature, proxies can assume conference leadership, in case the top server becomes off-line for any reason. This functionality is achieved through the use of a keep-alive mechanism that is active between the top server and its proxies.

Load balancing between top server and proxies can also be implemented in systems that use the proposed architecture, as a way to improve system performance.

Conferences also require a control information database, holding information on their members and applications. This database is distributed on all nodes and keeps track of all the occurrences in each conference, as for example the entrance of a new customer node or the start-up of applications in a given node. All the changes in each node are communicated to the direct hierarchic superior, and they will progress up the server tree until they arrive at the top server. When this happens, control information will be updated in the top server and it will be replicated back (in its totality or partially-*delta*) to all nodes down the hierarchy through the CC channels (*CORBA Event Service*).

In each node, the information base administration is performed locally by the CC, which has the advantage of having local information accesses and not accesses to the direct hierarchic superior.

The applications data are transferred directly by JSDT through public and private channels, with no need to effect their routing through the conference structure. Channels are managed by the server to which each participating node is connected. JSDT can use several transport protocols, namely reliable multicasting protocols such as LRMP – Light-Weight Reliable Multicast Protocol.

1.2. Conference Prototype Implementation

The CC service was developed in Java, with the reliable multicasting communication component implemented according to the CORBA Event Service model (Figure 4).

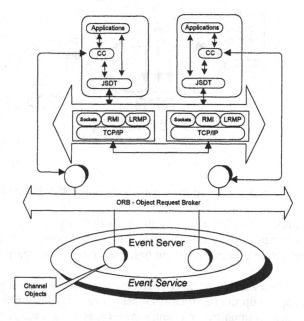

Fig. 4. Conference system communication architecture

The CORBA *Event Service* simplifies application development because it allows asynchronous communication (event messages) between consumers and suppliers, without the need to know in advance the communicating parties, which makes it very attractive for group distributed communication [11][12].

However, this service has some limitations that span the following important issues: multicast communication environments [13][14][15], communication reliability [14][15], event filtering and correlation [12], event ordering and prioritisation [14], and bulk data handling [16]. In what concerns this last case, CORBA Event Service was conceived for the diffusion of small messages, so it has many limitations for non-real-time bulk multicast communication (e.g. data replication).

Our CC service was developed having in mind the need to overcome some of the above mentioned limitations. For this, some additional features were introduced, namely mapping mechanisms for native IP Multicasting (for a more efficient multicast communication), reliability process optimisation, filtering and total ordering of events/messages and fragmenting/reassembling of large messages. Moreover, as the CORBA Event Service is based on a centralised architecture, a channel is not more than another CORBA object that introduces a single point of failure [16]. With the implementation of multicast communication using native IP multicasting, channels are mapped for a specific IP multicasting address, which does not introduce a single point of failure [16] and reduces the computation requirement on the Event Channel server, contributing for the increase of system availability.

With these refinements, we intend to provide high scalability and performance to the developed conferencing system prototype, maintaining compatibility with ITU T.120 [17] recommendations.

Figure 5 shows the considered object model and its relationships, in Unified Modelling Language (UML). The conference structure is initiated by the Node object. This object has the responsibility for creating the *Generic Controller* and the *ChatApplication* and the *WhiteBoardApplication*. The *Generic Controller* is the core of the model, having the tasks of conference management, and creation and management of all of theremaining objects that constitute the physical resources and support the data communication mechanisms. When a conference is initiated by a Top Provider the *Generic Controller* interacts with the *EventsSupplierA*, *EventsSupplierB*, *EventsConsumerA* and *EventsConsumerB* objects. These objects deal directly with our CORBA *Event Service Java* library. In addition to communicating via ORB, the conference system uses another type of communication that is based on JSDT objects. In this way, the *Generic Controller* also has the task to create a session (*JSDTSession*) and a channel (*JSDTChannel*) for the control of information exchange. When it is intended to have a private channel or when synchronisation is to be carried out in disordered channels, the system creates a *JSDToken* object. To be able to consume the available information in the channels, it is necessary to have one object that acts as a client (*JSDTClient*), which will be joined to *JSDTSession* object, and thus will be in position to *hang* one *JSDTConsumer* object for each channel that will be interested in listening. All the interactions with databases are executed by the *DBExecute* object.

The conference prototype was developed to operate with two different types of Java collaborative applications: WB (WhiteBoard) and Chat.

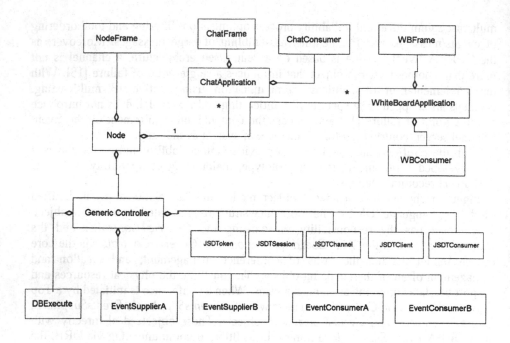

Fig. 5. Object model of the conference service

2 Extensions to the CORBA Event Service

As already mentioned, the standard OMG COS (CORBA Object Specification) Event Service has some limitations that span the following important issues: multicast communication environments [13][14][16], communication reliability [14][16], event filtering and correlation [15], event ordering and prioritisation [14], and bulk data handling [16]. In what concerns this last case, CORBA Event Service was conceived for the diffusion of small messages, so it has many limitations for non-real-time bulk multicast communication (e.g. data replication). The standard OMG COS Event Specification doesn't mandate that the Event Channel provide persistence. Therefore, if some problems occur and the host is shut down, it is possible for the Event Channels to lose events and connectivity information [18], which turns them inadequate for publish/subscribe environments.

The proposed extensions to the CORBA Event Service, which we name *CORBA Event Service Reliable Multicast Architecture* extends the standard OMG Corba Event Service specification with mechanisms for reliable IP multicast communication, with total ordering of the event transmission, filtering and fragmenting/reassembling of large events.

As in the original specifications made by OMG CORBA Event Service, Suppliers and Consumers are decouple, that is, they don't know each other's identities; so that the standard Event Service may still be provided by this extended service.

The approach on which we intend to work is the one called *Push* model, no work is being made on the *Pull model*. So, we intend to allow the consumer to be sent the event data as soon as it is produced. The canonical Push Model allows Suppliers of events to initiate the transfer of event data to Consumers.

In this phase of development, we do not address filtering, identity of sources and persistence limitations. These will be addressed in further stages of the work, after the stability of the architecture is achieved.

In order to overcome some of the above mentioned limitations and to achieve a powerful and scalable solution, we added new CORBA IDL Multicast interfaces to the modules *CosEventComm* and *CosEventChannelAdmin* of the OMG standard COS Event Service architecture (Figure 6). The CORBA IDL interfaces details are in Appendix.

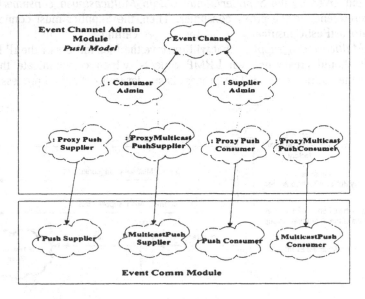

Fig. 6. Structure of IDL Interfaces in the CORBA Event Channel Reliable Multicast Architecture

2.1. The Event Channel

The Event Channel must keep the properties of the Event Service provided for standard use, and further provide interoperability between both services. Connections with the Event Channel use the same interfaces *SupplierAdmin* and *ConsumerAdmin*, each one providing new methods that return references to the respective proxies. In addition, new methods are provided to connect Suppliers and Consumers with their proxies, which also return the *InetAddress* (Figure 7). The *InetAddress* will be used either by the Supplier or the Consumer, to directly invoke Multicast communication of events. Consumers using the standard or multicast interfaces may receive events that are being pushed by standard or multicast Suppliers. This is one of the basic reasons that leaded us to implement one single proxy (*ProxyMulticastPushSupplier*) to deal with all multicast suppliers, and also, one single proxy (*ProxyMulticastPushConsumer*) to deal with all multicast consumers, using the same Multicast Group. As one can easily imagine, in this way redundancy of event communications can be avoided. This Event Channel provides a set of new interfaces for administrating Multicast Groups.

2.2. Suppliers

This service provides a *MulticastPushSupplier*, in the CosEventComm module, that applications can use for multicast communication. The multicast Supplier will register (which requires the steps shown in Figure 7) in the Event Channel by invoking the new method given by the *SupplierAdmin*: *obtain_Multicastpush_consumer()*, which gives the reference to the proxy consumer. Then, the Supplier must connect to the ProxyMulticastPushConsumer, by calling its method *connect_Multicastpush_supplier*, that will retrieve the InetAddress of the IP Multicast Group. It should create its own LRMP objects, which communicate the events directly to the multicast group. This process is similar to the standard process.

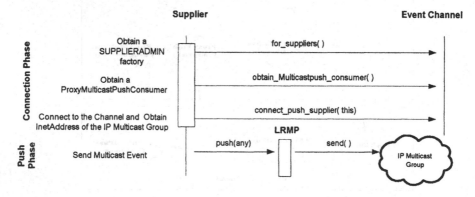

Fig. 7. Connecting a Multicast Supplier to an Event Channel

2.3. Consumers

Multicast Consumer is defined as a new interface: *MulticastPushConsumer* in the CosEventComm module. If applications are willing to receive multicast event communication, they may use an instance of this interface and connect to a ProxyMulticastPushSupplier on the Event Channel side. After receiving the InetAddress of the Multicast Group, they should create their own LRMP objects, that pick events directly from the group and push them to the Consumer. This process requires the steps shown in Figure 8. If applications are willing to use the standard communication, they may use an instance of the standard interface.

An application could have a Standard Consumer and a Multicast Consumer, for receiving events supplied by multicast Suppliers and supplied by standard Suppliers, which are dealt by different push interfaces.

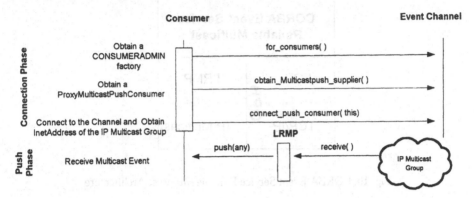

Fig. 8. Connecting a Multicast Consumer to an Event Channel

2.4. The Service Architecture

The reliable multicast extension can be seen as second way, or the other way to get the event across to the consumer. This assumption obliges the new service to keep all the standard interfaces with the same functionality (the same methods) that are defined by OMG, for a choice to be made by the supplier/consumer: IIOP or Reliable Multicast. Figure 9 shows a multicast supplier pushing events to multicast and standard consumers.

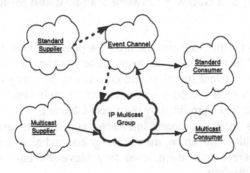

Fig. 9. High-level view of the CORBA Event Service Reliable Multicast Architecture

The reliable multicast solution, is based on the Light-weight Reliable Multicast Protocol (LRMP), which deals with IP Multicasting and provides the necessary reliability (Figure 10).

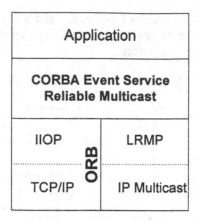

Fig. 10. CORBA Event Service Reliable Multicast Architecture

LRMP has been implemented in Java as a reusable library. It is designed to work in heterogeneous network environments with multiple data senders. A totally distributed control scheme is adopted for local recovery so that no prior configuration and no router support are required [10]. LRMP works in NACK-only mode to ensure reliability for the majority of receivers. Some congestion control mechanisms are included to fairly share network bandwidth with other data flows [10]. LRMP has been used with different applications and a number of measurements have been carried out, with convinced results [10]. Inadequate data fragmentation may introduce inefficiency in the network and should be generally avoided [10]. LRMP places a upper bound on the maximum length of packets that can be transmitted, in other words, the maximum transmission unit (MTU). Therefore proper data fragmentation/reassembly are left to our service (higher layer protocol) to perform.

2.5. Interoperability and Gateway Between Standard and Multicast Event Communications

On behalf of interoperability, our service must be provided with the ability to work over any Java Object Request Broker. The proposed service guarantees interoperability between event suppliers and consumers, in a way that is independent of the used communication facilities - standard or IP multicast. To achieve this, a transparent gateway must be provided by the Event Channel.

We described the most important interfaces and objects used in the service implementation to achieve the above mentioned goals, and also described the objects and methods that operate upon them, used to achieve the integration of the " *Any* Values" upon LRMP packets.

As can easily be understood, the interfaces that are provided by OMG must be maintained by our service in order to deal with *Any* type of data. In this way, we try to achieve interoperability between the proposed service and most of the commercially available ORBs.

In order to deal with all the values supported by the *Any* type, we created a new object *Any*, as an extension of the OMG class, to override the ORBs implementation.

The values to be passed to this service, in order to be sent via multicast, shall be created with the proprietary interfaces of the service, that provide a set of new methods to deal with the resulting buffers and fill the LRMP data packet with the data contained in those buffers.

So, by creating a method for extracting the sequence of bytes generated by the *Any* marshalling operation, the result is a stream of bytes that mean nothing to LRMP. Once they get across to the consumer LRMP object, it reassembles them to form an *Any*, and delivers them to the consumer as a valid *Any* that shall be extracted by the application interfaces.

To provide the inter-change of Multicast events and Standard events, two scenarios are considered (Figure 9):

- when the supplier is a *MulticastPushSupplier* and the consumer is a normal Consumer;
- when the supplier is a normal Supplier and consumer is a *MulticastPushConsumer*.

For the first scenario, and knowing that the model we deal with is the Push Model, the event is sent via LRMP, e.g. is sent to a well known multicast group, and is pushed to Multicast Consumers by one LRMP object that receives events directly (Figure 7 and 8). As has been explained before, the Event Channel holds one *ProxyMulticastPushConsumer* for all Suppliers. This proxy acts as a normal consumer to whom events are pushed. Each time this proxy is pushed a new event, it invokes the receive method on the Event Channel. This invokes a receive method on the *ConsumerAdmin*, that holds references to all *ProxyPushSuppliers* (standard) and shall invoke the receive method on all those proxies, that will then invoke the push method on the consumer they are attached to. The *ConsumerAdmin* does nothing on the *ProxyMulticastPushSupplier* or else events would be sent twice to the multicast group.

For the second scenario, the supplier calls the push method on its *ProxyPushConsumer*, that will call receive on the Event Channel, to get the event to standard consumers, and will also ask for any existent *ProxyMulticastPushSuppliers*. If there are any, the proxy shall call the receive method on that proxy. In this way, the event is sent to the multicast group, and all *MulticastPushConsumers* receive it. So, the Event Channel is responsible for doing all the necessary gateway work.

In both scenarios, the gateway operation implies changes to the *Any* Object that is being forwarded. The Event Channel will also take care of this detail, at the cost of an additional overhead caused by the de-marshalling of the proprietary *Any* and subsequent marshalling as ORB *Any*, and vice-versa.

3 Conclusions and Guidelines for Further Work

In the present paper, the authors presented a proposal for the extension of the CORBA *Event Service* with reliable multicasting communication capabilities, adjusting it to environments where data replication/updating actions are involved, as it is the case of conference control information according to ITU T.120 model.

This proposal is being evaluated in the context of a prototype development already operational [19]. This prototype provides a conference control service based on the CORBA Event Service, characterised by a scalability refinement of tightly-coupled conference models based on the ITU T.120 recommendations, and the use of reliable multicasting communication mechanisms (mapping of native IP multicasting, reliability process optimisation, filtering and total ordering, and fragmentation/reassembling).

Our following steps will be the fine tuning and further evaluation of the proposed extensions to the CORBA Event Service, as well as atomic multicast, filtering and a persistent service. In addition, we plan to investigate and explore the applicability of the proposed extensions to multimedia information distribution (VRML, XML, audio and video), and for environments with bulk data replication.

References

[1]   Object Management Group-OMG (1997): "CORBAservices: Common Object Services Specification", formal/97-07-04, July1997.
[2]   ITU-T Study Group 8, ITU-T Recommendation T.124 (Draft, March 1995): Generic Conference Control for Audiographic and Audiovisual Teleconference Applications.
[3]   IMTC (1996): IMTC GCC API, ftp://ftp.imtc-files.org/imtc-site/AP-AG/TECH/CD/GCC-API
[4]   IMTC (1996): Implementor's Guide, ftp://ftp.imtc-files.org/imtc-site/IMPGUIDE/impguid4.zip
[5]   Java Shared Data Toolkit (JSDT), SUN Microsystems, JavaSoft Division, http://java.sun.com/people/richb/jsdt
[6]   ITU-T Study Group 8, ITU-T Recommendation T.122 (1995): Multipoint Communication Service for Audio Graphics and Audiovisual Conferencing.
[7]   IMTC (1996): IMTC MCS API, ftp://ftp.imtc-files.org/imtc-site/AP-AG/TECH/CD/MCS-API
[8]   JavaSoft, www.javasoft.com
[9]   Tie Liao: "Light-weight Reliable Multicast Protocol Specification", Internet-Draf: draft-liao-lrmp-00.txt, 13 October, 1998.
[10] Tie Liao: "Light-weight Reliable Multicast Protocol", INRIA, France.
[11] S. Maffeis (1995): Adding Group Communication and Fault-Tolerance to CORBA, Proceedings of USEUNIX Conf. On Object Oriented Technologies, Monterey CA, June 1995.
[12] T. H. Harrison, D. L. Levine, Douglas C. Schmidt (1997): "The Design and Performance of a Real-time CORBA Object Event Service", Proceedings of the OOPSLA'97 conference, Atlanta, Georgia, October 1997.
[13] D. Trossen, K.H. Scharer (1998): "CCS: CORBA-based Conferencing Service", Proceedings of the International Workshop on Interactive Multimedia Systems ans Telecommunication Services (IDMS'98).
[14] S. Maffeis, D. C. Schmidt (1997): "Constructing Reliable Distributed Communication Systems with CORBA", on topic issue Distributed Object Computing in the IEEE Communications Magazine, Vol. 14, No. 2, February 1997.
[15] P. Felber, R. Guerraoui, A. Schiper (1997): "The CORBA Object Group Service", EPFL, Computer Science Department, Technical Report, 1997.
[16] K. Maad (1996): "Efficient Bulk Transfers over CORBA", Department of

Computer Systems, Uppsala University, Sweden, 1996.

[17]   ITU-T Study Group 8, ITU-T Recommendation T.120 (Draft, March 1995): Data Protocols for Multimedia Conferencing.

[18]   Douglas C. Schmidt, Steve Maffeis (1997): "Object Interconnections", SIGS C++ Report magazine, February, 1997.

[19]   João Orvalho, T. Andrade, Fernando Boavida (1999): "Extension to Corba Event Service for a Conference Control System", Proceedings IEEE International Conference on Multimedia Computing and Systems ICMCS'99, Florence, Italy, June 7-11, 1999.

Appendix: CORBA IDL Interfaces in the New Modules CosEventComm and CosEventChannelAdmin.

The following CORBA IDL interfaces are used in our implementation of the CORBA Event Service Reliable Multicast Architecture. The new module CosEventComm:

```
module CosEventComm
{

exception Disconnected {};

interface PushConsumer
{
    void push(in any data)
raises(Disconnected);
    void disconnect_push_consumer();
};

interface PushSupplier
{
    void disconnect_push_supplier();
};

// New interface to support multicast on the Supplier
interface MulticastPushSupplier
{
    void disconnect_push_supplier();
};

// New interface to support multicast on the Consumer
interface MulticastPushConsumer
{
    void push(in any data);   // 3
    void disconnect_push_consumer();
};
};


The new module CosEventChannelAdmin:

module CosEventChannelAdmin
{
```

---

3   No "exception disconnected" is thrown. This method will always be called locally by the object that provides Multicast event receptions.

```
   exception AlreadyConnected {};
   exception TypeError {};

   interface ProxyPushSupplier : CosEventComm::PushSupplier
   {
        void connect_push_consumer(in CosEventComm::PushConsumer
push_consumer)
   raises(AlreadyConnected, TypeError);
   };

   interface ProxyPushConsumer : CosEventComm::PushConsumer
   {
        void connect_push_supplier(in CosEventComm::PushSupplier
push_supplier)
   raises(AlreadyConnected);
   };

   // New (proxy) interface to handle Consumers using multicast,
   // this interface shall return the IPMulticast Group to be
   // used by Consumers

   interface ProxyMulticastPushSupplier :
CosEventComm::MulticastPushSupplier
   {
        // returns a long that corresponds
        // to the integer representation of the
        // IP address.

        long connect_Multicastpush_consumer(in
CosEventComm::MulticastPushConsumer Multicastpush_consumer);
   };

   // New (proxy) interface to handle Suppliers using multicast,
   // this interface shall return the IPMulticast Group to be
   // used by Consumers

   interface ProxyMulticastPushConsumer :
CosEventComm::MulticastPushConsumer
   {

        // returns a long that corresponds
        // to the integer representation of the
        // IP address.

        long connect_Multicastpush_supplier(in
CosEventComm::MulticastPushSupplier Multicastpush_supplier);
   };

   interface ConsumerAdmin
   {
      ProxyPushSupplier obtain_push_supplier();
        ProxyMulticastPushSupplier obtain_Multicastpush_supplier();
    };

   interface SupplierAdmin
   {
      ProxyPushConsumer obtain_push_consumer();
        ProxyMulticastPushConsumer obtain_Multicastpush_consumer();
   };

   interface EventChannel
   {
        ConsumerAdmin for_consumers();
        SupplierAdmin for_suppliers();
        void destroy();
   };
   };
```

MBone2Tel – Telephone Users Meeting the MBone

Ralf Ackermann[1], *Jörg Pommnitz*[2], *Lars C. Wolf*[1], *Ralf Steinmetz*[1,2]

[1]	[2]
KOM	GMD IPSI
Darmstadt University of Technology	Dolivostr. 15 • D-64293 Darmstadt • Germany
Merckstr. 25 • D-64283 Darmstadt • Germany	

{Ralf.Ackermann, Lars.Wolf, Ralf.Steinmetz}@kom.tu-darmstadt.de
Joerg.Pommnitz@darmstadt.gmd.de

Abstract: The integration of the Internet with the existing PSTN found much interest recently. While in many cases the aim is to use the Internet as a carrier medium for voice phone calls, we will present a proposal to extend classic Internet multimedia services to users on the PSTN.
We introduce an architecture and prototype implementation that allow users on the PSTN to actively participate in MBone audio conferences. We present a set of building blocks for applications using voice access via telephones and give an overview of how these can be combined to meet individual requirements. Finally, we discuss application scenarios, usage experiences and potential future enhancements to improve usability and to extend the range of available services.

Keywords: Multimedia Gateway Technology, MBone, Public Switched Telephone Network and Internet Internetworking.

1 Introduction

The MBone forms the IP Multicast Backbone [9] of the Internet. In operation since 1992, it uses multicast directly and connects networks capable of supporting it via unicast IP tunnels. Originally meant to be a research tool in the multicast area it quickly became a valuable resource for users outside the network research area as well.

Today it is widely used to distribute multimedia contents like audio and video to recipients in diverse environments ranging from research and education to business and entertainment all around the world. Though it has been used for some years now, IP multicast is still considered a somewhat uncommon feature and is not yet available to all users on all platforms. One of the technical reasons for this is the difficulty to send multicast traffic over point-to-point dial-up links to end nodes. Since many people access the Internet over point-to-point dial-up links, they have to establish IP-IP tunnels across their access link. This imposes an additional burden on the Internet Service Provider (ISP) and the end user.

Additionally, in many situations, for instance for mobile users, access bandwidth, hardware resources or the available software for the used devices are not sufficient for using attractive MBone services directly, though the user may have – at least temporarily – some sort of connectivity to the Internet. In these cases, even applications like mTunnel [10] which allow to easily build up and use multicast tunnels are no adequate means.

One of the most attractive MBone services is audio (and video) conferencing using tools such as the "Robust Audio Tool" rat [11] or vat. Our work was initially inspired by the lack of an adequate access facility for our research staff members. When they worked at home but tried to take part in a scientific conference transmitted via the MBone, we actually had to attach the handset of a conventional phone to a workstation loudspeaker manually.

The idea came up, that if it would be possible to access MBone audio conferences from standard Public Switched Telephone Network (PSTN) terminals (i.e. simple phones or video phones) in a convenient and comfortable way, the possible audience for MBone distributed content could be significantly increased. Hence, the MBone2Tel Gateway (Figure 1) described in this paper is designed to bridge the gap between the MBone and the PSTN.

After a discussion of related issues in the next section, we will describe the concept and architecture of the gateway in Section 3 before we focus on both the implementation of audio forwarding (Section 4) and the control facilities (Section 5). Finally, usage scenarios and possible extensions of the Gateway are shown.

2 Related Work

Over the last years many MBone based applications have been developed, covering the domain of audio and video conferencing including archival of MBone sessions and the reliable and efficient distribution of data to many customers.

Much of that work has been done as part of the MERCI (Multimedia European Research Conferencing Integration) [2] project which resulted in the availability of powerful applications like the *Robust-Audio Tool* (rat). These tools can cope with varying network conditions causing jitter, corruption or packet loss and still deliver a reasonable good quality to the end user and interoperate with other solutions. Their main drawback is that they are limited to a pure MBone environment and cannot be used by plain PSTN users.

The interworking between MBone tools and applications implementing the ITU recommendations of the T.12x and H.3xx series is regarded to be of high importance, as stated e.g. in [5]. The ITU recommendations describe the implementation of multipoint multimedia conferences both for circuit switched networks as well as for packet switched ones. Especially the H.323 standard [8] is one of the most important protocol candidates within the evolving field of IP telephony.

To provide interoperability of the MBone and the PSTN world, work has been carried out as part of research projects such as MERCI as well as by individual commercial initiatives [18]. These efforts resulted in the design, development and public availability of applications and services [19] which are capable of forwarding multimedia content between the MBone and the PSTN H.3xx domain while also providing means for multiple access and an adequate mapping of control semantics. While this already broadens the coverage of potential users, it still leaves those without access to H.3xx/T.12x equipment without support. These can be users who are either not attached via the necessary access lines, such as mobile users with a GSM phone or persons who – probably even more often – may not use the appropriate applications due to hard- or software restrictions. Therefore, our implementation does not compete with approaches that bridge into the T.12x or H.3xx world directly. Our work is orthogonal and covers a large and yet still mostly unsupported area.

Actually there are many efforts to use the Internet as a carrier medium for telephone calls and to integrate the usage of the PSTN with Internet applications. The IETF working group "Internet Telephony" (iptel) [12] concentrates on the engineering of protocols to be used for IP telephony applications and signalling as well as on the deployment and interaction of gateways between the Internet and the PSTN. This primarily focuses on the transport of telephone traffic via the IP infrastructure.

The "PSTN-Internet Internetworking" (pint) [13] working group describes application scenarios and frameworks such as "Click to dial" and "Voice access to Web content". Many of the building blocks we implemented can easily be adapted to be used for providing a selection of these services as well.

In summary, while these approaches provide access to MBone sessions from H.32x devices resp. use the Internet as transport medium for phone calls, our work adds the support for phone users to access MBone sessions.

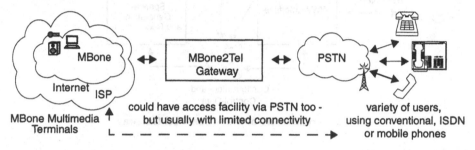

Fig. 1. MBone2Tel Gateway

3 The MBone2Tel Gateway

3.1 Intended Features and Design Considerations

The design of the MBone2Tel Gateway was driven by the following goals:

- implement the basic service of forwarding audio content bidirectionally between MBone audio conferencing sessions and the PSTN,
- do not restrict the intended audience by means of special hardware or software requirements
- keep the operational overhead for a regular service as small as possible while still retaining a maximum of configurability,
- develop components which can easily be adapted and combined to support further scenarios such as dial-up voice access to Email or active notifications on the occurrence of certain events.

There are various well established MBone and other audio conferencing tools. We decided to use already available and proven components and concentrate on enhancing and integrating them in a new way. By keeping the interfaces towards these applications as small and universal as possible, we can also benefit from advances in these base tools.

As shown in Figure 1, the MBone2Tel Gateway is located between the Internet and the PSTN and is connected to both of them. A regular phone user makes a standard PSTN call to the gateway and can participate in a MBone session. Figure 2 shows a more detailed view of the gateway and its building blocks as well as their interactions.

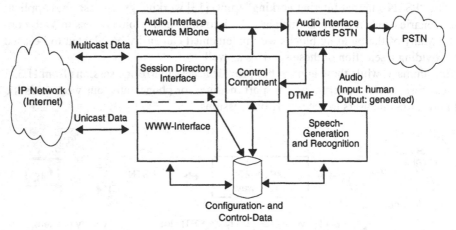

Fig. 2. Architecture of the MBone2Tel Gateway

3.2 Basic Operation

When a user with a conventional phone dials up the gateway, the Audio Interface towards the PSTN accepts the call and presents a short introduction message and a description of the available services is generated on the fly by a speech generation component. All the control information is locally stored and is dynamically updated either by the system's operator or automatically through interacting with the Session Directory Interface that permanently receives and decodes Session Announcements from the MBone. The caller is then guided through an audio menu and may select his further operation either by Dial Tone Multi-Frequency (DTMF) tones, by giving single word answers to questions or by naming a MBone session directly through its multicast address and port.

When a valid session is chosen, an enhanced "rat" program is started with the corresponding parameters and audio data is forwarded bidirectionally through the Audio Interface towards the MBone. After the connection is set up that way it can be used to take part in the selected MBone audio conference as passive listener or even as an active talker and the configuration database is updated to represent the current state of the gateway.

All the control and management information can also be accessed via a WWW and a Java based interface that is used locally as well as from the Internet via a unicast connection. The management of the system is also done this way.

4 Audio Forwarding

The prototype implementation of our MBone2Tel Gateway has been done on the Linux operating system and largely benefits from the features the system supports for accessing the PSTN via ISDN-cards or voice modems.

The part responsible for audio forwarding (see Figure 3) is basically split into two components. One component, the audio interface towards the MBone, receives and decodes the incoming data stream from the network in the receiving case or encodes and sends it when the phone user speaks. It is based on the popular MBone "Robust Audio Tool" (rat) which has been extended with a generic audio interface that is able to forward and receive audio data using the systems IPC mechanisms, namely Named Pipes. By using "rat" we profit from the work already done in the area of jitter compensation and redundant audio transmission as well as from its adaptivity to varying network conditions.

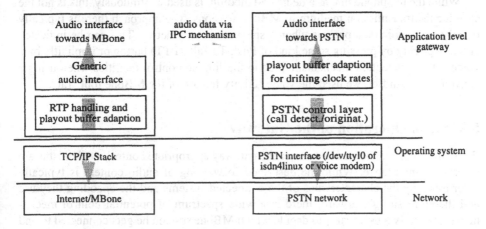

Fig. 3. Audio Forwarding – Components and Data Flow

The other component, the audio interface towards the PSTN, waits for incoming calls or originates them, establishes a connection and sends the audio data to the PSTN. Doing our experiments in Germany, where the ISDN system has a widespread deployment, we used the functionality provided by the ISDN kernel extension (isdn4linux) [14] of the Linux operating system. It provides a modem emulation accessible over a TTY-like interface. The code we implemented has been largely inspired by the "vbox" package [15], which implements a telephone answering machine on top of the isdn4linux facilities.

To keep compatibility with conventional analog modems and to have a means of providing the service also without ISDN, we decided to use the isdn4linux modem emulation interface and to strictly follow a layered approach. This modem emulation is fairly complete and provides extended commands for voice processing compatible with the popular ZyXEL voice command pseudo standard. It supports features like voice activity detection and real-time in-band dial tone (DTMF) decoding which can be used to trigger control operations.

Within the whole system – which also has interfaces towards a synthetic speech generation and a speaker independent single-word recognition module, that we will describe later on – we use a uniform 16 bit signed little endian audio format. Thus the components which all have to implement an interface towards this format can be combined in a very easy manner and form a kind of a tool box.

The independent timing used by the two components of the gateway system was a subtle problem. Forwarding audio data from the MBone to the PSTN layer and vice versa is typically application driven. In our scenario data has to be sent at a rate that corresponds to the sampling rate of the PSTN side (e.g. 8 kHz in the ISDN system). This rate should be the same for both parts, but the sampling clocks may drift apart. If the "producer" is too slow, the audio stream gets interrupted which is audible by annoying clicks. When too much data is arriving, buffers may overflow and audio information gets lost. We cope with these problems by using a playout buffer adaption algorithm.

While the telephone line as a dedicated medium is used continuously, this is not the case for the transmission medium in MBone sessions. The approach there is typically to only emit audio data packets when a speaker is really active. This can be indicated either explicitly by pressing some kind of Push To Talk (PTT) button or implicitly by a voice activity detection mechanism. Due to the limited control facilities when using a conventional handset, we used the voice activity feature of the MBone tool "rat".

5 Control Operation of the Gateway

For widespread and convenient use of the gateway appropriate control mechanisms are of crucial importance. While the described forwarding of audio content is typically determined by the characteristics of the connected systems and the encoding they use and therefore straightforward, there is a wide spectrum of potential control mechanisms. Basically a user wants to decide which MBone session he gets connected to and whether the gateway should play a passive role by just accepting calls or an active one by calling him itself.

Typically, there is a varying number of different MBone sessions that can be received at the gateway location at a time and might be of potential interest for a caller. There are multiple ways for a user to select one of these sessions. The various possibilities to control the operation of the gateway can be categorized along the following parameters:

- the direction in which descriptive data or control information is transferred,
- the media and encoding used for the descriptive and control data,
- the time when the information is presented or the control operation takes place.

This leads to the matrix of usage scenarios shown in Table 1. We show the third dimension (time) by distinguishing between white (if the information flow or control operation takes place before the actual session) and shaded (if the operation takes place during the session) table fields.

white – before the session	shaded – during the session

		data presented to the user	descriptive and control data originated by the user
in-band (using the session channel)		(1) Audio description of available sessions and navigation support in a voice guided menu – but in a call before the actual session itself.	(2) Control managed via an initial call to a dedicated service telephone number (which could even be operated by a human), while the gateway's MBone audio data service is provided by another telephone number.
		(3) Equivalent to (1) but as part of the session.	(4) Navigation through a menu by means of DTMF touchtones generated either by the telephone itself or an adequate external dialer. Additionally speaker independent single word recognition can be used to choose alternatives presented by the voice menu.
out-of-band		(5) Calls can be originated by the system whenever they are scheduled and the MBone session actually starts. Additionally all the features described under (7) may be used.	(6) MBone sessions can be initiated and advertised via a proxy interface to a session directory tool, thus not requiring direct MBone connectivity. Additionally all the features described under (8) may be used.
		(7) The announcement of available services (dial-up number according to a certain MBone session) can be done by means of Electronic Mail or a WWW-Interface.	(8) User may actively manipulate the assignment of a dial-up number to a MBone session, may give additional information (such as his real name) and may change transmissions parameters or activation of its channel using a WWW browser interface enhanced with a Java Applet.

Tab. 1. Systematization of Control Modes

All of the described cases are supported by the gateway and have been evaluated. We now describe the most common ones in more detail and give an overview of their implications and the software components that have been developed to support them.

5.1 Static Mapping between MBone Sessions and PSTN Numbers

In the case of a static mapping from MBone sessions to PSTN numbers, the session to be accessed through the gateway is fully determined by the telephone number the user dials up to and the time he does so. In a very basic implementation the provider of the gateway may thus statically assign a MBone session to a certain number and – as a first additional service – reject calls or present a corresponding informative audio message during the time the session is not yet or no longer active. This configuration, while usable and implementable without much additional effort, still has a considerable management overhead and an even more serious problem: How do potential users learn about this mapping? In order to solve this problem it is desirable to give users additional information and let them choose the session themselves.

5.2 User Controlled Session Mapping

With a user controlled session mapping the particular session which a user receives can either be predetermined before the call is done or may be chosen or even changed while the telephone connection is already established. In this approach descriptive information and control data may either be handled in-band during the connection or out-of-band through a call in advance or via a computer based interface. All variants will now be explained in more detail. We show that despite of our initial concentration on access by means of a simple telephone also the latter mode where a computer is used in addition to a phone may be very useful.

5.2.1 Integration of a Text-to-Speech, DTMF- and Single-Word-Recognition Component

For a regular service, it is desirable to

- have a means to inform the user about the session he will be attached to when he dials up a certain number,
- to guide him through navigation menus, where he can choose between several options
- or to inform him about errors that have occurred.

While some of this information is static and could be prerecorded and replayed whenever this is necessary, problems may arise for data that changes frequently such as, for instance, the MBone session announcements. Session announcements are encoded according to the rules of the Session Description Protocol SDP [4] and transmitted using the Session Announcement Protocol SAP [6]. A tools like the Session directory tool sdr [7] can be used to send and receive announcements.

We integrated a Text-to-Speech component which gives us the opportunity to generate audio information on the fly just by means of accessing text input files or commands. Since it is freely available and supports a rich set of languages and speaker voices, we use the Festival Speech Synthesis System [3] here. Due to its command-line interface, powerful Scheme-based scripting language and the free availability of all sources for research and educational purposes, it can be augmented to use the IPC mechanisms used in our gateway to attach various audio sources.

In our scenario, the speech synthesis component is used to generate audio from english text input and has been proven to work quite satisfying. It supports an operation mode that – after a short introduction message – guides a caller through a menu of available MBone sessions from which he may choose by signalling through DTMF tones.

Since DTMF tone generation is not always available, we incorporated the speaker-independent Single-Word-Recognition component "EARS" [16] into our system. It has been enhanced with the generic audio interface and may thus be trained and used via a telephone line. In the interest of a high recognition rate we limited the vocabulary to the answers "Yes" or "No" for decision questions, the ordinal numbers 0 to 9 and the word "dot". This way a user may choose to select a MBone session – that must be known to him – directly by naming its associated multicast address and port. If the recognition result is valid within the current context we trigger the selected operation, otherwise the question and its possible answers are presented to the user again. Doing both the training and usage experiments with a couple of different users we got an acceptable behavior of the recognition component though its utilization needs very disciplined callers and takes some time to get accustomed to.

After a session is chosen the gateway starts up the "rat" component and connects the audio input and output streams. This mode is suitable for users with a conventional phone and no further means of control. However, a selection-only mode requires the operator of the gateway to do a preselection of the MBone sessions presented, otherwise the number of alternatives is usually unacceptable large.

5.2.2 Management via a WWW-Interface

As stated before the number of MBone sessions available simultaneously at a site makes handling them just by means of DTMF or voice navigation quite uncomfortable or even impossible and typically requires a considerable human configuration effort for preselecting a reasonable subset. Therefore, in order to provide a regular service, we incorporated a WWW interface that allows to manage the system in an easy and yet very efficient way.

It reads session announcements that are received and decoded by the session directory tool sdr and enables a remote or local user to assign dial-in or dial-out connections to a certain MBone session by means of a Java applet.

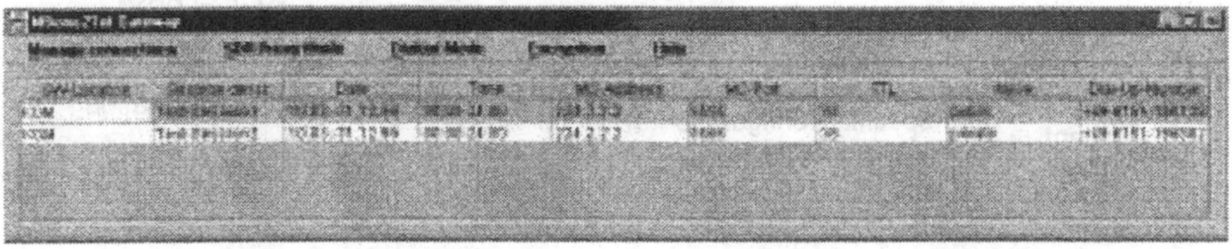

Fig. 4. Controlling the Gateway via a Java Interface

Though it may not be obvious from the very beginning, why a user who is directly connected to the Internet – which is a necessary prerequisite for using this management facility – should use the gateway, there are several reasonable scenarios, where this is quite desirable. Examples are mobile users, who scheduled the configuration in advance or are using a device such as a personal communicator via a GSM data connection, but have no means for taking part in the audio conference otherwise. The interface also allows the initiation of MBone sessions by means of a proxy session directory mode. This is important for users who want to originate a MBone session but have no means of accessing the MBone directly. Users may also be called by the Gateway at a preconfigured time when using the "Dialout Mode".

6 Conclusion and Future Work

The gateway has been in experimental operation for a few months at our site and has proven to work stable. It is mainly used in the operation mode incorporating the WWW control component that allows to configure it using the information that is dynamically generated by the session directory tool sdr.

We also envision a couple of other interesting application scenarios. Universities or local ISPs could establish dial-up nodes that allow their students or customers to follow lessons distributed over the MBone. This would be useful for tele-learning applications.

With an adequate deployment of gateways it is also possible to arrange ad-hoc multiparty telephone conferences by just using conventional phones and having the MBone build a kind of virtual Multipoint Control Unit (MCU) (Figure 5). Such a scenario is otherwise still difficult to set up since it usually needs to be planned and arranged with a MCU provider in advance.

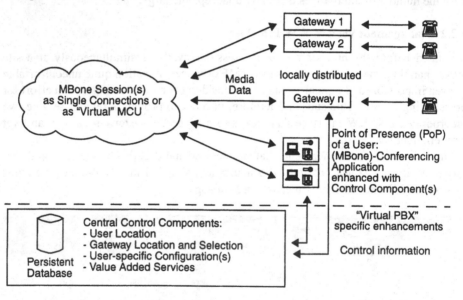

Fig. 5. Gateways Forming a Virtual MCU for Multiparty Telephone Conferences

The configuration can even be adapted to be used as a "Virtual PBX" [1], which in combination with an existing conventional telephony exchange provides additional services such as local user mobility.

When using a packet switched network for transmitting telephone calls, security considerations play an inherently important role for ensuring the callers privacy. This applies even more in our scenario where any user or malicious attacker can subscribe for receiving the corresponding multicast packets by just sending adequate IGMP messages. The audio tool "rat" supports closed group sessions by means of encrypting its audio data streams, whereas session keys have to be agreed upon, exchanged and typed in externally and all communication partners in a session have to use the same key.

Therefore, we suggest to incorporate public key mechanisms for determining and exchanging session keys to authorized components which have to prove their authenticity in advance.

We consider parts of the implementation as appropriate building blocks for future applications spanning a wide range and covering for instance audio access to mailboxes or web content. With the emerging use of IP telephony, the realization of value-added applications becomes a very reasonable challenge. So it is quite reasonable to extent the accessibility of the gateway to users of IP telephony applications as well.

The MBone2Tel Gateway is a convenient way to extend the range of potential participants in MBone audio conferences. We identified a number of additional usage scenarios where the gateway has experimentally been proven to be of great value. Currently we try to extend the possible range of interaction by incorporating an implementation of the H.323 protocol [17] developed as part of the Linux VOXILLA telecom project.

References

[1]   R. Ackermann, J. Pommnitz, L. Wolf, R. Steinmetz. "Eine Virtuelle PBX", 1. GI-Workshop "Multicast-Protokolle und Anwendungen", Braunschweig, Mai 1999, S. 187-197

[2]   R. Bennett, P. T. Kirstein. "Technical Innovations Deployed by the MERCI Project", Proc Networkshop 25, Belfast, March 1997, pages 181-189

[3]   A. Black and P. Taylor. "Festival Speech Synthesis System: system documentation (1.1.1)" Human Communication Research Centre, Technical Report HCRC/TR-83, 1997, http://www.cstr.ed.ac.uk/projects/festival/festival.html

[4]   A. Black and P. Taylor. "Festival Speech Synthesis System: system documentation (1.1.1)" Human Communication Research Centre, Technical Report HCRC/TR-83, 1997, http://www.cstr.ed.ac.uk/projects/festival/festival.html

[5]   S. Clayman, B. Hestne, P. T. Kirstein. "The Interworking of Internet and ISDN Networks for Multimedia Conferencing", IOS Press 1995

[6]   Mark Handley. "SAP: Session announcement protocol", Internet Draft, Internet Engineering Task Force, Nov. 1996. Work in progress

[7]   Mark Handley. "The sdr Session Directory: An Mbone Conference Scheduling and Booking System", Department of Computer Science University College London Draft 1.1, 14th April 1996, http://www-mice.cs.ucl.ac.uk/mice-nsc/tools/sdr.html

[8]   International Telecommunication Union. "Visual telephone systems and equipment for local area networks which provide a non-guaranteed quality of service", Recommendation H.323, Telecommunication Standardization Sector of ITU, Geneva, Switzerland, May 1996

[9]   Hans Erikson. "MBONE: The Multicast Backbone", Communications of the ACM, August 1994, Vol. 37, No. 8, pp. 54-60

[10]   P. Parnes, K. Synnes, D. Schefström. "mTunnel: A Multicast Tunneling System with a User Based Quality-of-Service Model", 4th International Workshop, IDMS '97, Darmstadt, Sept. 97, pages 87-96

[11]   Angela Sasse, Vicky Hardman, Isidor Kouvelas, Colin Perkins, Orion Hodson, Anna Watson, Mark Handley, Jon Crowcroft, Darren Harris, Anna Bouch, Marcus Iken, Kris Hasler, Socrates Varakliotis and Dimitrios Miras. "Rat (robust-audio tool)", 1995
http://www-mice.cs.ucl.ac.uk/multimedia/software/rat/

[12]   "Charter of the IETF Working Group IP Telephony (iptel)"
http://www.ietf.org/html.charters/iptel-charter.html

[13]   "Charter of the IETF Working Group PSTN and Internet Internetworking (pint)"
http://www.ietf.org/html.charters/pint-charter.html

[14]   "FAQ for isdn4linux" – Version pre-1.0.5
http://www.lrz-muenchen.de/~ui161ab/www/isdn/

[15]   "vbox – Anrufbeantworter für Linux"
ftp://ftp.franken.de/pub/isdn4linux/contributions/vbox-1.1.tgz

[16]   "EARS: Single Word Recognition Package"
http://robotweb.ri.cmu.edu/comp.speech/Section6/Recognition/ears.html
(no longer maintained or supported by the original author)

[17]   "OpenH323", Part of the Linux VOXILLA Telecom Project
http://www.openh323.org/

[18]   "RADVision Products for IP Telephony and Multimedia Conferencing"
http://www.radvision.com/

[19]   "The JANET Videoconferencing Switching Service"
http://www.ja.net/video/service/pilot_service.html

Quality of Service Management for Teleteaching Applications Using the MPEG-4/DMIF

Gregor v. Bochmann and Zhen Yang

School of Information Technology and Engineering (SITE), University of Ottawa, Canada
Bochmann@site.uottawa.ca, zyang@site.uottawa.ca

Abstract. In the context of distributed multimedia applications involving multicast to a large number of users, a single quality of service level may not be appropriate for all participants. It is necessary to distribute part of the QoS management process and allow each user process to make certain QoS decisions based on its local context. In order to allow for different QoS options, we assume that each source provides, for each logical multimedia stream, several different *stream variants*, representing different choices of user-level QoS parameters. The paper presents the design of a teleteaching system which uses this paradigm for QoS negotiation, and explains how the Delivery Multimedia Integration Framework (DMIF) of MPEG-4 can be adapted as a session protocol for such an application. In fact, it appears that this DMIF protocol, which is now being extended by ISO (DMIF Version 2) to the context of multicasting, provides some general session management functions which are quite useful for many distributed multimedia applications using broadcasting.

1. Introduction

Over the past several years, there has been a large amount of research focussed on the issue of QoS management techniques. The topics range from end-to-end QoS specification, adaptive QoS architecture, to QoS management agents, etc. The research covered both architecture and implementation issues of QoS management functions, such as QoS negotiation, QoS renegotiation, QoS adaptation, QoS mapping, resource reservation, and QoS monitoring. For instance, [1] investigated the problems related to providing applications with a negotiated QoS on an end to end basis. [2] illustrated some examples of applications that adapt to a certain situation in which the available QoS may be severely limited. In our previous CITR project, "Quality of Service negotiation and adaptation", solutions were developed for applications involving access to remote multimedia database [3], [4].

The concept of multicasting arose years ago [8]. It is playing an important role in distributed multimedia applications, such as remote-education, tele-conferencing, computer supported collaborative work, etc. However, QoS management in the context of multicast has only been addressed recently. Most work in this field has been done for providing QoS-based multicast routing schemes. [9] proposed a multicast protocol that considers QoS in the routing phase and can create a multicast tree that better meets the needs of QoS sensitive applications. Another study was done to provide QoS control for an existing core-based multicast routing protocol [10]. Stefan Fischer and other researchers proposed a new scheme for cooperative QoS management, which takes a different approach. They considered an adaptive application that suits different QoS requirements of different users in the context of multicast applications [5] [6] [7]. Our project has been greatly inspired by this approach.

Quality of Service management for distributed multimedia applications becomes more complex when a large number of users are involved. A single quality of service level may not be appropriate for all participants, since some of the users may participate with a very limited workstation, which cannot provide for the quality that is adopted by the other users. It is necessary to distribute part of the QoS management process and allow each user process to make certain QoS decisions based on its local context.

We have developed a framework for QoS management of tele-teaching applications. It was assumed that different variants (with the identical content, but different QoS characteristics) of mono-media streams are available to the students' workstations throughout the network by means of multicasting. A so-called QoS agent is installed on each student's workstation. The QoS agent may select the stream that is most appropriate based on the student's preference.

2. Providing QoS Alternatives for Multicast Applications

2.1. General Assumptions

We consider in this paper distributed multimedia applications where most data originates in one system node, called the sender, and is multicast to a large number of receiver nodes. A typical application would be teleteaching where the teacher's node is the sender and the students' nodes are receivers. We note that in such an application, there may also be some real-time data going from a receiver node to the sender and the other receiver nodes. For instance, a student may ask a question which is broadcast to all participants in the teaching session. However, we will ignore this aspect in our discussion.

Figure 1 shows the overall system architecture which we consider for a single instance of our multimedia application. The sender and receiver nodes are connected to one another through a network which supports a multicast service. The video and audio streams originating from the sender are therefore broadcast to the different receiver nodes that participate in the application session.

We assume that the users at the different receiver workstations have different quality of service requirements. These requirements may be due to the following reasons:

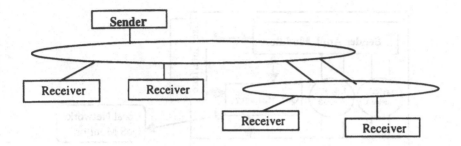

Fig. 1. Overall System Architecture

1) different *hardware and/or software resources* available in the workstation,
2) different *user-level QoS parameters* (that is, user preferences), such as requiring low-cost network service or high reception quality (which may imply higher costs); there may be different priorities for different aspects of quality, such as frame rate, color, resolution, disturbances through packet losses, etc.
3) different *transmission-level QoS parameters* (provided by the network) for the different receivers due to the specific network architecture and interconnection structure.

In order to accommodate these different QoS requirements, we assume that the sender node provides for each *logical multimedia stream* (such as for instance "video of teacher", "video of demonstration", and "audio of teacher") different *stream variants*, each representing a specific choice of user-level QoS parameters. Actually, some stream variant may represent several user-level qualities in the case that some form of scalable encoding is used; however, in most cases scalable encoding implies several elementary streams which can be combined in order to obtain a specific quality.

2.2. General System Architecture

Figure 2 shows a general system architecture for multicast applications with QoS alternatives. The figure includes a sender node and two receiver nodes. The receiver nodes communicate with a user profile manager that contains information about the user's QoS preferences and may also be used for user authentification. The sender node may also communicates with a local network QoS monitor which, in the case of best-effort networks, provides information about the transmission quality that is presently available.

The architecture of the sender and that of the receiver nodes are similar. They contain the application module that performs the application-specific functions, including all streams processing (e.g. video capture, coding in several stream variants, transmission, reception, decoding and display, etc.). The transport layer provides an end-to-end data transmission service with multicasting. In our prototype implementation, we assume that this includes the protocols IP (with MBone multicasting), UDP and RTP/RTCP. The session management layer looks after the

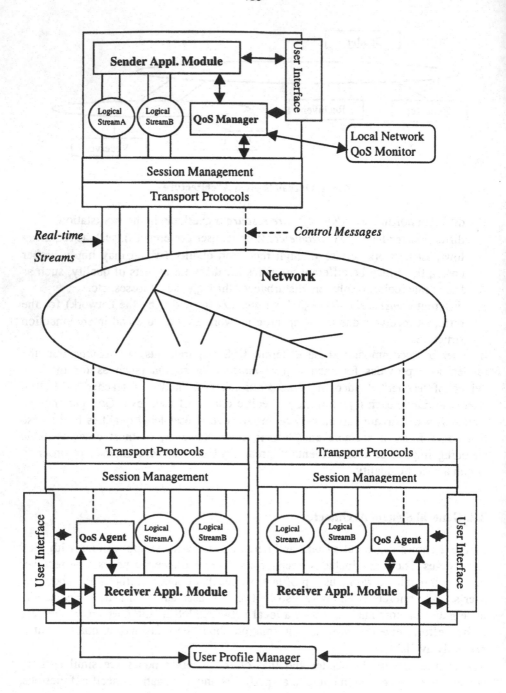

Fig. 2. General System Architecture

management of the application session, including the management of the transport channels for the different multimedia streams and the knowledge about the participating users.

The QoS manager in the sender node determines the list of potential stream variants for each logical multimedia stream. However, not all of these variants will actually be transmitted. A stream variant become active (and is transmitted) if the QoS manager considers that there are enough users that receive that stream, or that have requested that the stream be activated. For the selection of the potential stream variants and the activation of some of these variants, the QoS manager may take into account the information about the presently available network transmission quality that can be obtained from the local network QoS monitor and through the monitoring of the active transport channels. It may also take into account specific requests sent by the users participating in the application. An example of potential stream variants is shown in the table of Figure 3.

Video Stream A			
	Ch1	...	Chn
Frame Rate	10		30
Color	Grey		Color
Resolution	640*480		640*480
Coding Scheme	h.261		jpeg
Cost	10		20
Active Flag	Yes		No

Audio Stream B			
	Ch1	...	Chn
Quality	CD		Phone
Coding Scheme	PCM		PCM
Cost	2		1
Active Flag	Yes		No

Fig. 3. QoS Table for Logical Multimedia Streams

The QoS agent in the receiver node obtains information about the user QoS preferences (either directly through the user interface, or by retrieving the user's QoS profile) and selects for each logical multimedia stream a specific stream variants which best fits the QoS preferences of the user. In case that the most suitable stream variant is not active, it sends an activation request to the QoS manager at the sender node.

An example of a possible interaction scenario is shown in Figure 4. Point [1] shows the QoS Manager of the sender that initiates the session and broadcast the

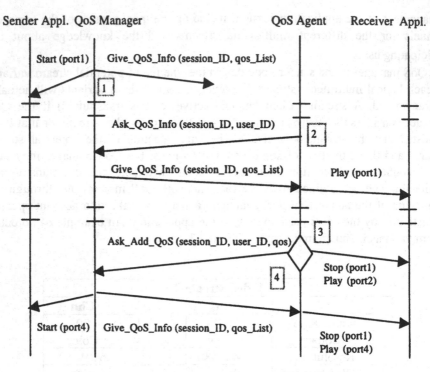

Fig. 4. Example of an Application-Level Interaction Scenario

Notes:

Point [1]: When the QoS manager initiates a session, it should broadcast the information about all potential stream variants. The parameter *qosList* contains the QoS table (see Figure 3) plus the IP addresses and port numbers of all active channels.

Point [2]: Some time later, a new receiver wants to join the session. First, its QoS agent asks the QoS manager for the QoS information (*qosList*) of the session. Then it selects an appropriate stream variant, and asks the Receiver Application Module to start the application with the selected stream.

Point [3]: After some time, QoS violation occurs. If there is another currently active stream that is acceptable, the QoS agent simply asks the Receiver Application Module to switch to that stream.

Point [4]: If no currently active stream is acceptable, the QoS agent asks the QoS manager to activate one of the other potential streams. The scenario above assumes that the QoS manager activates the requested stream; when the receiver's QoS agent is informed, it switches to the new stream. -- If none of the variants that the QoS Manager offers, whether active or not, satisfies the user's expectations, the QoS agent should inform the user about this situation and check whether the user wants to change the preferences in his/her profile.

information about the available streams. When a new receiver joins the session (point [2]), its QoS Agent sends first an *Ask_QoS_Info (session_ID, user_ID)* to the QoS Manager. The QoS Manager then sends *Give_QoS_Info (session_ID, qos_List)* to the Agent, providing information for all potential stream variants, as shown in Figure 3. According to the user's profile, the QoS Agent then selects the best stream variant (*port1* in our example).

When a QoS violation is detected by the QoS agent, it first checks whether one of the other currently active streams is acceptable. If so, it switches to it (point [3]), otherwise (point [4]) the QoS Agent will ask the QoS Manager to activate one of the other potential stream by sending an *Ask_Add_QoS (session_ID, user_ID, qos)*. If none of the streams, whether active or not, satisfies the user's expectations, the QoS Agent should inform the user about this situation and check whether the user wants to change the preferences in his/her profile.

Whenever there is a change in table of potential streams, the QoS manager should broadcast this information to all the receivers by sending a *Give_QoS_Info (session_ID, qos_List)* message.

3. The MPEG-4 / DMIF

MPEG-4 [11] is a new ISO standard on multimedia stream coding which goes beyond the previous MPEG-1 and MPEG-2 standards by providing adaptation to for low-data-rate transmission and support for multiple video and audio streams which can be useful for teleconferencing applications and applications involving virtual environments. In contrast to traditional coding standards, MPEG-4 allows the definition of video streams which represents objects with arbitrary contours, not only the rectangular screens of TV or film presentations.

The suite of MPEG-4 standards include a so-called Delivery Multimedia Integration Framework (DMIF) which is functionally located between the MPEG-4 application and the transport service. The main purposes of DMIF are to define a session level protocol for the management of real time, QoS sensitive streams, to hide the delivery technology details from the DMIF user, and to ensure interoperability between end-systems in the control plane.

DMIF defines an interface called DMIF application interface (DAI), in order to hide the delivery technology details from applications. Also, by using media related QoS metrics at the DAI interface, applications are able to express their needs for QoS without the knowledge of the delivery technology. It satisfies the requirements of broadcast, local storage and remote interactive scenarios. In addition, in case of interactive operation across a network, it ensures interoperability between end systems through a common DMIF protocol and network interface (DNI), which is mapped into the corresponding native network signaling messages. The basic DMIF concepts are defined in MPEG-4 version 1 [12]. Version 2, now being developed, specifies extensions for multicast scenarios [13]. Our project is based on these extensions.

The DAI provides primitives for an application to attach itself to a given session (or create a new session), called *DA_ServiceAttach*, primitives for adding or deleting a transport channel for the session (*DA_ChannelAdd* and *DA_ChannelDelete* primitives). It also includes a *DA_UserCommand* primitive that provides a means for transmitting, between the applications. Similar primitives exist at the network interface (DNI), which includes in addition so-called *SyncSource* and *SyncState* primitives which allow the distribution of the information about the multimedia streams provided by the different senders in the application session. The *AddChannel* primitives at the DNI also include a parameter for requesting different options of QoS monitoring for the transmission service provided by the channel.

In a multicast session, a DMIF terminal can be either a Data Producer DMIF Terminal (DPDT) that is a information source, or/and a Data Consumer DMIF Terminal (DCDT) that is a information receiver. A DMIF multicast session consists of a DMIF multicast signaling channel (C-plane) to distribute the state information of the session, and one or more multicast transport channels (U-plane) to deliver the multimedia data. We will use the DMIF C-plane in our project since it meets our need for message exchange between QoS manager and QoS agents.

The following are some details on the establishment of a multicast session between data produces and consumer DMIF terminals:

- A DMIF multicast session is identified by an DMIF-URL. DMIF-URL is a URL string, whose basic format was defined in DMIF version 1. It is used to identify the location of a remote DMIF instance. In the multicast scenario, this URL is extended with the role of the DMIF terminal in the multicast session (DPDT / DCDT). So that the local DMIF layer is capable of recognizing the role of the application (sender or receiver) in the multicast session.
- The session's signaling channel address (IP address and port number) is available to the interested DMIF terminals by mean of a session directory or through e-mail, etc.
- Each DPDT must explicitly join and leave the DMIF multicast session by sending messages over the signaling channel.
- When a DCDT joins a session, in order to reduce the number of signaling messages, it should listen on the signaling channel to collect the state information from all DPDTs participating in the session. If the DCDT does not acquire the information within a given time period, it should ask the DPDTs for their state information.

4. Using DMIF for Session Management in Tele-teaching Applications

Although the DMIF of MPEG-4 has been designed for use with MPEG-4 applications, it appears that its functionality can be used in a much wider context. It appears to be a quite general session protocol that can be useful for any application using a number of concurrent multimedia streams in a distributed environment.

Looking at it from this perspective, we asked ourselves whether it could be used for the teleteaching application described in Section 2. The answer is Yes, and the modalities are explained below.

We consider our teleteaching application in the context of the Internet. Therefore we assume that the underlying transport is provided by the IP/UDP protocols complemented with some multicasting facility. For our prototype implementation, we plan to use the multicasting facility provided by MBone. Each multicast channel, including the signaling channel, is identified by a broadcast IP address plus a port number.

We assume that RTP (and an associated RTCP) is used for each multimedia stream for the purpose of synchronization and for monitoring of the network-level QoS parameters. The DMIF network interface is therefore mapped onto the multicast transport service provided by RTP/RTCP. General guidelines are given in the DMIF specifications.

Over the DMIF layer lies the application. The abstract interactions shown in Figure 4 can be mapped to the DMIF primitives provided to the application through the DAI. A typical example is shown in Figure 5, which corresponds to the abstract scenario given in Figure 4. Specifically, the figure shows the following interactions:

Point [1]: The QoS manager initiates a session. The information on the potential stream variants (see Figure 3) is encoded in the user data field *uuData* of the *DA_ServiceAttach* primitive and forwarded by the DMIF through the *SyncSource* and *SyncState* messages which are broadcast over the network through the signaling channel.

Point [2]: The QoS agent sends a *DA_ServieAttach* primitive to inform the local DMIF protocol entity that a receiver wants to join the session. To avoid requesting the information that other receivers just requested, the DMIF entity listens on the signaling channel for a reasonable time period, and collects the *SyncSource* and *SyncState* messages from the sender side. If these messages are not gotten during a given time period, the DMIF entity should request them. The received information is passed to the QoS agent in the *uuData* parameter of the *Attach* confirmation.

Point [2']: The QoS agent selects an appropriate stream variant, and sends a *DA_ChannelAddRsp* primitive. This primitive is originally used to respond to the DA_ChannelAdd primitive, but it can also be used at any time during a session when the application wants to connect to a channel. The DMIF will perform a group join operation at the DNI in order to join the MBone multicast group for the selected multimedia stream. The primitive *DA_ChannelMonitor* is invoked in order to inform the DMIF to start monitoring the selected channel.

Point [3]: The DMIF entity indicates a QoS violation, with detailed QoS information in *qosReport* paramter. If one of the other currently active streams is acceptable, the QoS agent simply decides to switch to that stream.

Point [4]: If none of the currently active streams is acceptable, but an inactive streams is acceptable, the QoS agent sends a *DA_UserCommand* primitive to ask for the activation of that stream, as specified in the *uuData* parameter. If neither active nor inactive streams are acceptable, the QoS agent should ask the user whether he/she wants to change his/her QoS profile or abandon the session.

142

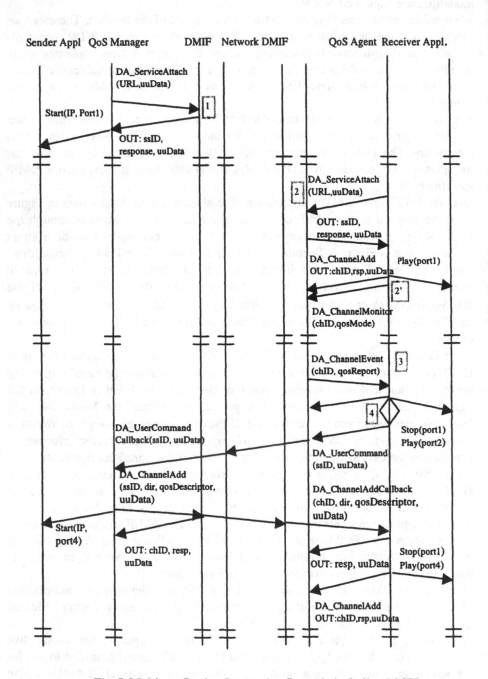

Fig. 5. Multicast Session Interaction Scenario including DMIF

We found that the DMIF specifications fit well with the session and QoS management functions that were required by our teleteaching application. However, there are certain points where the extensions for multicast applications described in ISO Working Draft of DMIF Version 2 do not fit completely our requirements. We mention in particular the following points:

- **SyncState message parameters:** The information about the potential stream variants (see Figure 3) is encoded in the user data field *uudata* of the *DA_ServiceAttach* primitive invoked by the QoS and session manager in the sender node. When a receiver node joins the application session, this information is passed along in the *SyncState* message from the sender to the receiver DMIF protocol entity. This requires certain changes to the *SyncState* message parameters, as specified in the Working Draft. More specifically, the SyncState message, as defined, contains information about the active channels, however, the QoS information included provides only information about the transmission-level QoS parameters of the channels, but not the user-level parameters contained in the table of Figure 3. These parameters must be added to the *SyncState* message. In addition, the information for the inactive channels must be added.

- **Requesting new stream variants:** Another issue is the question how a QoS agent in a receiver node could request that an inactive stream variant be activated. We have adopted the use of the *DA_UserCommand* primitives for this purpose. This primitive allows the transfer of user information from application to application, in our case from the QoS agent to the QoS manager in the sender node. If the QoS manager decides to activate a new stream variant, it will invoke a *DA_ChannelAdd* that leads to a multicast control message to all participating receiver nodes and a *DA_ChannelAdd* indication to all QoS agents.

5. Prototype Implementation

Our implementation environment is based on MBone. A single sender (teacher) node and several receiver (student) nodes communicate over the Internet via RTP/RTCP, UDP/IP and MBone protocols. At each side, there are two major parts: real-time video transmission and session control message transmission. For the first part, we did not develop a completely new multimedia application, but used an existed MBone video conferencing tool 'vic' at the teacher side, and a Java application 'RTP player' at the student side. 'RTP player' is part of the Java Media Framework (JMF), which specifies a programming interface for time-based media playback. It enables multimedia content in Java applets and applications, and allows for Web-based multimedia solutions that run in any Java compatible environment. However, these tools have to be modified in order to be suitable for multicast application and QoS management. The Java interfaces representing the abstract DAI primitives are specified in the DMIF (version 2) specification. We implemented these interfaces in the sender and receiver applications with the necessary modifications (mentioned above) for realizing the session and QoS management functions. The overall system is described in [14].

6. Conclusions

Session management for multimedia applications with multicasting is a complex task, especially when a large number of users are involved. It appears that the Delivery Multimedia Integration Framework (DMIF) for MPEG-4, which is presently extended for multicast applications, provides interesting session management functions for distributed multimedia applications in general, independently of the question whether MPEG-4 encoding is used.

We have shown how these DMIF functions can be used for session management of a teleteaching application including different QoS alternatives for the participating users. In such an application, the sender node of the teacher provides different stream variants (with different QoS attributes) for each logical multimedia stream. Each user participating as a student may then select one of these variants according to its QoS preferences.

The detailed analysis of the DMIF protocol in the multicasting context has identified certain generalizations that would be useful in order to make it more generally usable for various distributed multimedia applications. This includes general means for distributing user information from the stream producers to the stream consumers, and some means for sending general user requests from a consumer to a particular producer.

A prototype implementation of a teleteaching application with QoS alternatives is in progress and includes a variant of the MPEG-4/DMIF for the management of the application session and associated QoS management. A demonstration should be available in October 1999.

Acknowledgements

The authors would like to thank Khalil El-Khatib and Qing Zhu for some interesting discussions. This work is supported by research grants from Communications and Information Technology Ontario (CITO) and Nortel-Networks.

References

[1] D. Hutchinson, G. Coulson, A. Campbell, G. Blair. "Quality of Service Management in Distributed Systems", Network and Distributed Systems Management, page 273-303, 1994.
[2] J. Gecsei, "Adaptation in Distributed Multimedia Systems". IEEE Multimedia, 1997.
[3] G. v. Bochmann, A. Hafid, "Some Principles for Quality of Service Management", Distributed Systems Engineering Journal, 4:16-27, 1997.

[4] A. Hafid, G. v. Bochmann, "Quality of Service Adaptation in Distributed Multimedia Applications", ACM Multimedia Multimedia Systems Journal, volume 6, issue 5, 1998.

[5] S. Fischer, A. Hafid, G. v. Bochmann, H. d. Meer, "Cooperative Quality of Service Manaement for Multimedia Applications", proceedings of the 4th IEEE Internatinal Conference on Multimedia Computing and Systems, Ottawa, Canada, june 1997, pp. 303-310.

[6] S. Fischer, M. v. Salem, G. v. Bochmann, "Application Design for Cooperative QoS Management", proceedings of the IFIP 5th International Workshop on QoS (IWQoS'97), New York, May 1997, pp 191-194.

[7] A. Hafid, S. Fischer, "A Multi-Agent Architecture for Cooperative Quality of Service Management", Proceedings of IFIP/IEEE International Conference on Management of Multimedia Networks and Services (MMNS'97), Montreal, Canada.

[8] S. Deering, "Host Extensions for IP Multicasting", IETF RFC 1112.

[9] A. Banerjea, M. Faloutsos, R. Pankaj, " Designing QoSMIC: A Quality of Service sensitive Multicast Internet protoCol", IETF Internet Draft: draft-ietf-idmr-qosmic-00.txt.

[10] J. Hou, H. Y. Tyan, B. Wang, "QoS Extension to CBT", IETF Internet Draft: draft-hou-cbt-qos-00.txt.

[11] "Overview of MPEG-4 Standard", (N2725), http://drogo.cselt.stet.it/mpeg/standards/mpeg-4/mpeg-4.htm

[12] "Information technology – very-low bit-rate audio-visual coding – Part 6: Delivery Multimedia Integration Framework (DMIF)", (N2506), http://drogo.cselt.stet.it/mpeg

[13] "Information technology – very-low bit-rate audio-visual coding – Part 6: Delivery Multimedia Integration Framework (DMIF)", (N2720), http://drogo.cselt.stet.it/mpeg

[14] Z. Yang, "A tele-teaching application with quality of service management", MSc thesis, University of Ottawa, expected 1999.

IP Services Deployment: A US Carrier Strategy

Christophe Diot

Sprint labs ATL

Abstract. This talk will describe the Sprint approach to providing new and integrated services to the customers. The problem will be addressed from various standpoint, from backbone and access loop technologies to internet services and applications. We will start from the current technology deployed by Sprint (including the ION service) and see what is the expected evolution middle term and long term. This talk will focus more on the technological aspects than on marketing ones, even if customer requirements will be the central focus point justifying described technologies. Among the discussed technologies, we will pay a particular attention to multicast deployment, quality of service with regard to enabling new applications.

Network-Diffused Media Scaling for Multimedia Content Services

Omid E. Kia[1], Jaakko J. Sauvola[2], David S. Doermann[3]

[1]National Institute of Standards and Technology, Information Technology Laboratory
Gaithersburg, MD 20899
omid.kia@nist.gov
[2]Machine Vision and Media Processing Group, Infotech Oulu, University of Oulu,
90570, Oulu Finland
jjs@ee.oulu.fi
[3]Language and Media Processing Laboratory, Institute for Advanced Computing Studies
University of Maryland, College park, MD 20742
doermann@cfar.umd.edu

Abstract. In this paper we propose a new approach to adaptation of content-based processing, media preparation and presentation. We address multimedia consumption terminals having a variable amount of resources. The proposed technique adapts the required service data contained within the multimedia to the abilities of the hybrid network elements and application requirements. In our approach the data representations are transformed and converted efficiently to reduce delivery requirements and to emphasize quality of service ratio. An experimental system is implemented in a distributed environment providing only low processing requirements from the data processing nodes. We perform media profiling based on cross-media translation to achieve an efficient scaling and fit the content with terminal capabilities. Our approach aims to build an intelligent content service infrastructure in a way that the servers, nodes, and terminals are aware of their capabilities along with the capabilities of their surroundings. We demonstrate with an example service with scalable multimedia delivery over hybrid network elements.

1. Introduction

Service quality, transmission requirements and capabilities range widely across multimedia information services. Creating custom services for specific requirement and capability entities is difficult, and porting them efficiently to different delivery domains requires significant effort. Thus, it is beneficial to design a service infrastructure that automatically adapts to network availability and terminal capability. Media content adaptation technology installed in programmable gateways that embed content-based intelligence allows for implementation of such services. In a multicasting environment, it is desired to submit multimedia messages that will then be reprocessed for current network conditions and terminal capabilities. This will distribute the processing requirements progressively and allow for a more efficient information transmission.

Current "value added services" integrate circuit-switched telephony with packet-based networks to deliver multimedia data and services. Due to the large infiltration of circuit-switched telephony there exists a large number of innovative techniques and standards to implement this, such as ISDN and H.323 [1]. Current developments in this area focus on multimedia services and media telephony with advances in network gateways, distributed processing and media processing technologies. Due to varying amounts of network bandwidth and processing capability, a viable system should match resources efficiently and timely. In our approach we intend to implement a comprehensive and scalable multimedia presentation system which can adapt to a large array of network resources. Our underlying task is to adapt content with semantic invariance.

The majority of the literature that addresses adaptation or scaling problems has dealt with system level issues related to network- and client-driven adaptation, without emphasis on the information content of the media. In particular, research has focused on reducing the amount of data to be transmitted by trading CPU cycles for bandwidth usage [10, 13] and bandwidth control in a multicast environment [11]. Most approaches do not address difficulties in delivering data via wireless, modem lines, and even dedicated lines because of the demand for multimedia services and almost all approaches do not allow for wide deviation in quality of service (QoS) parameters. These network parameters along with client resources such as screen size, resolution and pixel depth, and the resource consumption, such as available memory, CPU and I/O capabilities, dictate the types of applications, which the client can support, as well as the types of media objects that can be consumed.

In our approach we incorporate media understanding techniques with the transmission resources to design an efficient multimedia service delivery scenario. Our approach is to design a set of distributed database servers that multicast the media to be delivered. The network transmission nodes then process and route the components and scale them to match the network and terminal conditions.

This paper is organized as follows. In Section 2 we present our approach to media adaptation and QoS control. In Section 3 we discuss scaling of multimedia services. In Section 4 we cover media transmission, routing and diffused scaling. In Section 5 we present actual adaptation schemes and present example services and test results for different media and delivery conditions.

2 Content-Based Media Adaptation

Data contained within media components of a multimedia document is usually designed to deliver a message. The adaptation applied to the different media components needs to be done coherently to preserve that message and its planned presentation. Semantic content and presentational features allow a vectored adaptation so that a specific schedule is implemented across different media types. Combining progressive representation and semantic relationships between media components allows for the media adaptation. This approach is process intensive, especially when multicasting the content to various terminal types, thus motivating a distributed processing paradigm.

Using network routers as scaling agents provides a unique approach for joint distributing and routing paradigm. Such nodes are increasingly used in forwarding multimedia data and have processing power to determine routing and switching conditions. The processing ability can be extended to include small amounts of media processing which is realistic with the availibility of cheap processing resource. Distributed tasks that can be implemented are not general purpose tasks and they need to be designed on a per-service-basis. Furthermore, due to large amount of parameters that drives scaling, a monitoring agent needs to forward scaling profiles to these nodes. These agents take into account network conditions along with client/server characteristics. We then use these nodes to 'tune' the contents of a message by a small amount in a prescribed manner (possibly included in transmission stream by the server) *without* full decompression or composition of the document. Specific indexes need to be placed in the media to denote start and stop of the specific sub-component along with the knowledge of how to perform manipulation in compressed domain. Some of the benefits of our approach follows.

- Processing requirements are offloaded from the server.
- Multicast by progressively scaling the content and dealing with network characteristics as they are intercepted.
- Achieve faster presentation by absorbing the processing delay in the network.
- Improve network utilization in places where bottlenecks might arise.
- Can use numerical analysis tools to synchronize the adaptation parameters to data and routing decision in each node.

Figure 1 shows the multimedia service adaptation control flow from the initialization of a session by a user (service) event to delivery of a multimedia presentation session by the delivery infrastructure. Our multimedia adaptation dialogue performs processing that ranges from content-based adaptation to case scripting optimization according to underlying network infrastructure, terminal properties and user (process, application, human) profile: (1) After initialization of a service event, the service profile is defined in a presentation service broker. (2) The profile of each event is analyzed by a service distribution agent, which performs the decomposition into catalogued media sources and allocates content. (3) The content structure is coordinated from each media type index or presentation domain, and physical media is assigned federated distribution, whose main function is to provide efficient media type-dedicated service. (4) While needed objects are collected, their presentation is synchronized to facilitate the event to assure the correct pre-processing, avoid jitter and perform an optimized buffering. (5) The presentation is synchronized through available media objects, and resolution for content-based adaptation is determined. (6) The indexes are attached to content of a presentation and follow each media object to presentation. (7) The presentation broker utilizes object push techniques that synchronize with the presentation script. Control agents finally perform the execution of a script.

As an example, Figure 2 shows a multimedia document containing a scenic image surrounded by binary textual content and embedded audio annotation. The image and text roughly require the same amount of storage space, about 800 Kbytes. The color image can be compressed to about 200 Kbytes and the binary portion can be compressed to 100 Kbytes without any significant loss in quality. These components can be scaled down to 10K for the color image and 30K for the document image before showing a large amount of degradation. After that, an iconic representation can be done to reduce storage requirements to 1500 bytes to contain only ASCII characters and a minor format information, and to 500 bytes to show title and author in ASCII and the rest of the text in gray shade.

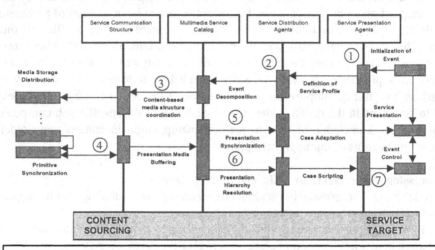

Fig. 1. Schematics of applying media adaptation over allocation and delivery services.

SAMPLE OF SCALING TWO DIFFERENT IMAGE CONTENTS

Omid E. Kia, Jaakko Sauvola, David S. Doermann

ABSTRACT

This paper proposes a new approach to address different multimedia consumption criteria in hybrid network and terminal constructions. In our model same media can be used for conferencing, content-based retrieval, remote storage and transmission. We describe techniques needed to perform efficient media understanding based task profiling and execution over the different elements. Techniques for content-based media preparation for multimedia presentations are introduced. We demonstrate media scaling with two different multimedia services.

Keywords: multimedia, scaling, distributed processing

1. INTRODUCTION

Range of the transmission requirements and capabilities across information delivery services are great. Creating custom services for specific requirement/capability pair is extremely difficult and porting such services to different transmission paradigms requires a lot of work. Designing a service that automatically adapts to network availability and terminal capability can be bene-

Figure 1: A color textured image mixed with a binary structured document.

Fig. 2. Partial sample of a multimedia document containing two media types

3 Adaptation of Value Added Services

In previous work [2], the most important aspect of CTI integration for multimedia applications is the delivery of messages. A Unified message allows different media types to appear in a single stream with the ability to route and process in unison. Integrated messaging provides a technique to arrange the information body that can be accessed with different communication devices utilizing suitable transmission Codecs. Figure 3 illustrates the principle of an integrated messaging system that contains media types where the presentation is in the form of messages.

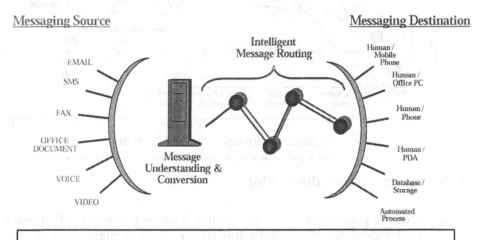

Fig. 3. Integrated multimedia messaging as a value added telephony service.

The adoption of value added telephony features, such as advanced call control, party profile, intelligent routing and usage of caller/process ID provides a number of benefits, which vary from interactive session control to advanced synchronization services. In our system, in addition to associating user data with calls that control the media, advanced call management enables the exploitation of user, terminal or service profiles, and thus efficient distribution and adaptation of media to presentation for each terminal. Figure 4 shows the variety of available terminals along with their capabilities.

Multimedia services set typically high demands for underlying transmission and processing. We pursue an agent-based service network principle, where different steps of a progressive multimedia service event are described. Typically, from the services and network point of view, users connecting to the network can be classified as having a wireless or fixed connection and as local or remote access. Our distributed multimedia services (DMS) network uses network nodes capable of performing processing and re-routing decisions based on agent functionality. While the presentation system is actually located inside the DMS network, the user adapter agents take care of session and content over terminal service event. Thus, a terminal can freely perform its own processing, or ask DMS network to provide refined multimedia and functional service object directly.

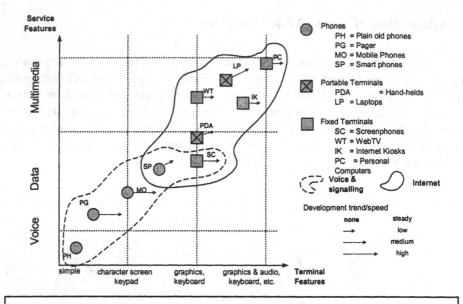

Fig. 4. Segmentation of multimedia service capability for different terminal types.

4 Techniques for Media Scaling

In our system we propose scaling methods to match data requirements with network and terminal capabilities. We consider lossless and lossy compression tools that operate on a wide array of media data; however, when such tools do not match the given resources we proceed to distill the data to a format that can match the resources. Figure 5 illustrates a scaling schedule for a single media type where after a certain amount of scaling a media conversion takes place, M_a to M_b. The vertical scale represents the introduced distortion and the horizontal scale represents the amount of scaling analogous to the amount of transmission or storage. The scaling schedule for M_a is different than for M_b and the point of discontinuity at (K) denotes the cross-media translation point. We use a simple threshold, based on media type and content, to determine the point of cross-media translation.

Popular compression methods allow lossy and lossless data representation for in-media scaling such as vector quantization [6], transform coding [7], Shannon-Fano coding [Shannon48], Huffman coding [5, 9, 14, 15], and run-length coding. Some are inherently lossy and others need to be adapted to achieve lossy representation. Some methods not only allow compression but allow self-indexing mechanisms. Previous work [8] has demonstrated this on document images. Compression methods for spatial (images) and temporal data (audio) are relatively mature, however, spatio-temporal data such as video presents several new twists. Advances in video coding, mostly implemented in the MPEG standards, consider similar scaling functionality. MPEG-4 promises to be content scalable, where objects can be selected, extracted, transmitted and generated from the original data.

155

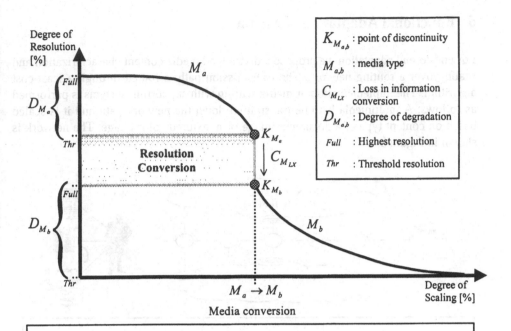

Fig. 5. Principle of intra- and cross-media adaptation.

"Distillation" is a term in the literature that is reserved for the process of reducing a media object to a minimal representation via a lossy data-type-specific compression to significantly limit the amount of data that must be transmitted over low-bandwidth networks, and subsequently improves the perceived latency for the user. In [3, 4], Fox et al describe a system for providing network and client adaptable on demand dynamic distillation for WWW pages. In order to provide an autonomous system, they implement the system as a proxy, removed from both the server and the clients. Similarly, McCanne [11] describes a multicast environment for delivery of data to light-weight sessions through their MBone network.

There are available tools for upscaling of media content such as graphics packages that allow us to convert tabular data to visual graphics, rendering text to images, and overlaying them on videos along with visualization packages that can render synthetic scenes. However, for downscaling, some conversions are needed that require recognition or analysis of media objects. For example, with document images, OCR can be used to provide an electronic text version of the document that can be accessed. Voice recognition can be used to provide textual transcripts of an audio track. Video segmentation can be used to abstract based on key-frames. A more complete analysis is required to actually "describe" the semantics of image and video. Currently, such analysis capabilities work in very limited domains.

In our system we perform content-based adaptation, which refers to the ability to analyze and provide appropriate media scaling, indexing and representation based on feature extracted from content. Based on previous work [12], presentation layout provides a rich set of features that allows selective scaling based on content. This allows us to implement a semantic invariant scaling paradigm. This has been used in improved multimedia database organization and indexing.

5 Functional Adaptation of Media

For end-to-end adaptation we propose a distributed media content characterization and scaling over a routing network. The transmission path is routed through a least-cost algorithm. The implementation of media scaling with supporting systems is performed as follows: A multimedia is to be transmitted down the network path and it is scaled based on content type of a document image or a textured color image. The network is shown in Figure 6.

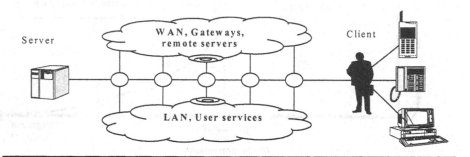

Fig. 6. Simulated network used to show network diffused media scaling.

An example network construction is shown with five nodes (Figure 6). Each node in effect is a router that accepts data from a specifiable port and transmits it along another port based on its programming and routing schedule. Then, for each node several factors exist which can decide on the level of media scaling. Such factors are incoming bandwidth (B^{in}), incoming load (L^{in}), local cache (C), outgoing bandwidth (B^{out}), and outgoing load (L^{out}). The overall scale function per node is defined as:

$$S_i = K_i \times \frac{B_i^{in}}{L_i^{in}} \times \frac{1}{C_i} \times \frac{L_i^{out}}{B_i^{out}} \tag{1}$$

Where K_i is a normalizing constant. It is hard to justify scaling based on input bandwidth and input load since functions of this type measure cost of delivery and not a cost of reception. In such cases, a handshaking algorithm is used and local queues cache the data. The scaling equation is written as:

$$S_i = \frac{K_i}{C_i} \times \frac{L_i^{out}}{B_i^{out}} \in F \tag{2}$$

where C_i denotes the amount of local cache the node is willing to set aside per handshake cycle. Also note that the Load and Bandwidth ratio is a measure of effective bandwidth. In essence, the scale function is computed as:

$$S_i = \frac{K_i}{C_i \times B_i^{eff}} \tag{3}$$

where the scaling is limited to a range $F=[1,10]$.

In our tests the range of available bandwidth(s), load(s) and cache for three different scenarios is considered. The range of bandwidths through various service nodes fluctuate between 10-100 Mbps for LANs, 1-10 Gbps for fiber-based FDDI, 0.1-10 Mbps for delivery of data to wireless antennas, 150 Kbps for GPRS terminals, and 9.6 Kbps (effective 0.2-5) for GSM/PCS (cellular phones). The coefficients used for scaling are shown in Tables 1-3.

Table 1 reflects transmission between a server and a high-end workstation. The server is connected to a 10 MB LAN having only 10 simultaneous sessions on average, which in turn is connected to an enterprise 100 MB enterprise backbone, and a service provider's 1 GB Fiber-optic backbone that has a load of 10,000 sessions. The receiving workstation is similarly connected to a 10 MB LAN and a 100 MB enterprise backbone with 10 and 500 sessions, respectively. The enterprise link nodes cache an amount of data reflective of an index of 5. However, the service provider caches twice as much because of the higher throughput.

MM station	Node 1	Node 2	Node 3	Node 4	Node 5
K_i	1 M	1 M	1 M	1 M	1 M
B_i	10 M	100 M	1 G	100 M	10 M
L_i	10	500	10000	500	10
C_i	5	5	10	5	5
S_i	1	1.5	5	1.5	1

Table 1. Transmission schedule for a multimedia workstation.

Table 2 denotes a connection to a PDA device with a bandwidth of 10Kbps. The server in this case is connected directly to a Fiber backbone and a high-speed fiber backbone. The effective bandwidth is still low due to a large amount of load. The 10K normalizing factor takes into account that the handshake signaling for a slow connection is anticipated along with a long transmission delay. The amount of caching is similar to the motivation for the previous example.

PDA Device	Node 1	Node 2	Node 3	Node 4	Node 5
K_i	1 M	1 M	1 M	1 M	10 K
B_i	10 M	1 G	100 G	100M	10 K
L_i	100	50000	1000000	5000	5
C_i	2	10	100	25	2
S_i	1	2.5	1	2	2.5

Table 2. Transmission schedule for a Personal Data Assistant.

Table 3 denotes connection to a cellular phone, which has a connection of 5 Kbps. Similar to the PDA device, the bandwidth and load are represented adequately. The normalizing factor and cache capability is analogous to similar network conditions. Due to the low transmission speed, the last scaling factor is high.

GSM/PCS phone	Node 1	Node 2	Node 3	Node 4	Node 5
Ki	1 M	1 M	1 M	1 M	1 K
Bi	10 M	100 M	1 G	10 M	50 K
Li	10	100	10000	100	100
C$_i$	5	5	10	5	5
Si	1	1	1	2	4

Table 3. Transmission schedule for a cellular phone.

We analyzed the overall QoS and relationships between the main features over the different network paths. First one considers the information transfer from a workstation to the other workstation and a second scenario considers the transfer of information from a high-end server to a cellular PDA device. For the first scenario, the path from one workstation includes two 10 Mbps LAN nodes to three 100 Mbps LAN nodes and five 1 Gbps Fiber nodes and to nine high speed 100 Gbps nodes. The path reverses itself to five 1 Gbps nodes, three 100 Mbps nodes, and finally to two 10 Mbps nodes. During the transmission session, the packets experience rises in load respective to the transmission speed while the other half experience a symmetric reduction in load, respectively. The plot of effective bandwidth versus required cache size for constant scaling is shown in Figure 7. In the second scenario, a server is connected to a high speed 100 Gbps node and the information travels down five 1 Gbps nodes, ten 100 Mbps nodes, ten 10 Mbps nodes, and finally to the 10 Kbps PDA device. The load decreases substantially for each network. The scaling function for this network and cache size is shown in Figure 8. For constant scaling of the first scenario, the cache size decreases once a higher network speed is reached, but the size increases gradually as load is increased.

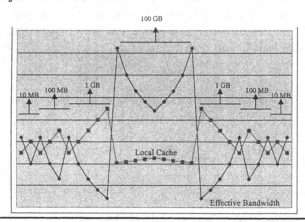

Fig. 7. Plot of effective bandwidth along a long network transmission path (In logarithmic scale) and the required local cache.

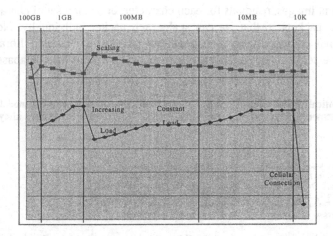

Fig. 8. Simulation of content delivery to a cellular PDA device showing effective bandwidth (In logarithmic scale) and amount of scaling.

For our experiments, we used a wavelet-based representation for the transmission of textured content and pattern-matching and substitution for textual content. In wavelet transformation, we used a Haar wavelet with decomposition shown in Figure 9. For the textual components we decomposed the image into character shape clusters and residuals as shown in Figure 10.

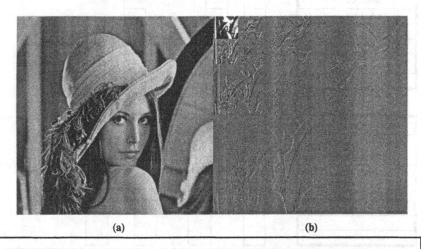

(a)   (b)

Fig. 9. Wavelet decomposition of a color textured image. The original image (a) is decomposed into three sub-band levels.

The wavelet representation is designed to organize the pixels in order of importance. The processing node would only need to access that part of the data and ignore as many bits as needed. Similar to the document image decompositions, the majority of the data is in the residual and effective scaling is done to only that part of the data. Data packets are organized as shown in Figure 11. In wavelet decomposition, all subbands may be present in each packet allowing for scaling on a per-packet basis.

For document images, residuals for each character, or a small set of characters, may be formed into packets so that we can scale each packet individually. Notice that the computational complexity of such a scheme is only to access a library of packet formats. In essence, node intelligence is required in a form of a database information access.

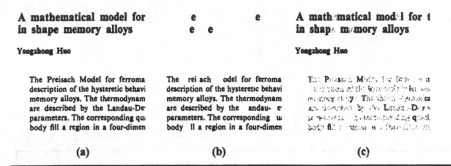

(a)   (b)   (c)

Fig. 10. Decomposition of a binary document image. The original image (a) is decomposed into patter clusters (b) and residuals (c).

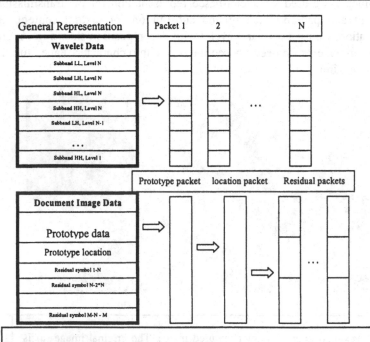

Fig. 11. Packetizing wavelet compressed data and document image data. In each packet, scaling can be done so as to reduce transmission requirements.

A sample scaling is shown for a multimedia document. The scaling on the portions of the image are done separately as shown in Figure 12. In this, the original image (Figure 12a) is scaled to that of Figure 12b. When the textured region starts to show large pixels, the textual regions begin to lose their font information, and contain some common errors such as replacement of 'b' with 'h', 'a' with 'e' or 'o'. Further scaling requires cross media translation. That is denoted by the image shown in Figure 12c.

The textured image is replaced by a local icon (or text), the title and authorship still contain textual content and the other textual content is replaced with gray boxes. The amount of information to convey the entire page is in the order of 12:1 and 120:1 for the scaled versions.

SAMPLE OF SCALING TWO DIFFERENT IMAGE CONTENTS
Omid E. Kia, Jaakko Sauvola, David S. Doermann

ABSTRACT
This paper proposes a new approach to address different multimedia consumption criteria in hybrid network and terminal constructions. In our model same media can be used for conferencing, content-based retrieval, remote storage and transmission. We describe techniques needed to perform efficient media understanding based task profiling and execution over the different elements. Techniques for content-based media preparation for multimedia presentations are introduced. We demonstrate media scaling with two different multimedia services.

Keywords: multimedia, scaling, distributed processing

1. INTRODUCTION
Range of the transmission requirements and capabilities across information delivery services are great. Cre-
ating custom services for specific requirement/capability

(a)

Sample of Scaling Two Different Image Contents
Omid E. Kia, Jaakko Sauvola, David S. Doermann

abstract: This paper proposes a new approach to address different multimedia consumption criteria in hybrid network and terminal constructions. In our model same media can be used for conferencing, content-based retrieval, remote storage and transmission. We describe techniques needed to perform efficient media understanding based task profiling and execution over the different elements. Techniques for content-based media preparation for multimedia presentations are introduced. We demonstrate media scaling with two different multimedia services.
1. Introduction
Range of the transmission requirements and capabilities across information delivery services are great. Creating custom services for specific requirement/capability pair is extremely difficult and porting such services to different transmission paradigms requires a lot of work. Designing a service that automatically adapts to network availability and terminal capability can be bene-

(b)

Sample of Scaling Two Different Contents

Omid E. Kia, Jaakko Sauvola, David S. Doermann

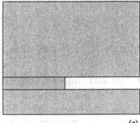

(c)

Fig. 12. Example of scaling where (a) is the original image, (b) is scaled but bounded to the same media types and (c) is translated to alternate media.

6   Conclusions

In this paper we presented a new technique for adaptation of multimedia content to different environmental conditions. In this approach we use network distributed processing to diffuse the processing requirements while simultaneously transmitting the content. The processing requirements needed at each node are relatively low with respect to the amount of a local cache that needs to be maintained. While our scaling approach has been motivated by transmission speeds and amount of loads on these mediums, another factor is terminal capability and quality of service. Content adaptation at each node serves a difficult task for scaling under conditions of varying terminal capability. Also of interest is the ability to perform cross-media adaptation in a distributed manner. Such techniques are now under consideration and existing distributed techniques are applied, still, further network diffusion needs to be studied. Our system also enables up-scaling by media conversion. However, this is considered to be a presentation problem rather than a scaling problem. Since up-scaling is done on terminals of relatively higher resources, it is possible to perform that within a server or a client.

Acknowledgement

The funding from the Academy of Finland under the contract 'Media Processing and Hybrid Networks' and the Technology Development Centre of Finland is gratefully acknowledged.

7 References

1. International Telecommunications Union (1998) H.323 standards specification, version 2.0.
2. M. Palola, R. Kaksonen, M. Heikkinen, S. Kaukonen, T. Ojala, T. Tervo & J. Sauvola (1998) CTI State-of-the-art. A technical report, CTI project, 64 pages.
3. Armando Fox, Steven D. Gribble, Eric A. Brewer, and Elan Amir. Adapting to network and client variability via on-demand dynamic distillation. In Proc. Seventh Intl. Conf. on Arch. Support for Prog. Lang. and Oper. Sys. (ASPLOS-VII), 1997.
4. Armando Fox, Steven D. Gribble, Yatin Chawathe, and Eric A. Brewer. Cluster-based scalable network services. In Proc. 1997 Symposium on Operating Systems Principles (SOSP-16), St-Malo, France, 1997.
5. R. Gallager. Variations on a theme by Huffman. IEEE Transactions on Information Theory, 24:668--674, 1978.
6. A. Gersho and R. Gray. Vector Quantization and Signal Compression. Kluwer Academic Publishers, Boston, 1992.
7. A. Jain. Fundamentals of Digital Image Processing. Prentice-Hall, 1989.
8. O. Kia. Document Image Compression and Analysis. PhD Thesis, University of Maryland at College Park, 1997.
9. D. Knuth. Optimal binary search trees. Acta Informatica, 1:14--25, 1971.
10. J. Li and C.C.J. Kuo. Progressive Coding of 3-D Graphic Models. IEEE Proceedings, 86(6), pp 1052-1063, 1998.
11. S. McCanne. Scalable Compression and Transmission of Internet Multicast Video. PhD Thesis, University of California, Berkeley, 1996.
12. J. Sauvola. Document Analysis Techniques and System Components with Applications in Image Retrieval. PhD Thesis, University of Oulu, Finland, 1997.
13. K. Sayood. Introduction to data compression, Morgan Kaufman Publishers, 1996.
14. J. Vitter. Design and analysis of dynamic Huffman codes. Journal of the Association for Computing Machinary, 34:825--845, 1987.
15. J. Ziv and A. Lempel. A universal algorithm for sequential data compression. IEEE Transactions on Information Theory, 23:337--343, 1977.

Extended Package-Segment Model and Adaptable Applications

Yuki Wakita[12], Takayuki Kunieda[12], Nozomu Takahashi[2],
Takako Hashimoto[12], and Junichi Kuboki[1]

[1] Information Broadcasting Laboratories, Inc ***
{yuki, kunieda, takako, kuboki}@ibl.co.jp
[2] Ricoh Company, Ltd.
{yuki, kunieda, nozomu, takako}@src.ricoh.co.jp

Abstract. This paper aims to present and verify the Extended Package-Segment Model. It provides a more flexible representation of logical structure of multimedia content than the Package-Segment Model (PS-Model) which we have already proposed. The PS-Model structure represents recursively ways to divide content. It is suitable to represent logical structure of content generated by content-based analysis.

However, there are also some restrictions in generating the PS-Model structure. To solve these problems, we have reconsidered the generation rules and here propose a new data model called the Extended Package-Segment Model (EPS-Model). The EPS-Model structure can represent scenes and cuts that are focused. Moreover the EPS-Model inherits the advantages of the PS-Model framework such as indexing, browsing, and retrieval mechanisms.

We introduce two experimental systems based on the EPS-Model structure. We verified that this EPS-Model has flexibility and suitability of representation of the structure for multimedia content in any application.

1 Introduction

Multimedia data continue to grow rapidly and there is a huge demand for multimedia archiving systems. The imminent age of digital satellite broadcasting will bring about an explosion of TV programs. Along with the information, people appear to want to use multimedia content. However, before any content can be used, it will have to be located, and this is currently not easily possible for audiovisual content since most existing audiovisual sequences have no structured information. However, there are now several attempts to standardize content-based retrieval schemes [1].

We have already proposed a Package-Segment Model (PS-Model) to represent multimedia content in a logical structure [2]. Here we explain the PS-Model

---

*** Wakita, Kunieda, and Hashimoto have been temporarily transferred to Information Broadcasting Laboratories, Inc. from Ricoh Company, Ltd. Kuboki has been temporarily transferred to Information Broadcasting Laboratories, Inc. from Fuji Television Network, Inc.

briefly. The PS-Model consists of various objects where each object corresponds to a node in a movie content structure. The PS-Model is adaptable and can represent the logical structure, attributes, and supplements that are associated with the target content. The logical structure that is represented by using the PS-Model is composed by segmentation of the sequence, detection of still image features, and extraction of key frames. However, there are some implicit rules in generating the PS-Model structure. Therefore, a flexible representation is required in order to be applied to various applications. In this paper, we explain our solution to these problems. As a result, we extended the generation rules of the PS-Model structure. Thus our extended data model is called the *Extended Package-Segment Model* (EPS-Model). In the following sections we discuss the PS-Model's restrictions, and the EPS-Model's merit in detail. Finally, the experimental applications of the EPS-Model are introduced.

2 Package-Segment Model

We suggested the PS-Model structure as a structural formula and method for video and/or audio sequences. The PS-Model represents the logical structure of multimedia content. This structure is used as an index of retrieval or editing scheme for TV programs. Indexing, Browsing, and Retrieval mechanisms are provided in this framework.

2.1 The PS-Model Structure

The PS-Model is based on a tree structure and each node is composed of the following objects:

MovieTree	Root node of the structure
MoviePackage	Definition of the segment set
MovieSegment	Part of the logical structure in the content
MovieFrame	Frame in a sequence
MovieSound	Sound part in a sequence
MovieAttribute	Additional attribute or supplement
MovieFeature	Features extracted from the content

A MovieSegment object represents a cut, scene or logical duration in the sequence. The MoviePackage object plays a role in the aggregation of a segment set. Some structures and construction schemes of multimedia content have already been proposed[3][4]. The PS-Model is characterized by the intervention of the MoviePackage object, which enables multiple interpretations of a target content to be represented in a logical structure.

Two sample sequences are shown in Fig.1. Assume those two sequences are news programs. In this example, the content of a news program is represented by using four layers. The first layer represents the whole sequence. In the second layer, each program is partitioned into "News Topics", "Domestic news", "World news", and "Weather report" where each item is further partitioned into such items as "politics", "economy," and "the other" in the next layer. In the bottom layer, the content is partitioned into "live announcement part" and "recorded news part".

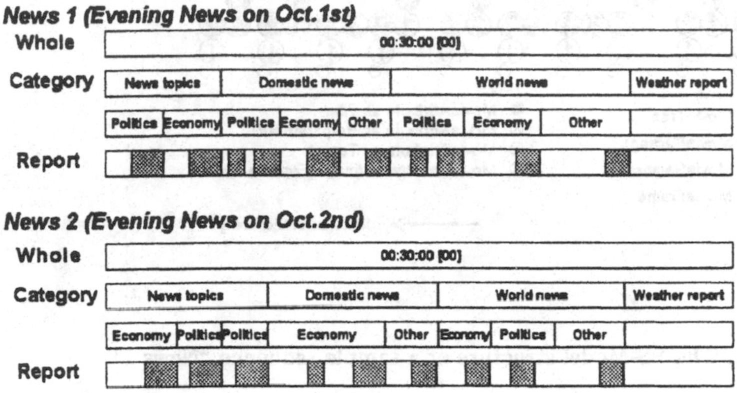

News 1 (Evening News on Oct.1st)

Whole	00:30:00 [00]					

Category	News topics	Domestic news		World news		Weather report

	Politics	Economy	Politics	Economy	Other	Politics	Economy	Other

Report

News 2 (Evening News on Oct.2nd)

Whole	00:30:00 [00]					

Category	News topics	Domestic news		World news		Weather report

	Economy	Politics	Politics	Economy	Other	Economy	Politics	Other

Report

Fig. 1. Sample sequences: "News1" and "News2"

Fig.2 shows the PS-Model structure of an example sequence "News1". The intervention of the MoviePackage object makes it possible to represent multiple scenarios that can be contained in one structure. The sequence can be divided into segments in multiple ways depending on different points of view. The MoviePackage objects can therefore represent the segmentation methods. MovieFrame objects that are extracted as key frames from a "recorded part" have some image features defined as a MovieAttribute object.

This sample represents the content "News1" above using the PS-Model. The MovieTree object is at the root and the layer is constructed with the MoviePackage object that expresses the whole content and a MovieSegment object. The second layer consists of an aggregation of MovieSegment objects that are classified by topics and a MoviePackage object which manages it. In this layer, each MovieSegment object is classified by genre and the aggregation of the MovieSegment objects is managed by a MoviePackage object. In this example, each node maintains specialized objects for the MovieAttribute object for describing additional attribute information according to its needs. In the bottom layer each MovieSegment object maintains a MovieFrame object representing a key frame while the MovieFrame object maintains a MovieFeature object that represents still image features. This example uses information, added manually, for its seg-

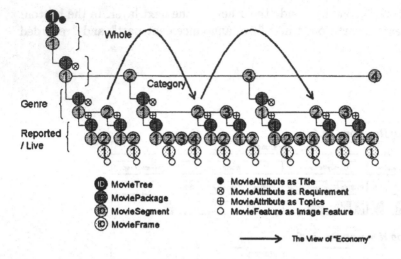

Fig. 2. The PS-Model structure of a sample sequence "News 1"

mentation criteria but content-based information can be manipulated just as flexibly as that manipulated in the PS-Model.

The PS-Model provides ways to browse the content that are compatible with structured representations. Browsing is required to preview a search result where several candidates have been retrieved. The MovieView object performs this function. The MovieView object has a list of MovieLink objects that point to a node inside or outside and also has a supplement that enables representation of the behavior of playing contents. In the Fig.2, the arrow shows how to trace nodes that are represented by a MovieView object.

3 Extended Package Segment Model

We applied the PS-Model structure to represent logical structure of content that is generated by content-based analysis. The PS-Model structure is generated automatically in the following processes:

1. Trace content of a sequence
2. Extraction of features and detection of cut change points
3. The segment range is decided

The PS-Model is adaptable to these processes. However, it has some restrictions in representing the content. The PS-Model structure can not be composed by selected parts only since it is designed for the representation of continuous media without any missing segments.

3.1 The Restrictions of the PS-Model Structure

The composition rules of the PS-Model structure is normally as follows:

- A segment is not allowed to overlap neighboring ones.
- Deleting segments from the parent segment is prohibited. That is, the duration of an aggregation of segments which are held in a MoviePackage object must be the same as the duration of the whole parent segment. These rules are shown in Fig.3.

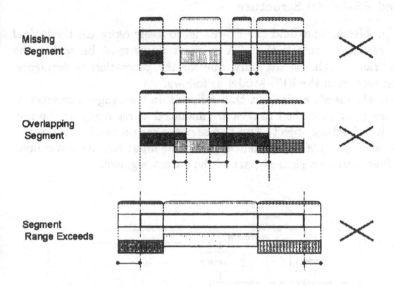

Fig. 3. The composition rules of the PS-Model structure

The PS-Model structure represents recursively ways to divide the content. For example, in Fig.1, both of the "recorded part" segments and the "live part" segments exist in the bottom layer. If an author wants to represent "recorded part" only, what should he/she do to represent the logical composition using the PS-Model ? Unfortunately, he/she is not able to delete the "live part" segments because the PS-Model prohibits missing segments. If a "live part" segment is deleted, the PS-Model structure is not complete. The mechanism of dividing is as if a cake were divided by a knife. Any missing or overlapping segment is not allowed in the PS-Model. This is an implicit rule, but this rule is a restriction for composing the structure. The MovieView class enables the representation of only selected nodes in the structure. Therefore, he/she should use a MovieView object to represent only the "recorded part", and the "live part" segments remain in the structure as dummy segments.

These restriction rules are effective for analyzing continuous data such as an audio or video sequence. Thus this structure can represent the content completely where a segment is always covered by child segments which are in fact refined parent segments. However, it is not suitable to represent the logical structure which is edited and arranged for a specific purpose. Since the PS-Model structure assigns a segment to a disused part of a sequence, it is lengthy and wastes resources. Although a MovieView object can represent the selected segments, it can not maintain the refined selected segments or packages in the lower layer. A MovieView object can not represent a hierarchical structure.

Therefore, flexible representation mechanisms are required for the PS-Model.

3.2 Extended PS-Model Structure

To solve those problems, we extend the PS-Model to freely represent the logical structure. The restriction rule is that "A segment range must be within the parent segment range". This is the only rule for the generation of segments. Segments are generated in the EPS-Model as follows.

A user can freely specify a segment that falls within the range of interest at the parent segment. The specified ranges are managed as segments by the mediation of the MoviePackage object. Any two segment ranges can be overlapped as needed. Undesired parts of the sequence will not be input into the lower layer since it is considered to be a disused part of the parent segment.

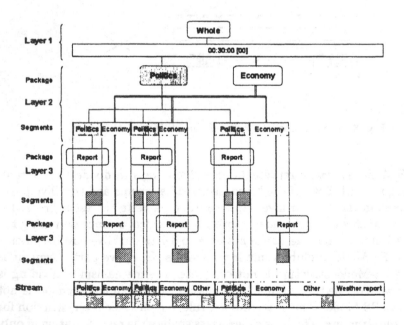

Fig. 4. The EPS-Model representation of sample sequence "News 1"

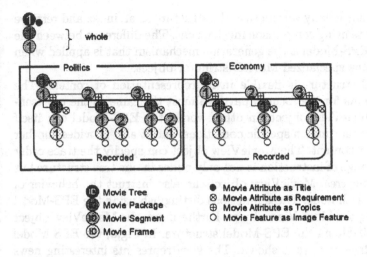

Fig. 5. The EPS-Model structure of sample sequence "News 1"

The sample content "News1" is represented by the EPS-Model structure in Figs.4 and 5. The same content is also represented by the PS-Model in Figs.1 and 2. In Fig.4, the content is shown on the bottom. The position and length of segments are mapped from the corresponding parts of content. In this case, we focus on the categories of news "politics" and "economy". We must arrange the content logically. In Layer 2, two packages are generated. The packages show the news that is categorized as "politics", and "economy", respectively. Each part of news content categorized as "politics" is mapped into a segment in the "Politics" package. This package does not contain any other type of news. The "economy" package is composed as freely as "politics". Although the news parts that are categorized into "Weather report" or "Other" are contained in the whole segment, they are ignored and not described in any packages because they are out of focus. In Layer 3, the recorded parts are focused. The "recorded" packages are generated under the each segment, and the segments which are mapped as the "recorded" part are generated. The "recorded" segment range must be required to be within the parent segment range. The "live" parts do not appear as any segment in this layer. A "recorded" package that represents all "recorded" parts irrespective of genre can be generated in Layer 2 as required.

Fig.5 shows the tree structure of the sample above as a representation of object nodes. This structure is very simple, and it is easy to understand. The structure arranged by editors can be represented clearly using the EPS-Model. Also the role of package is more important in the EPS-Model to describe an aggregation of segments conceptually. But the EPS-Model still has the same advantages as the PS-Model framework. We can freely define and specialize attribute classes. Additionally we can save the structure as an index and retrieve segments by attribute using comparison mechanisms. The difference between the PS-Model and the EPS-Model is the generation mechanism that is applied when using or not using the specialized MovieDistributor object.

The EPS-Model structure is flexible in its representation of content. The generation of segments for the focused part is almost the same as the representation using the MovieView object. In other words, the EPS-Model can itself represent the scenes that match specific conditions, and it also provides another way to easily browse content. The MovieView object can specify the trace order of the nodes. The range of instruction is not only nodes in one tree structure but also nodes in another tree. MovieView object can also instruct the behavior of playing content for each node. These functions distinguish using the EPS-Model structure from using the MovieView object. Furthermore, the MovieView object for browsing is available on the EPS-Model structure. In Fig.6, the EPS-Model structure for two days of news is shown. The view represents interesting news topics collected from two days of news as a digest which appears like a trace from node to node in a couple of sequences.

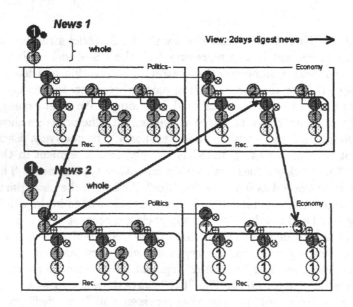

Fig. 6. The View representation in the EPS-Model

The EPS-Model structure makes full use of the functions of the MoviePackage object. There are many ways to represent the logical structure of content. The EPS-Model provides the function of keeping multiple aggregations of segments under a parent segment using the MoviePackage object. In Sect.4, we present some experimental applications that are based upon the EPS-Model structure.

3.3 The EPS-Model Framework

The functions of the PS-Model are a subset of the EPS-Model. Indexing, Browsing, and Retrieval mechanisms are also provided in the EPS-Model framework. The EPS-Model structure consists of a MovieTree object that serves as a root of the tree structure and the MoviePackage object that manages the aggregation of MovieSegment objects. The MovieSegment object can manage a MovieSound object as a key sound, a MovieFrame object as a key frame, and a MovieMotion object as a key motion. All of these objects can have MovieAttribute objects representing attributes and the content of the MovieAttribute object can be freely defined. The MovieFrame, MovieSound, and MovieMotion objects can have MovieFeature objects by mediation of the MovieAttribute objects. This MovieFeature object can maintain features that are extracted from content.

Object Composition and Functions. The EPS-Model is composed of various classes that are classified by their functions:

- Structure definition class
- Content mediation class
- Operation class
- View definition class
- Type definition class

In Fig.7, object relations are represented by a class diagram. The functions of these classes are the same as the PS-Model.

Operation classes are base classes and need to be specialized as required in order to compose the EPS-Model structure and extract feature information. In the PS-Model, a MovieDivider class is provided. The MovieDivider object divides a segment and generates new segments under the target segment. The structure generated by the MovieDivider object is restricted by the implicit rules of the PS-Model. Now, we add a new class called MovieDistributor to form the EPS-Model structure. The MovieDistributor object selects parts in a segment and generates segments sets which correspond to the selected parts under the target segment. It enables the composition of structure by selected parts only.

Data Conversion. The EPS-Model structure can be converted into the following various data.

- EventList (input):
  This is a text-based list of events in the content. It is provided for real-time event marking using an easy interface.
- Description by XML (input/output):
  This is an eXtended Markup Language(XML) formatted text data. It describes the EPS-Model structure that is created by an analysis module. This data can also be edited manually and reloaded into the application.

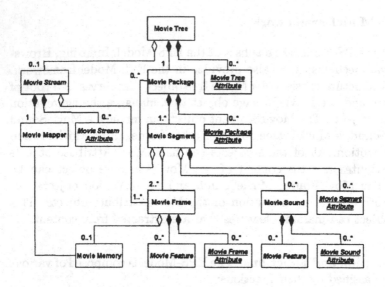

Fig. 7. Class diagram for the EPS/PS Model structure

- Retrieval Index (input/output):
  This index is referred to by the retrieval engine and specifies the content sequence and scene position that a user requests.
- EDL (output):
  This data format is in an Editing Definition Language(EDL) format for non-linear editing system so that new sequences can be created. This format depends on the type of editing machine.
- The program index for broadcasting (output): The program index has been standardized in the Association of Radio Industries and Businesses in Japan[5]. It is a Meta-data for TV programs for new broadcasting services such as hierarchical programs. It enables the specification of content inside an index for broadcasting.

4   Adaptable Applications

We can still apply the PS-Model to represent the structure of content generated by content-based analysis, and apply the EPS-Model to the logical structure of the content that is arranged or reconstructed by hand.

Two applications are presented in this section. They parse an EventList file which describes many events and times. These systems select event types as required and construct the EPS-Model structure that is focused on selected events. This EPS-Model structure composes a new sequence and is converted into a Meta-data index for broadcasting.

4.1 Authoring System for Information Broadcasting Program

This is the TV program authoring system[6] which is impremented based on the EPS-Model. This system generates the program related information which is structured sheme for information broadcasting. The TV program authoring system processes by the following procedures. Here the EPS-Model is utilized for specification, feature extraction, and the editing of the structural and additional information in each scene and key frame.

1. The events which is happen on the TV program are recorded manually. This unit defines a event profile according to the type of each program where the event that fits is input into the profile. The events of the content roughly are described for a program and the EventList file is created. An event description consists of an event type, a starting time code, any attribute, and a duration.
2. After inputting the EventList file into the editor, which is constructed by the EPS-Model framework, the focused events are specified as an aggregation of segments. This aggregation is represented by the MoviePackage object. An EPS-Model structure is generated by selected events. There is usually some error in the manual designation of a scene start/end point, automatic extraction is used to determine precisely the cut turning points. The key frame of each scene is then selected and features are extracted making the EPS-Model structure complete.
3. The desired scenes are selected for use in browsing. A new EPS-Model structure or a MovieView object can be created to describe this browsing scheme.
4. This completed browsing scheme is converted into EDL and the important scenes can be reconfigured as one new complete sequence by non-linear video editing system. The list of both starting points and corresponding durations is recorded for the selected parts in EDL.
5. For the application of information broadcasting, all of the EPS-Model structure is converted into the program index and then sent to the transmitter as content for broadcasting. These received data can be utilized for watching a TV program at home. A user can select the points by his/her taste and directly seek them from the program index has indicated.

This system is applied to sports and news programs on TV where they are live programs. In live programs, producers do not know exactly what events will occur in the next moment. Thus structure of live programs can not be constructed in advance. Real-time marking is required to represent the logical structure of the content in this area. Currently, humans have to check and note all events manually while watching a game in real time to generate a digest. So after generating a digest program, event information is then discarded, and a logical program structure is not constructed.

This system constructs the logical structure of content by selected events in the EventList file. These events are classified into two types - structural and actual. The structural events are available to be used to construct the content

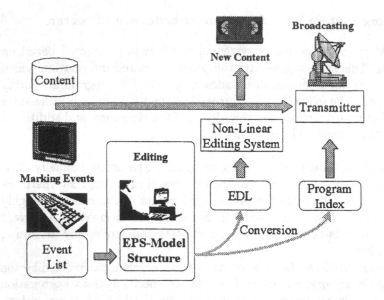

Fig. 8. Authoring System for Information Broadcasting Program

structure. The actual events are further classified into some types that are associated with the event profile. The actual events describe everything that occurs and they are essential for understanding the content and for semantically reconstructing the structure. However, many actual events are marked in content and some of their durations overlap. If content is divided by all of the actual events using the rules which are based on the PS-Model, a segment is too fine and it makes no sense. If the content is divided by the events with the same event type, the part in which nothing occurs must be described as a padding segment. The EPS-Model structure solves these problems by selecting the desired events, creating a package for each meaning, and reconstructing the structure.

Fig.9 shows events for a football game. For example, the system selects the events whose type is "Shoot" and creates a MoviePackage object and MovieSegment objects. These MovieSegment objects have, associated with the information of the events, MovieAttribute objects. This associated information is important and utilized in order to retrieve. Also this information generates a view or another EPS-Model structure as a digest. It produces the best logical representation of the content semantically.

4.2 Digest Generation System

The Authoring System for Information Broadcasting Program is utilized by TV program producers, while the Digest Generation System[7], [8] is used at the home.

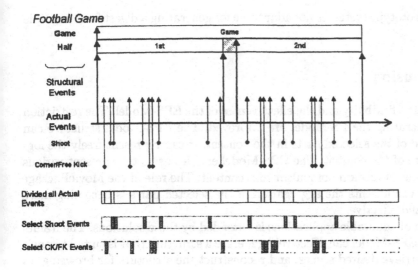

Fig. 9. A sample of events for a football game program

The EPS-Model structure is reconstructed from the program index and the digest of a user's choice is generated automatically and dynamically in the set top box (STB) of each home. The Digest Generation System processes the following procedures.

1. The program index is generated from the EPS-Model structure in a TV broadcasting station, and broadcasts with the content.
2. The program index and content stream are received at home. The EPS-Model structure to represent the content is reconstructed in the STB.
3. A user can input his/her own taste to the system as a rule. It is stored in the user's database. The user can specify temporal parameters to generate his/her own digest. For example, the total time for digest, focused sports players, selected events, and interesting genres are given. The rules on how to select scenes are decided according to these parameters.
4. The system selects important scenes, according to the rule, dynamically. Searching is based upon attribute information and the features of the EPS-Model structure.
5. The system reconstructs a new EPS-Model structure as a digest by the selected scenes and associated information. This must be his/her own digest and satisfy the user's taste requirement. If a user changes the setting of parameters, a new different digest is generated.

The MovieView object is more adaptable to represent a digest. Currently, however, this system applies the EPS-Model structure to represent a digest because this prototype system is not adaptable for generating a digest from multiple contents.

5 Conclusion

We verified the flexibility and the effectiveness of the EPS-Model. The restriction rules in generating the PS-Model are improved. The EPS-Model structure can be composed of the selected parts in the content. It can represent freely the logical structure of the content. The EPS-Model enables users to represent various interpretations of any kind of multimedia content. The role of the MoviePackage object which represents the aggregation of the selected parts is becoming more important and effective.

The area of adaptable applications is extended by these solutions. The framework of the EPS-Model can effectively generate a structure that is suited to one's purposes, retrieve desired scenes, and reconstruct the structure for browsing. In the next step, we must implement various applications using this EPS-Model and increasingly reconfirm the effectiveness.

References

1. MPEG Requirements group, "MPEG-7 Requirements Document V.7", Doc. ISO/MPEG N2461, Atlantic City, Oct 1998.
2. Kunieda, T. and Wakita, Y., "Package-Segment Model for Movie Retrieval System and Adaptable Applications," Proc.of IEEE International Conference on Multimedia Computing and Systems'99, Vol.2, Florence, pp 944-948.
3. H. J. Zhang, C. Y. Low, S. W. Smoliar and J. H. Wu, "Video Parsing, Retrieval and Browsing: An Integrated and Content-Based Solution", Proc. ACM Multimedia '95, San Francisco, pp 15-24.
4. H. Ueda, T. Miyatake, "Automatic Scene Separation and Tree Structure GUI for Video Editing", Proc. ACM Multimedia '96, Boston, pp 405-406.
5. "Service Information for Digital Broadcasting System", ARIB STD-B10 V1.2, Part.3, Association of Radio industries and Businesses Japan(1999)
6. J. Kuboki, T. Hashimoto, T. Kimura, "Method of Creation of Meta-Data for TV Production Using General Event List (GEL)", The Institute of Image Information and Television Engineers Technical Report, Vol.23, No.28, 1999, pp 1-6.
7. T. Hashimoto, Y. Shirota, and T. Kimura, "Digested TV Program Viewing Application Using Program Index", The Institute of Image Information and Television Engineers Technical Report, Vol.23, No.28, 1999, pp 7-12.
8. T. Hashimoto, Y. Shirota, J. Kuboki, T. Kunieda, A. Iizawa, "A Prototype of Digest Making Method Using the Program Index", Proc. of IEICE Data Engineering Workshop '99 CD-ROM.

Tailoring Protocols for Dynamic Network Conditions and User Requirements

R. De Silva† and A. Seneviratne‡

†rdesilva@cisco.com
CISCO Development Labs
Sydney, Australia

‡a.seneviratne@unsw.edu.au
University of NSW
Sydney, Australia

Abstract. This paper shows the use of protocols dynamically generated for a particular network environment and an application's requirements. We have developed a novel system called PNUT (Protocols configured for Network and User Transmissions), which can be used to dynamically generate protocols. A distributed MPEG player has also been developed using PNUT to demonstrate the need for adaptive protocols that can reconfigure their functionality as the operating environment changes. One of the key features of PNUT is that it allows for the dynamic reconfiguration of the protocol stack, which we believe is necessary when developing protocols for mobile computing environments. The paper also shows how intermediate servers can be used to provide better services to the mobile application. Furthermore, we show how intermediate servers can be used to allow mobile applications using PNUT to interact with servers based around traditional applications

1. Introduction

The Internet protocol suite was developed for network environments and applications, which are static and support a single media type. However, now its evident that the Internet of the future will be very different and will consist of heterogeneous networks, and the applications that utilize multiple media types. Moreover, the applications will require different levels of service from the underlying communication system. It has been shown that these traditional protocols in their original form perform poorly under these conditions. In order to cope, communication system designers classified applications into classes and developed different protocols [5,7,17], and extended the internet protocol suite in numerous ways [10,11]. Firstly, the development of systems to a set of classes results in sub optimal solutions to all but a small set of applications. Secondly, the acceptance of new protocols, and extensions or modifications to existing protocols requires long lead times. Finally, none of these schemes take into account the dynamic nature of both the network environments, especially when considering wireless networks, and the user requirements/preferences of a session.

In order to efficiently deal with specific application requirements, it is necessary develop the required mechanisms, i.e. tailor the communication subsystem. To cope with the dynamic changes in the operating environment and user requirements/preferences, it is necessary enable the communication subsystem to adapt. The development of application specific protocols has been addressed by a number of researchers. This work has shown that protocols tailored to their operating environment produce better performance over traditional protocols and other protocols not tailored to the current environment [3]. Furthermore, new protocols have been developed which provide the necessary support for a wide class of applications [17,6]. The main weaknesses of this approach is the complexity associated with designing and implementing new communication sub-systems for each application, and its inability to deal with dynamic changes in operating environments.

To cope with the heterogeneity, the applications in the future will attempt to adapt their behavior to maximize the resources available to them in the different operating environments [9]. In addition, firstly, the users will change their preferences during a session for a number of reasons, such as the use of the application, location and cost. Secondly, the operating environment may also change during a session. For example, the computer may be moved from a wired network segment to a wireless network segment. Consequently, the services required from the communication subsystem will change. A number of projects have investigated the development of adaptive protocols [13,16,18]. These projects have shown the advantages of adaptive communication systems, but have used specialized hardware or run-time environments.

In this paper we present method of generating application specific protocols which can be dynamically re-configured which can be used in general purpose systems. Then the viability of the proposed method is demonstrated through a prototype implementation called PNUT (Protocols configured for Network and User Transmissions), and its use in a distributed MPEG player. The paper is organized as follows. The next section describes the protocol generation and dynamic configuration framework. We then discuss the development of a distributed MPEG player that was developed to operate in a mobile environment using PNUT in Section 3. Section 4 outlines experiments and the results obtained using the distributed MPEG Player. Section 5 provides a discussion on the finding, section 6 reviews related work, and section 7 presents the conclusions.

2. PNUT Framework

The PNUT framework consists of 4 phases, namely
  1. The requirements specification,
  2. Determination of the status of the operating environment and user preferences,
  3. Selection of the required communication subsystem functionality, and
  4. Re-configuration

In the first phase, the application designer writes a set of application requirements. This will be done by the application designer, in parallel with the development of the application code. The requirements can be written using a formal language as described in [4] or selected from a list of available options. We have used the latter scheme, and have provided the designer the option of specifying requirements of reliability, flow control, and sequencing. The choice of this set of options is discussed below.

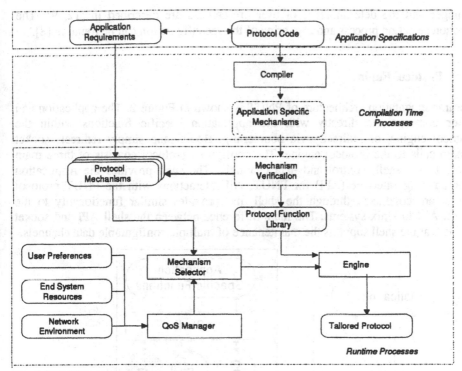

Fig. 1. PNUT Framework

User preferences and the status of the operating environment will be determined via a simple user interface and a set of resource monitors. In our overall architecture, the user interface and monitors are part of the Quality of Service (QoS) Manager as described in [12]. The status of the operational environment, and the user preferences will then be used by the protocol engine to select the necessary functions from a protocol function library, and will be linked together to generate the communication subsystem. For example, error control functionality can be supported using a go-back-n, a selective retransmission, or using a forward error correction mechanism. Thus, when reliable functionality is required, and if the system is in power saving mode, the engine will choose go-back-n.

In addition to the functions provided with the library, application developers have the freedom to define application specific protocol mechanisms that will allow for a higher level of tailoring to the application's requirements. This process is indicated as the outcome of the compilation process. These application specific mechanisms can be included in the current library of mechanisms and later be used when configuring the tailored protocol.

Changes in the user preferences, or operating environment, e.g. end system resources or the network conditions are handled by the QoS manager. If the user indicates that change in the current configuration is warranted, this is achieved by repeating steps 2 to 4.

Figure 1 summarizes the PNUT framework. In this paper, we will only focus on the synthesis of the protocol and its dynamic configuration. Details about the QoS manager and the determination of user preferences are described in [12, 9]. The function selection process and triggering of the reconfiguration can be found in [8].

2.1 Protocol Engine

The implementation architecture of PNUT is shown in Figure 2. The application can either communicate directly with the application specific functions within the protocol engine, or indirectly via boomerang. Boomerang simply redirects socket system calls to the protocol engine. The configured protocol consists of three main elements - a shell, control and data channels. The shell provides the Application Programming Interface (API) for PNUT. All interactions with the PNUT protocol engine are conducted through the shell that provides similar functionality to the socket API in Unix systems. The main difference between the shell API and socket API is that the shell supports the maintenance of multiple configurable data channels.

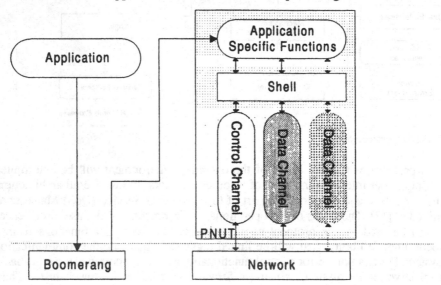

Fig. 2. The PNUT Engine

The data channels transfer the user data and are configured individually, using the protocol function library. The control channel is responsible for the management of the data channels. It is used to create, configure and close each of the data channels. Thus, unlike the data channels, the control channel uses a standard reliable protocol, ie TCP, thus enabling initial communication with all peer entities

2.2 PNUT Protocol Library

PNUT defines four groups of functionality, namely sequencing, error control, flow control and timing. Two of these groups can be supported by a number of mechanisms, and all of them can be enabled or disabled. Table 1 shows the mechanisms that are currently being used to support these functionality groups.

Table 1. Functionality & Mechanisms of PNUT

Functionality	Mechanisms
Sequencing	None, Sequence No.
Error Control	None, Go-back-N, Selective Repeat
Flow Control	None, Credit, Rate
Timing	None, Time Stamp

Sequencing defines the ordering constraints on the packets being transmitted and received on a data channel. Error control functionality defines how the protocol will behave when errors occur during the data transfer phase. Flow control functionality allows mechanisms to be specified that will allow flow and congestion control functionality to be included in the protocol. Finally, the timing functionality provides timing information for real time data streams.

By using the out-of-band signaling through the control channel, PNUT protocol engine is able to support the dynamic reconfiguration of data channels while minimizing the disruptions to the user data transfer.

2.3 Protocol Configuration

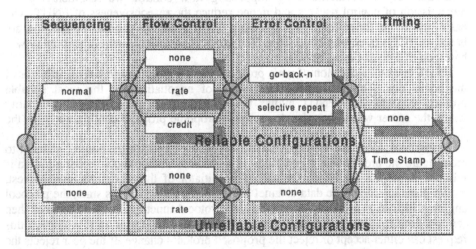

Fig. 3. Protocol Configurations

A variety of different configurations can be realized using the four groups of functionality described above. However, there are dependencies. For example, the error control functions - go-back-n and selective repeat - and the flow control function - credit control - require sequencing and hence can not be used with no sequencing. The different combinations of functionality of the current implementation of PNUT are shown in Figure 3. Configurations can be divided into reliable configurations, that guarantee the delivery of data and unreliable configurations that does not guarantee of delivery of data packets.

The functionality and mechanisms defined in this prototype of PNUT are not exhaustive but are designed to show the feasibility of the proposed framework.

However, although there may be many of different mechanisms that can be added to allow for a greater flexibility, we do not believe that there need to be many more functional groups.

2.4 Dynamic Configuration of Data Channels

Similar to other operations, the negotiation of the new protocol configuration takes place across the control channel. However, because we use out-of-band signaling it is necessary to provide synchronization to co-ordinate the change of functionality within the data channel. Therefore, the reconfiguration of the data channel requires two processes. The negotiation of the new protocol via the control channel and the synchronization of the change of the configuration of the data channel. In this section, we will briefly discuss both of these processes. A detailed description is provided in [8].

Since the control channel is required for the management of the data channels, it needs to be set up before any requests regarding data channels are handled. The main requirement of the control channel protocol is that it must be reliable and responsive. Because of the transmissions on the control channel are small, flow control functionality is unnecessary, as the small amount of data will not overflow any receiving buffers. Furthermore by supporting flow control, we risk delaying the transmission of control packets and hence reduce the responsiveness of PNUT. The basic operations for setting up and tearing down a control channel connection are based on TCP's three way handshake. The connection management functionality was based on TCP to assure reliability.

PNUT leaves the decision of what protocol configuration to use, on the initiator of the connection. This simplifies the process of negotiating for the most suitable protocol configuration. PNUT allows the initiator to request default functionality where the server would then decide on the appropriate protocol configuration for the data channel.

In Figure 4, we show the handshaking that takes place via the control channel to reconfigure the protocol for a particular data channel. The protocol reconfiguration is initiated through a shell function call. The initiator of the reconfiguration request, would place its side of the data channel into a prepare state. In this state, the protocol engine is instructed to anticipate a change of protocol functionality. The initiator then sends a Protocol Change Request message to the peer. The peer on reception of this request can either accept or reject the proposed protocol change. If the peer rejects the change, it would send a reject message informing the initiator that the reconfiguration request has failed. Otherwise, it would acknowledge the change by sending a confirm message to the initiator, and inform the local end of the data channel of the protocol change. On reception of the confirmation message, the initiator would similarly inform its local data channel endpoint that the change has been accepted. This sequence of message exchanges completes the handshake procedure for reconfiguring a data channel across the control channel. Once the data channels have been informed that the protocol change has been negotiated, they will need to synchronize the change of the protocol mechanisms.

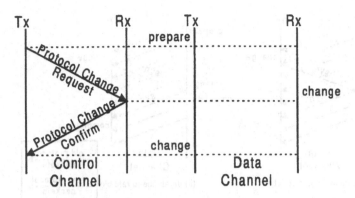

Fig. 4. Dynamic Protocol Reconfiguration

When changing mid stream, careful synchronization of the two end points are necessary. For example, changing from or to a reliable protocol configuration it is necessary to assure that all data being transferred by the reliable protocol is correctly received. Figure 5 shows the synchronization process for reconfiguring a protocol on the data channel with the four possible combinations of changing between reliable and unreliable protocols. The synchronization is achieved by each side indicating when they will begin to transfer data using the newly negotiated protocol. This is done by sending a packet with a SYN flag set in the header of the data packet. Once each side has sent and received a SYN flag, the synchronization process on the data channel is completed. Because the packet containing the SYN flag can be lost, when changing to an unreliable protocol, the first packet correctly received containing the new protocol ID within the packet header, is considered as the packet carrying the SYN flag.

A summary of the costs for the different cases of dynamic reconfiguration is shown in Table 2. For reconfiguration to be viable, the cost of the reconfiguration must be balanced by the gains achieved through the new protocol.

Table 2. Cost of Dynamic Protocol Reconfiguration of Data Channels

	Changing from an Unreliable Protocol	Changing from a Reliable Protocol
Irrespective of Data	None	Round trip time
Respective to Data		
Changeover Data arrives at Prepare signal	round trip time	Round trip time
Changeover Data arrives in between Prepare and Change signals	< round trip time	Round trip time
Changeover Data arrives after Change signals	None	Round trip time

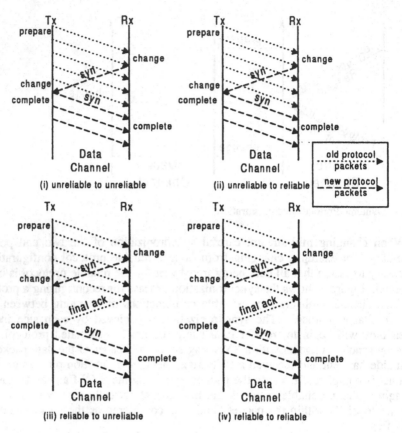

Fig. 5. Reconfiguration Process on the Data Channel

Reconfiguring a protocol, irrespective of the data, as shown in Table 2, there is only a cost of reconfiguration if we are changing from a reliable protocol to an unreliable protocol or to another reliable protocol.

2.4.3 Protocol Synthesis

PNUT, as described earlier, is specified in terms of the functionality it provides. Each functionality can be provided by a number of different mechanisms. Each of these mechanisms can be considered as an object consisting of a set of methods. A method is implemented as a C language function. A mechanism's methods correspond to the processing to be performed at certain events – transmitting data, receiving data and at time-outs. PNUT calls the appropriate method of a mechanism when one of these events occur. By indexing these sets of methods, configuration/re-configuration can be simply achieved by changing the reference pointer to the set of methods for the chosen mechanisms.

3. Prototype PNUT Implementation

The prototype PNUT framework was implemented in user-level over UDP under Linux. PNUT was implemented as an user-level protocol for a number of reasons. Firstly, it allowed for greater level of tailoring to the application requirements by allowing application specific mechanisms to be defined. Secondly, the user level protocol allowed for quicker development of PNUT when compared to it being developed within the kernel. Thirdly, by developing PNUT as an user-level protocol, we believe it will be easily distributed as part of applications rather than as part of the kernel. Finally, it has been shown that PNUT is able to provide high performance implementations even as an user-level implementation [8].

3.1 Design of a Distributed MPEG Player

To demonstrate the viability and the benefits of using the PNUT framework, a distributed MPEG application was developed. The MPEG player is essentially the MPEG player developed at the University of California, Berkeley [2] modified into a client and server application. The client requests a MPEG file from the server. The server transfers the requested MPEG file to the client, which processes and displays it. The distinguishing feature was the use of the application specific functions to handle changes in user requirements and network environment within the communication subsystem, thus minimizing the changes necessary in the application.

To highlight the usefulness of adaptive protocols, a mobile computing environment with an intermediate system was used for the experiments. Furthermore, a split the connection similar to work done in I-TCP [1] with separate communication subsystems for the wired and wireless sections of the network were used. PNUT in the intermediate also could filter the incoming traffic to be suitable for the end system in two ways. Firstly, by dropping P or B frames [14] at the intermediate system and only transmitting I frames. Secondly, by using unreliable channels for the transmission of the P and B frames, thus allowing them to be dropped when congestion occurs at either the intermediate system or the receiver. The experimental environment thus consisted of a fixed server, an intermediate server and a mobile client. TCP was used for the connection between the server and the intermediate system, and a protocol specifically configured for the varying network between the intermediate server and the client. This configuration is shown in Figure 6.

The communication system between the intermediate server and the MPEG client uses three individually configured data channels to transfer the different MPEG frame types, namely I, P and B frames as shown in Figure 6. The channel containing the I-frames was specified as a reliable, as I frames are essential for decoding. By defining the remaining data channels as either reliable or unreliable, it was possible to control the quality of the decoded video. For example if all three data channels are defined reliable, then the user will view the entire MPEG video clip as all the frames will be delivered and displayed. If the latter two channels are defined as unreliable, then when the network gets congested or the application is slow processing the MPEG clip, P and B frames will be lost, reducing the level of detail viewed by the user.

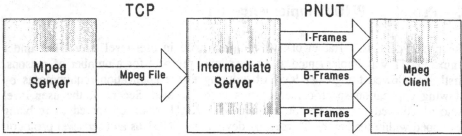

Fig. 6. Transfer of the MPEG file

The three channels are synchronized through framing services available via the PNUT shell. These synchronization services allow for the creation of synchronization groups where data channels can be added. When transmitting data on a synchronized group, a shell call similar to the socket call write() is made containing information on which data channel the data is to be sent and defining any dependencies. On the receiver side, the application would read data from the synchronized group and not any particular data channel. The PNUT shell passes the data sent to the client application in the same order it was transmitted but takes into account the possibility that data transferred on unreliable channels may be lost.

We are also able to define, through these synchronization services, the dependencies between the frames sent through different data channels. A packet with dependencies not met is not delivered to the user. For example, if a B-frame arrives at the MPEG Client, but the previous P-frame on which this B-frame is dependent on has been lost, this B-frame would be dropped.

We also added extra functionality to PNUT to provide dynamic filtering at the source of a data channel. Therefore in this scenario, we could dynamically prevent the transmission of any P and B frames, by having these frames dropped at the source. The advantage of this feature is that it prevents excess data being transmitted that will not be processed at the destination, and thereby freeing up bandwidth on the underlying network.

Two buttons on the MPEG player's interface were used to allow the users to dynamically reconfigure PNUT's functionality via a simple QoS. The first toggled whether the P and B data channels would use reliable or unreliable functionality while the second toggled whether filtering of these channels were activated or deactivated.

4. Experiments and Results

The experiments were conducted on PCs (the server was a 486dx-66, the intermediate Pentium-120, mobile computer was a Pentium-150MMX) running the Linux operating system (version 2.0.32). As PNUT is implemented in user space, and to facilitate the support of legacy applications, it is necessary to boomerang to direct the data. However, this can lead sever performance degradation. To evaluate the effects of boomerang, a dummy PNUT module, which simply transferred coming data to the intermediate system, was tested. The results shown in Table 3 confirm that the overhead of indirection is minimal

Table 3. Overhead of Boomerang

	TCP
With	5.473 Mbps
Without	5.352 Mbps

Then we firstly conducted experiments with a direct TCP connection over Ethernet and recorded the time to play out the clip and the number of frames processed by the MPEG player. We then conducted experiments with the indirect PNUT version of the MPEG player, recording the play out times for the full video clip, which was achieved by setting all the data channels to reliable protocol configurations. Similar experiments were done when playing out a lower quality video clip, by setting the higher detail channels to an unreliable protocol configuration. Thus using this setting we achieved a partial video affect where the level of quality displayed would be determined by the amount of packets dropped by either the underlying network or the end system resources. Finally by filtering the lower channels, we display a low bandwidth, low quality video clip. We also recorded the number of frames processed in each of these cases. These results are shown in Table 4.

The results show firstly that PNUT performances are very similar to TCP for the full video clip and there are no extra costs for using PNUT or the intermediate server. Secondly the processing time for the partial video clip is smaller and this is explained by the smaller number of frames processed. This is due to the dropping of frames by the MPEG player. The MPEG player is the main bottleneck in this distributed system and as packets on the unreliable channel are delayed waiting to get processed they are dropped. These losses are limited because of the nature of the framing services provided by the shell. Since the I-frames, are reliable, they will block shell calls to the I-frame data channel when the buffers are full, thereby blocking subsequent calls to send data on the unreliable channels.

Table 4. MPEG Player Results over Ethernet

	Play out Time (sec)	Frames Processed/Total
TCP		
Full video	77.23	960/960
PNUT		
I, P & B Reliable	77.75	960/960
I Reliable; P & B unreliable	67.25	832/960
P & B filtered	45.63	162/960

We conducted similar experiments over a congested Wavelan network and the results are shown in Table 5. These results show the benefits of PNUT. Firstly, PNUT performs slightly better than TCP when transferring the full video clip, this would be mostly due to the absence of flow control or congestion control mechanisms in the process model of PNUT that was used for these experiments. This resulted in more aggressive behavior and resulted in quicker recovery from errors.

Table 5. MPEG Player Results over Congested Wavelan

	Play out Time (sec)	Frames Processed/Total
TCP		
Full video	96.33	960/960
PNUT		
I, P & B Reliable	89.41	960/960
I Reliable; P & B unreliable	74.04	803/960
P & B filtered	45.74	162/960

When we use partial video through PNUT, we can see that the processing time has not greatly increased when on the congested network. By using unreliable protocols on the P and B data channels, we now only provide retransmissions for the I channel thus we reduce to the amount of packets we have to provide retransmissions for. This reduces the amount of delays, caused by retransmission, that the user views when observing the video clip. Secondly, as losses occur on the unreliable channels we prevent further retransmissions that will increase congestion on the underlying network.

The final case, of filtering, provides very similar performance to the case of the Ethernet. The main reason, for this is that by introducing filtering on the P and B channels we reduce the amount of data being transmitting and thereby reducing the congestion and the possibility of further data loss.

5. Discussion

It should be noted that during these experiments we did not move the base station as has been done in work on other indirect protocols (I-TCP and Snoop). This was outside our immediate area of concern. We believe that PNUT is able to migrate data channels using the control channel although there will have to be extra support added to the intermediate server to allow migration of application level filtering.

Any delays or interruptions caused during dynamic reconfiguration of protocols were unnoticeable to the user, mainly because of buffering delays within the application. Similarly, due to this buffering, it normally took a couple of seconds before the change of protocol functionality was noticeable. By allowing dynamic reconfiguration, the user is able to change the level of detail they view as their environment changes. Furthermore by placing services like filtering within the communication protocol rather at the end system, we can not only improve the performance of the client but also provide better usage of network resources.

Finally, using an intermediate server, we have shown that we can gain the benefits of tailored applications and protocols without having to modify currently available services. From these experiments, clearly demonstrate the benefits of using PNUT and tailored communication protocols.

6. Related Work

Over the last few years there have been a number of research projects that have investigated the development of adaptive protocols [13, 16, 18]. However, the work presented in this paper differs from this earlier work in a number of ways. Firstly, PNUT does not require specialist hardware or software system, this enabling it to be deployed in existing systems. Secondly, the PNUT implementation architecture makes it possible to be used with existing applications with no modifications. Finally, its user centric design makes it far less complex than any of the proposed systems to date. A detailed evaluation of these and a number of others can be found in [8].

[15] provides similar work on improving the transmission of MPEG videos over multiple partial-order transport channels. The main differences from our work is that PNUT is designed to provide a variety of functionality and mechanisms while they provide a transport protocol dedicated to providing partial-order service. They also provide a method for formalizing of the MPEG application's requirements that we have not done in this paper.

7. Conclusion

In conclusion, this paper has discussed the use of automatically generated communication protocols to support mobile computing environments. The paper discussed how the use of the PNUT model could be used to create protocols that not only provide services suitable to the application but can also benefit the underlying network environment.

Furthermore, this paper has shown how we can use an intermediate server not only to separate the protocols for the wired and wireless networks but also to provide intermediate application services like filtering. We also demonstrate that we are able to use the intermediate server to allow applications using tailored protocols to communicate with servers using traditional protocols.

Acknowledgements

This work was done while Ranil De Silva was working at the School of Computing Sciences, University Technology Sydney and was funded by the Australian Research Council. It does not represent the views of his current employer.

References

[1] Bakre, A., Badrinath, B., I-TCP : Indirect TCP for mobile hosts, in Proceedings of 15th Intl. Conf. On Distributed Computing Systems, May 1995.
[2] Berkeley Player, Berkeley Multimedia Research Center (BMRC), University of California, Berkeley, 1996. Http://bmrc.berkeley.edu/projects/mpeg/mpeg_play.html

[3] Bestavaros, A., Kim, G., Exploiting redundancy for timeliness in TCP Boston, in Proceedings of 3rd IEEE Real-Time Technology and Applications Symposium, Los Alamitos CA, USA, p184-90, 1997.

[4] Castelluccia, C., Chrisment, I., Dabbous, W., Diot, C., Huitema, C., Siegel, E., De Simone, R., Tailored protocol development using ESTEREL, Research report 2374, INRIA, France, Oct. 1994.

[5] Chen, Z., Tan, S.M., Campbell, R.H., Li, Y., Real Time Video and Audio in the World Wide Web, Fourth International World Wide Web Conference, Massachusetts, December 1995.

[6] Clark, D., Lambert, M., Zhang, L., Neblt: A high throughput transport protocol, in Proceedings of ACM SIGCOMM Conference (Communication Architecture and Protocols), USA, 1990, pp. 200-208.

[7] Clark, D., Tennenhouse, D., Architectural considerations for a new generation of protocols, in Proceedings of ACM SIGCOMM Conference (Communication Architecture and Protocols), USA, 1990, pp.200-208.

[8] De Silva, R., PNUT - Protocols configured for Network and User Transmissions, PhD Thesis, University of Technology, Sydney, Jan. 1998.

[9] Diot, C., Seneviratne, A., Quality of Service in Heterogeneous Distributed Systems, Proceedings of the 30th Hawaii Internation Conference on System Sciences HICSS-30, Hawaii, Jan. 1997.

[10] Jacobson, V., Modified TCP for Congestion Avoidence, end-to-end mailing list, April 1 1990.

[11] Jacobson, V., Braden, R.T., Borman, D.A., TCP Extensions for High Performance, RFC 1323, May 1992

[12] Landfeldt, B., Seneviratne, A., Diot, C., USA: User Service Agent, a New QoS Management Framework, Int. Workshop in Quality of Service Management (IWQOS98), USA, May 1998.

[13] O'Mailey, S.W., Peterson, L. L., A Highly Layered Architecture for High-Speed Networks, In Protocol for High-Speed Networks, IFIP, 1991.

[14] Rao, K. R., Hwang, J. J., Techniques & Standards for Image Video & Audio Coding, Prentice Hall, 1996.

[15] Rojas-Cardenas, L., Chaput, E., Dairaine, L., Senac, P., Diaz, M., Transport of Video over Partial Order Connections, Proceedings of the HIPPARCH 98 Workshop, London, pp158-171, 1998.

[16] Schmidt, D., Box, D., Suda, T., ADAPTIVE: A flexible and adaptive transport system architecture to support lightweight protocols for multimedia applications on high speed networks, in Proceedings of the Symposium on High Performance Distributed Computing Conference, Amsterdam, Sept.1992.

[17] Schulzrinne, Casner, Frederick, Jacobson, RTP : A transport protocol for real-time applications, Technical report, Internet Engineering Task Force - Audio-Video Transport WG, 1994.

[18] Zitterbart, M., Stiller, B., Tantawy, A., A model for flexible high performance communication subsystem, IEEE Journal on Selected Areas in Communications, 1993, pp.507-518.

Design of an Integrated Environment for Adaptive Multimedia Document Presentation Through Real Time Monitoring

Eduardo Carneiro da Cunha,
Luiz Fernando Rust da Costa Carmo,
and Luci Pirmez
E-mail: educc, rust, luci@nce.ufrj.br

Núcleo de Computação Eletrônica - UFRJ
P.O.Box 2324, 20001-970
Rio de Janeiro – RJ – Brazil

Abstract. The retrieval of multimedia objects is influenced by factor such as throughput and maximum delay offered by the network, and has to be carried out in accordance with the specification of object relationships. Many current network architectures address QoS from a provider's point of view and analyze network performance, failing to comprehensively address the quality needs of applications. The work presented in this paper concerns the development of an integrated environment for creation and retrieval of multimedia documents, that intends to preserve the coherence between the different media, even when the process is confronted with a temporary lack of communication resources. This environment implements a communication system that address QoS from the application's point of view and can help in handling variations in network resources availability through a real-time monitoring over these object relationships.

1 Introduction

Multimedia systems are computer systems that manipulate, in an integrated way, several types of media information. Frequently, multimedia systems are distributed; that is, its components are located at different processing nodes in a local or wide LAN. Quality of service in this context can be intuitively defined as a measure of how satisfied the user is with regard to a service rendered by a multimedia distributed system (DMS). Although the notion of QoS is intuitive, a series of measurable parameters can be established to define such concept objectively. These parameters are divided in two levels: system and user.

Most of the existing DMS architectures treats the quality of service from the system's point of view, that is, from the provider's point of view, and they use effective monitoring politics and resource management to provide quality of service support [1]. However, due to the heterogeneous nature and varying capabilities of today's end-systems and global network infrastructures, conventional resource reservation and admission techniques cannot guarantee QoS without considerable over-booking and inefficient resource utilization. Furthermore, these architectures fail to raise the QoS notion up to the user level, causing the QoS specification and management to be made through the system level parameters.

To address this problem, and avoid any adverse impact to the end-user, distributed applications and their infrastructures need to be adaptive. This means that either applications must tolerate fluctuations in resource availability or that the supporting infrastructure can itself mould to the dynamically changing requirements of the applications.

The work presented in this paper is originated from a research project on distance teaching that is being developed at NCE/UFRJ. Basically, this project, named ServiMedia [6], deals with multimedia documents authoring and storage, and with the network infrastructure for the remote retrieval of those documents. The basic subject is how accommodate in the same client/server, distance teaching environment, users with substantial differences on communication resources availability. The same document, retrieved by different users, can be interpreted in a form completely different, in agreement with the quality of the presentation that is noticed by the user. The quality of the presentation, in its turn, is directly related to the readiness of resources found in the network along the path that reaches the user, since some of the multimedia flows can be lost or partially damaged during the process of document retrieval.

In this work, we investigate a strategy for composition, storage and retrieval of multimedia documents that allows us to generate different formats of presentations at the user's site, starting from just one multimedia document specification at the document server. These presentation formats are adapted from the original specification in agreement with the network resources availability verified on the routes to the respective users. The key point in this strategy is that each generated format must preserve the semantic properties of the media originally specified by the document's author. This means that the multimedia document should maintain its semantics even when it is adapted and suffers some degradation in relation to the original document specification.

This paper is structured as follows: the section 2 presents the authoring strategy developed in this project and shows how to implement the strategy through extensions to the synchronized multimedia integration language (SMIL). The section 3 presents the ServiMedia Architecture and how the adaptive retrieval mechanism works. Still in the section 3, we describe the real-time stream relationship monitoring and we present an application example. Finally, the section 4 relates some conclusions of this work.

2 Composing a Multimedia Document

Current authoring systems accomplish the specification of multimedia documents based on three fundamental aspects: the logical structuring of the presentation, the establishment of spatial positioning and temporal relationships between the multimedia objects.

It is important to highlight that the concern about the maintenance of presentation coherence and semantic, associated to its QoS degradation control, and the inclusion of this facility in the authoring phase, is not explored by the current DMS architectures, which are most of the time just focused on the temporal issues. In order to provide the specification of semantic relationships between the multimedia objects, we use a modified version of the causal synchronization model presented in [3]. Thus, we've defined an authoring strategy that combines:

1) mechanisms of logical structuring of the presentations: the logical structuring worries about establishing abstraction mechanisms, intending to obtain a wide and structured view of the presentation. We use a logical structure that is based on the concept of groups of clips (parallel and sequential), where the clips are the media objects that compose the document. These groups are represented in a hierarchical tree similar to a tree of directories.

2) a model of spatial synchronization based on the definition of playback areas.

3) a model of temporal synchronization based on timelines: this imposes rules on how the objects can be linked to each other. Several models have been proposed in the literature, obeying, as possible, to some basic requirements pointed in [2]. A flexible temporal specification is obtained through the establishment of margins of tolerance for the beginning of the presentation of each object (fig.1). The advantage of the use of flexible temporal specifications is that it facilitates the use of relaxation and acceleration techniques of the presentations with synchronization purposes, aiming in the derivation of a schedule of the presentation as discussed in [8].

4) a model of causal synchronization based on conditional dependencies between the involved objects.

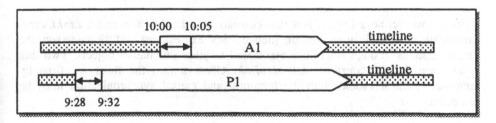

Fig. 1. Flexible temporal specification. Establishiment of a temporal interval in the beginning of an object indicates that its presentation can start at any instant within this interval.

2.1 Expressing Conditional Dependencies

Conditional dependencies have been proposed [4, 5] as a way of taking advantage of the knowledge on semantic relationships between different stream's objects. Conditional dependencies are causal relations associated with a stream's object aiming to express delivery constraints of that object, relative to the delivery of other objects belonging to either the same stream (intra-stream conditional dependencies) or distinct ones within the same bundle (inter-stream conditional dependencies).

To address the problem of different communication resource availability the author is allowed to specify the importance of each object that he inserts in the document and to delineate the QoS requirements and conditions that should be respected in order to preserve the consistency of the document.

Associated to each object, there is a *QoS descriptor* that defines the possible quality degradation. In the QoS descriptor the quality degradation is specified in terms of a range, within which the object can be manipulated to handle possible variations on the network resources.

Thus, the author can define whether an object is essential or just qualitative and also establish causal relationships between the qualitative ones. These relationships are specified through links that interconnect the qualitative objects forming a net of causality that describes the coherence wanted for the document [7].

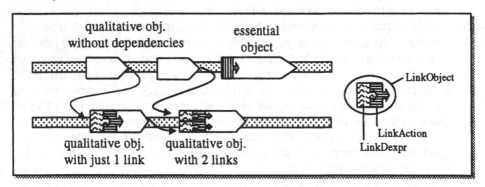

Fig. 2. Links representation. The Link objects are coupled to the respective qualitative objects.

In [7], we define a *Link object* that consists of a *LinkAction* and a *LinkDexpr*. LinkDexpr is a Boolean expression (dependency expression) that characterizes the causal relations through conditions associated to other qualitative objects. Two link types were defined: *startlink* and *stoplink*. Observe that the links are fired by temporal synchronization. Both the temporal and causal synchronization must be respected.

2.2 Implementing the Authoring Strategy

In relation to the multimedia presentation authoring, there is not a consent or standard widely accepted for the specification of multimedia documents that should be retrieval or presented through remote servers. The Synchronized Multimedia Integration Language (SMIL) [9] allows the integration of an independent multimedia group of objects in a synchronized multimedia presentation through a textual specification, with tags and very similar to HTML. In particular, SMIL is a format of multimedia data description for authoring tools and players.

SMIL introduces many valuable ideas that are similar to our authoring strategy and that can be used by the ServiMedia environment. This way, we've decided to adopt the language SMIL as reference for our ServiMedia authoring system. However, certain characteristics specific of our authoring strategy are not considered by SMIL 1.0. For example, the flexible temporal specification and the causal relationships through links are not considered. We have been working on this issue in order to describe the extensions to SMIL so as to implement our authoring strategy [7].

To allow a flexible temporal specification we have created the *can-begin* attribute that defines an interval of tolerance for the beginning of the presentation of any clip. To specify the links that describe the causal relationships between the clips of a document, we have created two attributes (one for each link type): *startlink* e *stoplink*.

```
<par>
    <audio id="A1" src="…" can-begin="5s"
                        startlink="(P1:started)"/>
    <img id="P1" src="…" can-begin="4s"
                        stoplink="(A1:concluded)"/>
</par>
```

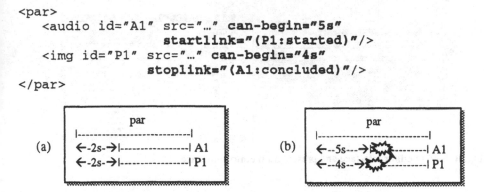

Fig. 3. (a) P1 and A1 begin synchronized at 2s and they finish synchronized when A1 finish; (b) it is not possible to begin P1 within the tolerance, P1 and A1 are discarded then.

The SMIL *switch* tag carries out the choice based on static variables that are configured or stored in the clients presentation tools (players). Using the causal relationships, that is, the *startlink* and *stoplink* attributes, we are testing dynamic variables that change their states during the presentation. Actually, we can even associate a causal relationship to the switch group, as shown in [7], obtaining a dynamic adaptation of the presentation. It depends only on the author of the document to specify the possible variations in the presentation formats.

2.3 The Authoring System Prototype

In this section we present a general view of the authoring tool developed in this project. It uses the language SMIL increased with the extensions that have been created and presented in the previous section. This tool was initially developed for the platform Windows98®.

Figure 4 illustrates one of the authoring tool interfaces. Through this interface, the author specifies the layout defining the playback areas where the clips must be presented. It is possible to create new areas, to move, resize and define its background colors. For each new area a new timeline is added where the objects (clips) can be placed. One timeline for the sound track of the presentation is always present.

In another interface the author specifies the logical structure of the presentation through a hierarchical tree. In this interface it is possible to create new groups (*par*, *seq* and *switch*), new clips, to organize them, and to define all its attributes and properties. All modifications done in the presentation structure, in the groups and its properties, are reflected in the visualization of the timeline of each area of the layout.

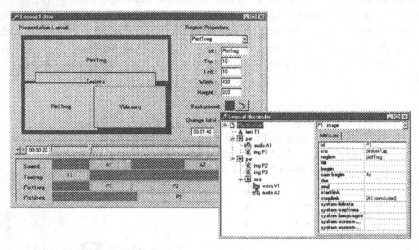

Fig. 4. Layout editor interface and presentation editor interface

3 Retrieving Multimedia Documents

A protocol for multimedia document retrieval control has been designed by the IETF recently. The real time stream protocol (RTSP) [13] is a client/server control protocol for multimedia presentation. RTSP was developed to deal with the needs of an efficient distribution of multimedia streams within IP networks. RTSP establishes and controls either a single or several time-synchronized streams of continuos media such as audio and video. In other words, RTSP acts as a network remote control for multimedia servers, and consists basically of request and response messages. The requests are issued sequentially on different connections.

3.1 Real-Time Delivery and Monitoring

Currently, there is an increasing demand for real-time applications that transfer continuous media. These new multimedia application, such as video conferencing and media-on-demand, impose new QoS requirements in terms of delay, error rate, jitter and throughput. The real-time applications must deliver data within an expected time frame and thus they cannot depend on TCP in the real-time transmission. Hence, in order to meet these new requirements, the Real-Time Transport Protocol (RTP) has been designed and it is used nowadays for real-time continuous media transmission over the Internet.

The real-time transport protocol (RTP) is both a IETF Proposed Standard [10] and a ITU Standard (H.225.0). RTP provides end-to-end delivery services for data with real-time characteristics, such as interactive audio and video. Those services include payload type identification, sequence numbering, timestamping and delivery monitoring. However, RTP itself does not provide any mechanism to ensure timely delivery or provide other QoS guarantees, but relies on lower-layer services to do so.

The data transport is augmented by a control protocol (RTCP) to allow monitoring of the data and to provide minimal control and identification functionality. The RTP control protocol (RTCP) is based on the periodic transmission of control packets to all participants in the session, using the same distribution mechanism as the data packets. RTCP provides feedback on the quality of data delivery. The feedback may be directly useful for control of adaptive multimedia applications. This feedback function is performed by the RTCP sender and receiver reports. The RTP functions require that all participants send RTCP packets, therefore the rate must be controlled in order for RTP to scale up to a large number of participants. The RTCP scales well for a small multicast group but a scalability problem arises when it comes to a group of thousands of users. Some of these problems are addressed in [11, 12].

Cumulative counts are used in both sender information and receiver report blocks so that differences may be calculated between any two reports to make measurements over both short and long time periods, and to provide resilience against the loss of a report. The difference between the last two reports received can be used to estimate the recent quality of the distribution. For example, we can calculate the number of packets lost during an interval, the number of packets expected during an interval, the packet loss fraction over an interval, the loss rate per second, the apparent throughput available to the receiver and the interarrival jitter. Packet loss tracks persistent congestion while the jitter measure tracks transient congestion.

3.2 Resource Reservation

Our communication architecture allows a resource reservation for the transmission of essential information. That is, for the traffic of essential information, a resource reservation protocol that communicates with a QoS routing algorithm should be used so as to select the best route capable to absorb the traffic and the resource reservation [16]. On the other hand, the transmission of qualitative information is accomplished without warranties through the routes statistically more favorable to the success of the transmission. This means that for the traffic of qualitative information the reservation is not carried out but the QoS routing algorithm is still used to choose the most favorable routes. That guarantees, at least, the retrieval and presentation will be in conformity with the basic QoS requirements specified by the author. The routing algorithm is based on the combination of two metrics: the bandwidth of each communication channel and the end-to-end delay variation generated by the route [17, 18].

RSVP [14] has been designed to support resource reservation in the Internet. However, it has two major problems: complexity and scalability. The former results in heavy message processing overhead at end-systems and routers. The latter implies that the amount of bandwidth consumed by refresh messages and the storage space that is needed to support a large number of flows at a router are too large, mainly in a backbone environment. Our next direction in the project is to investigate resource reservation associated with QoS routing. We are studying the RSVP complexity issues within the Integrated Services Model and the recent Differentiated Services Model. We intend to define a scheme that will use effectively the real-time protocols (RTP/RTCP and RTSP) not only to delivery continuous media but also to provide resource reservation and QoS-routing while addressing the complexity and scalability problems mentioned in [11, 12, 15, 19].

3.3 Adaptive Retrieval

QoS adaptation is the process of maintenance control, facilitated through alterations to either the balance and distribution of resources or to the application's level of service, on short time scales. Adaptation processes often occurs as a result of QoS notifications, usually emitted from QoS monitoring mechanisms, which indicate a change in the observed service affected through the availability of some element of the end-to-end resources. Notifications may indicate a imminent lack of resources and hence reduction in service quality (QoS degradation) or a failure to maintain service quality through a complete loss of resources (QoS failure). By QoS failure we mean either a complete route failure or an impossibility in keeping service quality inside the QoS requirement range provided by the QoS descriptor.

We can find in the literature a diversity of QoS adaptation mechanisms, such as QoS filters, sender rate adaptation, layered multicast, etc. However, all of them are concerned only with the traffic parameters and with the direct user perception of isolated media objects. The whole presentation, with all its media streams and their semantic relationships, is not treated as a complete documentation that has significance just due to the combination of several information objects. In the ServiMedia, starting from the document generated by the authoring system, a global QoS scenario is created. This scenario consists of the QoS requirements and conditional dependencies of the streams. An mechanism of dynamic path allocation and adaptive retrieval uses this scenario. This mechanism is implemented through two complementary modules: QDM (QoS descriptor monitor) and SRM (Stream relationship monitor).

This mechanism operates based on a real-time monitoring over the QoS scenario of qualitative streams, so as to achieve three main goals: 1) manage the delivery of qualitative streams, 2) verify whether the QoS requirements are being respected and adapt the QoS to fluctuations in the network resources, and 3) adapt the document structure based on the causal relations between the qualitative streams. Through this mechanism, we can release routes and resources that are being used for streams whose delivery, from the application point of view, became useless. We can also start delivering streams with less QoS requirements to replace other streams whose QoS requirements could not be respected.

An increasing number of current multicast applications prefer to use receiver-based adaptation schemes instead of sender-based adaptation schemes to adapt to congestion in the network. In sender-based adaptation, when congestion occurs, the sender decrease its rate of data transmission to suit the receiver with lowest capabilities.

Receiver-based adaptive applications have the advantage of accommodating to the heterogeneous capabilities and conflicting bandwidth requirements of different receivers in the same multicast group [20]. Besides, the real-time protocols represent a new style of protocol following the principles of application level framing (ALF) and integrated layer processing proposed in [21] and used in [22]. That is, these protocols intend to be malleable to provide the information required by a particular application and will often be integrated into the application processing rather than being implemented as a separate layer. Thus, the QDM and the SRM were implemented at the application level of the receiver side (fig.5).

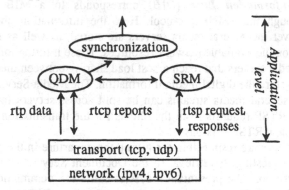

Fig. 5. Receiver-based adaptive structure. The QDM is responsible for monitoring the QoS feedback provided by the RTCP messages in order to preserve the QoS requirements as well as, when the QoS adaptation exceeds the QoS descriptor limits, to signal the SRM to adapt the document so as to preserve the causal relationships between the streams, in other words, to preserve the document consistency.

3.4 Overall Operation

An RTSP URL may identify each presentation and media stream. The overall presentation and the properties of the media the presentation is made up of are defined by a presentation description file. The presentation description file may be obtained by the client using HTTP or other means and may not necessarily be stored on the same server as the media streams.

The presentation description file contains a description of the media stream making up the presentation, including their encodings, language, QoS requirements, and other parameters that enable the receiver (client) to choose the most appropriate combination of media. In this presentation description, each media stream that is individually controllable by RTSP is identified by an RTSP URL, which points to the media server handling that particular media stream and names the stream stored on that server. Several media stream can be located on different server for load balancing purposes [23]. Following, we identified the components that participate in the architecture illustrated in the figure 6:

Document Server is the entity that receives the retrieval requests from the clients. The Document Server stores the specification of the multimedia document. An agent located in the document server receives the retrieval requests and, based on the information obtained from a media distribution information base (DIB), it find out how the information is replicated through the serves and verifies which media servers should be signaled to begin the transmission of the streams belonging to the document requested by the client. Then, it composes the presentation description file automatically inserting a list of URLs for each stream that points to different media copies.

Distribution Information Base (DIB) corresponds to a MIB extension for management through the SMNP protocol. Here the information about the media copies location over the several media servers are stored, as well as information on the topology and on the available network resources. These informations contribute to a choice of the media servers that reach a best load sharing between media servers.

Media Servers store the digital media information. The Media Servers may exist in any number and several media streams can be split across servers for load sharing. They receive the RTSP requests from the clients for the transmission of multimedia streams creating new RTSP sessions.

ServiMedia Clients are responsible for giving the departure in the processing of a document when requesting its retrieval to the Document Server. After receiving the presentation description, the presentation system starts the communication with the media servers. Afterwards, the QoS monitoring mechanisms (QDM and SRM) starts monitoring the communication to preserve the QoS requirements as well as the causal relationships specified by the author in the presentation description.

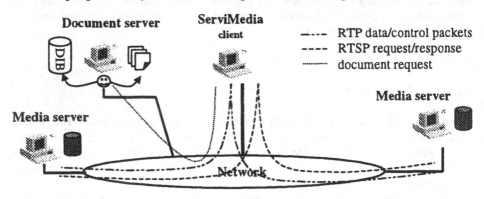

Fig. 6. Architecture of the ServiMedia Environment

In addition to the recent tendency in receiver-based adaptation scheme, there are at least three reasons for implementing the SRM as an integral part of the presentation system at the client site:

1. The presentation system possesses the information (contained on the presentation description) about the stream interdependencies. This way, it knows which causal relationships are to be monitored;
2. Once the presentation description, with the addresses of the media servers supplied by the document server, has been obtained, the RTSP sessions are

established between the client and the servers. Thus, the client can monitor the transfer of streams through the RTCP messages and send RTSP requests to stop, pause or start streams. If a client can start a stream, it must be able to stop and control the stream. Server should not start streaming to clients in such a way that clients cannot stop the stream.

3. Deciding that the monitoring should be made by the presentation system, we are adopting a decentralized control configuration where each client is responsible for monitoring the sessions (streams) that were established between it and the servers. One could think about a centralized configuration, where the SRM module could be implemented for example, in the document server. In spite of the document server also possessing the description of the stream interdependencies, it would have to participate in all RTSP sessions established between clients and servers for each one of the presentations. Besides, for the case of a multicast session with receiver diversity, it is much more suitable that the client makes the decision of abandoning or not the session while other clients stay connected receiving the corresponding flow.

3.5 Application Example

In this section, a simple example in the area of distance and interactive training is described. The application is assumed to be distributed over three nodes, the document server and two media servers. The training application comprises the following two parts: an introduction (the servers send information to the student in order to present the authors and copyrights, and give two option of training subject) and a training part (the servers send information so as to present the training material).

In the *introduction*, an opening video stream (IV1) presents the company providing the training and the credits, the authors are also presented. In parallel with the video there is a background sound track (IA1). An optional picture (IP1) is presented if the video cannot be presented with the specified QoS requirement. Following, a picture with two icons (IP2), representing the two possible trainings is presented. These two icons link the introduction to other specific presentation that contains the selected training (*training1* or *training2*). When the client request the document via http, the document server, after consulting the DIB, decides to stream IV1 from server 1 and IA1 from server 2.

Let's assume that the student has selected the *training2*. At this point, the process repeats, that is, the client request the document *"training2.smil"* from the document server, who consults the DIB and decides which severs should the streams be retrieved from. In the *training* phase, the document contains a video stream (TV1) that presents the training content, an audio stream (TA1) related to TV1, and a picture stream (TP1) that presents diagrams and other static images. In this phase there is a strong requirement for the delivery of the video stream. In other words, the delivery of only the audio and picture information is regarded as useless from an application point of view. If only the delivery of the audio information cannot be performed, it will be replaced by delivering a text information (TT1) to the student.

Figure 7 summarizes the communication process between client and servers for the training phase of the application. It shows the case where the audio TA1 cannot be delivered from the mediaserver2 within the QoS descriptor limits causing QDM to notify a QoS failure to SRM. SRM, in turn, examines the presentation description and decides to deliver the text TT1 from mediaserver2 to replace TA1. After that, SRM updates QDM with information about the new presentation structure (new streams) so that QDM can continue monitoring the quality of data delivery.

We consider that in a distance teaching environment, where the presentations (training) are most of the time relatively long, the receiver will not be bother by a delay caused by adaptation mechanisms, since we think the significance and consistency of the document are more critical than a initial delay. We assume that the receiver (student) will prefer to watch a coherent presentation with some casual delays than a continuous presentation with an unsatisfactory quality and that does not make sense. This situation comes to emphasized the importance of a flexible temporal specifications mentioned in the section 2.

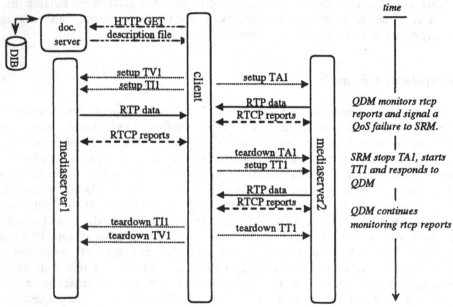

Fig. 7. Communication process

4 Conclusions

The authoring strategy presented in [7] promotes a strong integration between the systems that compose the ServiMedia environment. The establishment of a net of causal relationships in a document generates subsidies that allow the other systems (communication and presentation) to decide when adapting some information and how this adaptation must be carried in order to preserve the document semantic. That characteristic has been of great value, mainly in an integrated distance teaching environment.

An authoring system was developed based on the Synchronized Multimedia Integration Language (SMIL). A presentation system is under development in order to carry out the QoS monitoring and adaptation through the utilization of the QoS descriptor and the stream relationships. It is used the feedback provided by the RTCP reports and the real-time stream protocol to control the presentation. This real-time monitoring mechanism (QDM + SRM) allows the client to manage the QoS requirements as well as to preserve the causal relationships specified by the author. The whole presentation, with all its media streams and their semantic relationships, is considered as a complete documentation that must preserve its significance, which has been shown to be of great value in our distance teaching environment.

6 Reference

1. Aurrecoechea C, Campbell AT, Hauw L, "A survey of QoS architectures", ACM Multimedia Systems 6:138-151, 1998.
2. G. Blakowski, R. Steinmetz, "A Media synchronization Survey: Reference Model, Specification, and Case Studies", IEEE Journal on Selected Areas in Communications, Vol. 14, N. 1, pp. 5-35, January, 1996.
3. J.P. Courtiat, L.F.R.C. Carmo, R.C. de Oliveira, "A General-purpose Multimedia Synchronization Mechanism Based on Causal Relations", IEEE Journal on Selected Areas in Communications, Vol. 14, N. 1, pp. 185-195, January, 1996.
4. J.P. Courtiat, L.F.R.C. Carmo, R.C. de Oliveira, "A new mechanism for achieving inter-stream synchronization in multimedia communication systems", in Proc. IEEE Int. Conf. on Multimedia Comput. Systems, Boston, , pp.173-182, May 1994.
5. J.P. Courtiat, R.C. de Oliveira , L.F.R.C. Carmo, "Toward a new multimedia synchronization mechanism and its formal specification", in Proc. ACM Multimedia'94, San Francisco, pp. 133-140, October 1994.
6. L.F.R.C. Carmo, L.Pirmez, "ServiMedia: An Integrated System for Multimedia Document Creation and Retrieval with Adaptive QoS Control", 2° Franco-Brasilian Symposium on Distributed Computer Systems, Brazil, November, 1997.
7. E. C. da Cunha, L.F.R.C. Carmo, L. Pirmez, "A Multimedia Document Authoring Strategy for Adaptive Retrieval in a Distributed System", 6th International Conference on Distributed Multimedia Systems, University of Aizu-Japan, July 1999.
8. Candan KS, Prabhakaran B, Subrahmanian VS, "Retrieval schedules based on resource availability and flexible presentation specifications", ACM Multimedia Systems 6: 232-250, 1998.
9. Synchronized Multimedia Working Group of the W3 Consortium, "Synchronized Multimedia Integration Language (SMIL) 1.0 Specification", W3C Recommendation, June, 1998.
10. H. Schulzrine, S. Casner, R. Frederick, V. Jacobson, "RTP: A Transport Protocol for Real-Time Applications", RFC1889, January, 1996.
11. J. Rosenberg, H. Schulzrine, "Timer Reconsideration for Enhanced RTP Scalability", draft-ietf-avt-reconsider-00.ps, July 1997.
12. J. Rosenberg, H. Schulzrine, "New Results in RTP Scalability", draft-ietf-avt-byerecon-00.ps, November 1997.
13. H. Schulzrine, A. Rao, R. Lanphier, "Real Time Streaming Protocol (RTSP)", RFC2326, April, 1998.
14. R. Braden, L. Zhang, S. Berson, S. Herzog, S. Jamin, "Resource ReSerVation Protocol (RSVP) -- Version 1 Functional Spec. RFC2205, September, 1997.

15. P. Pan, H. Schulzrine, "YESSIR: A Simple Reservation Mechanism for the Internet", August 1997

16. E. Crawley, R. Nair, B. Rajagopalan, and H. Sandick, "A Framework for QoS-Based Routing in the Internet", RFC2386, Agosto 1998.

17. Z. Wang and J. Crowcroft, "Quality of Service Routing for Supporting Multimedia Applications", IEEE J. Selected Areas Communication, 14 (7), 1996.

18. R. Vogel, R. Herrtwich, W. Kalfa, H. Wittig, and L. Wolf, "QoS-Based Routing of Multimedia Streams in Computer Networks", IEEE J. Selected Areas Communication, 14(7), 1996.

19. R. El-Marakby, D. Hutchison, "A Scalability Scheme for the Real-time Control Protocol", Proceedings of the 8th IFIP International Conference on High Performance Networking (HPN'98), pp.147-162, Vienna, Austria, September 1998.

20. D. Waddington, D. Hutchison, "End-to-End QoS Provisioning through Resource Adaptation", Distributed Multimedia Research Group, Lancaster University, Internal report, "ftp://ftp.comp.lancs.ac.uk/pub/mpg/MPG-98-10.ps.gz" , May 1998.

21. D. Clark and D. Tennenhouse, "Architecture considerations for a new generation of protocols", in ACM SIGCOMM'90, pp. 200-208.

22. I. Chrisment, D. Kaplan, C. Diot, "An ALF Communication Architecture: Desing and Automated Implementation", IEEE Journal on Selected Areas in Communications, Vol.16, N°.3, pp.332-344, April 1998.

23. D. Pegler, N. Yeadon, D. Hutchison, D. Shepherd, "Incorporating Scalability into Networked Multimedia Storage Systems", technical report, Department of Computing, Lancaster University, 1996.

Authoring of Teletext Applications for Digital Television Broadcast

Christian Fuhrhop, Klaus Hofrichter, Andreas Kraft

GMD FOKUS - Kaiserin-Augusta-Allee 31 - 10589 Berlin - Germany
fuhrhop@fokus.gmd.de, hofrichter@fokus.gmd.de, kraft@fokus.gmd.de

Abstract. Teletext will be one of the first and most prominent multimedia services offered in digital television networks. Teletext is well known and has already been available for many years in analogue television in Europe and elsewhere as an additional service provided by television operators. Digital Broadcast technology allows to enhance this kind of service in quality and functionality.

Digital Television technology provides many advantages in comparison to the analogue distribution methods, such as possible reuse of content from Internet-based services, improved navigation, better layout capabilities and support of animated content. These capabilities give more freedom to the application designers. However, it is yet undecided which application format is best suited for Digital Television Networks. In order to develop a large market it is required to consider standardised formats to ensure interoperability between different service providers. Unfortunately, various industrial consortia such as DVB/MHP, ATSC/DASE and DigiTAG are working on different solutions. The formats have to be supported by the designers of the applications, the network operators and the end systems, and specific tools have to be developed to enable application production taking the advantages of the new technology into account. This paper discusses various aspects of the production of Teletext applications for Digital Television Broadcast. The authoring system presented here is based on a generic approach to describe the applications´ layout and behaviour utilising XML data representation, templates and server-side scripting. The system is independent of the final interchange format and can generate several different representations such as MHEG-5 and XHTML to meet the requirements of upcoming broadcast environments and to enable publication of applications in the Internet using the same content source. Navigation style-guides are covered and techniques to create applications and the associated content via server-side scripts are discussed.

Introduction

Digital Television Broadcast technology expands the potential of today's analogue broadcast services. Today's services are mainly based on the use of the Vertical Blanking Interval (VBI), where additional data is transmitted to the receivers along with the television signal. This method provides a convenient way to add application specific data without causing problems for legacy television receivers which are not aware of the service. Typical services are Teletext, mainly deployed in Europe and integrated into television sets, and some data services such as Intercast, which require a PC for viewing. This paper concentrates on Teletext-like services, and does not consider Intercast technology since it relies on PC clients with a different user community compared to a television-oriented audience.

The most relevant advantages of Digital Broadcast technology with respect to the application format are covered in the list below:

- **Richer Presentation Design**
  Digital content formats allow to create applications which make use of full multimedia functionality. Compared with analogue technology for Teletext this results in more colours, higher display resolution, advanced navigation and improved user interaction.

- **More Bandwidth**
  Digital Television Broadcast provides the option to distribute the available bandwidth between the television service and data services. The bandwidth is assigned by the service provider and may range between 6 Mbit per second for a data-only service and some hundred KBbit per second for Teletext-like applications.

- **Reuse of Content**
  The format for the analogue Teletext is fairly complex to generate and optimised for the target resources, e.g. the number of available colours for images, the font and the very restricted resolution. In contrast, content formats for Digital Broadcast allow to reuse existing content, such as images or text, without or with limited loss of quality. Moreover, the content formats discussed in this paper are capable of keeping different media separated from each other, which has significant advantages for the authoring process.

- **Service Convergence**
  Since Digital Broadcast Services and Online Services (which rely on bi-directional communication networks such as the Internet) are complementary to each other, it is possible to establish related services, which share information and content, or link to each other. Digital Television Broadcast technology eases this option, because content data such as images or text can be reused with little or no conversion required.

The Teletext Application

This section describes a design rationale for Teletext applications intended for the Digital Television Broadcast. The main focus of the experimental work carried out at GMD FOKUS is in the first step to recreate and enhance the functionality of the analogue Teletext. As a design study a Teletext-type application was developed in the course of the EC funded ACTS IMMP project. This application is designed around a tree-structure and features a specific area for the placement of advertisements.

Application Description

The application is developed to demonstrate and test an advanced Teletext service in a broadcast environment. It was designed to provide visual and attractive improvements over classic Teletext services while offering similar functionality. The target environment is a TV that is able to receive terrestrial, cable or satellite digital video. This implies limited bandwidth, a simple remote control unit as the control device and the absence of a back-channel. The application consists of a number of pages that are gathered in thematic groups like 'News', 'TV Programme', 'Weather' or 'Travel'. Every thematic group has an index page that lists all the pages in this group. The structure of the application is a simple tree structure, which is easy to conceptualise for the end user, easy to maintain by the author and easy to parse and present by the end system.

The following section describes the criteria that influenced the application design in detail.

Application Design Criteria

In the design of the Teletext application there were a number of criteria that had to be addressed. The navigation needed to be obvious and consistent, the authoring of the pages had to be simple, there needed to be a visible improvement over classical Teletext applications, the application needed to be appealing on 16:9 and on 4:3 displays without too much overhead, advertising space had to be provided and bandwidth considerations had to be taken into account.

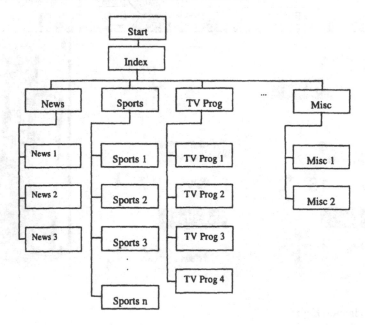

Fig. 1. Application Structure

Obvious and Consistent Navigation

For Teletext applications the input device is usually a simple remote control unit. Although some modern remote controls offer advanced input methods that work mouse-like and allow free cursor movement on the TV screen, only the keys '0'-'9' and the red, green, yellow and blue keys can be expected to be present on every TV remote control.

So the navigation had to be restricted to those keys, while structuring the pages of the application in such a way that more advanced TV sets can build their own meta-navigation system, similar to current TV sets that cache the Teletext structure and build their own navigation system around them. The structure of the application is a simple tree (see figure 1).

Navigation is based on these simple rules:

- Wherever you are, RED always returns you to the main index.
- When you are within a group, GREEN always brings you to the group index.
- YELLOW brings you to the previous page in a group.
- BLUE moves to the next page in a group.
- Pages within a group are numbered, the corresponding key acts as a shortcut to that page.

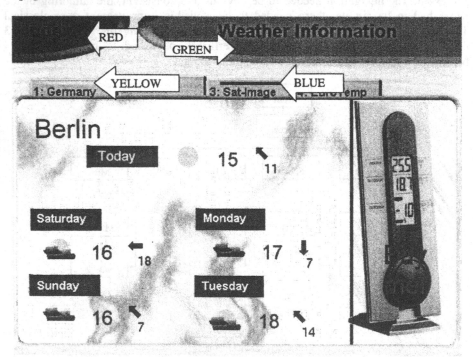

Fig. 2. Navigation Keys

These are the complete navigation rules. To aid the user in navigation, all available choices are always displayed on screen. There are no cases where a button that is not displayed on the screen has a function within the application.

The structure of the application is deliberately simple. As meta-information, each page contains the list of all pages within a group, the title of the group and its own title. Since this information is available to the displaying system, an additional navigation layer can be provided by the end system to make use of advanced navigation controls. For example, often or most recent used pages can be displayed in a spatial arrangement that is convenient for mouse-style navigation. Special features of the display can also be utilised, for example the list of pages within a group can be 'zoomed in' when the cursor passes over the group title.

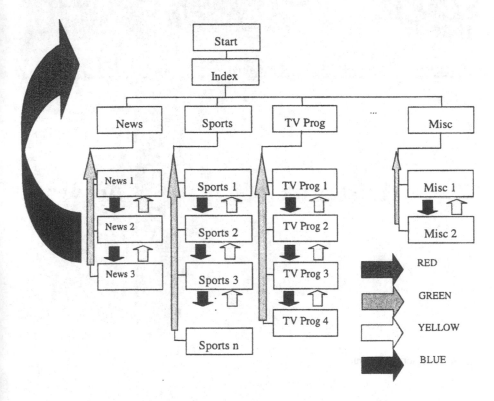

Fig. 3. Application tree with colour key navigation structure

Layout Considerations

Frame

There is a clear separation between the navigational frame and the area for the actual content and advertising. This ensures that all the information that is necessary to navigate between pages is visible on all pages, regardless of content. It also ensures a familiar environment for the user and reduces confusion that is caused if every page brings its own layout.

The frame contains the information where the user is in relation to other pages and which control elements lead where. Since the frame remains basically the same for all pages, the basic structure needs only to be done once and specific instances can be easily produced by automatic methods and for different platforms (for example for MHEG-5 as well as XHTML).

The actual authoring needs only to be concerned with the information area and the advertising area.

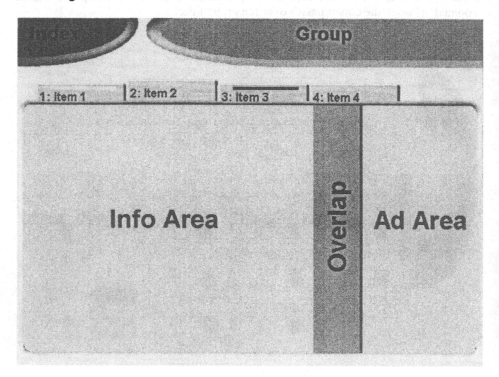

Fig. 4. Basic Screen Layout

Information Area

The information area contains the actual information for the audience. In this example only two variations are allowed for this area: Either this area contains an image or it contains a text. Other content types are possible and mixed forms might be desirable in some cases, but this is not covered here.

The content is generated by either creating a text file (for information that has to be provided fast or automatically, like news or stock exchange information) or a fixed sized bitmap for data that requires graphical presentation or is likely to live long enough to justify the additional effort of graphic design.

Advertising Area

End users are not expected to pay for Teletext services. This is unlikely to change in the near future. The only practical way to create revenue from Teletext services is the sale of advertising space. For this reason it is necessary to reserve an area specifically for advertising and separate this area logically from the information area to ensure that both areas can be changed independently from each other.

The advertisement is a bitmap of a fixed size. It overlaps the information area to a certain extend to allow advertising that laps over into that area. Usually this part of the advertising bitmap should be transparent, except for the overlapping image or logo. It is the responsibility of the content designer to place no important content in that area.

Advertisements that overlap into the information area are desirable for advertisers, since it helps to break down the conceptual border between information and advertising and makes it harder for the customer to mentally 'cut out' the advertising.

While the information area can be graphic or text based, the advertising area is always graphic based. Purely text based advertisements are of no interest to advertisers. Even though graphic advertisements consume more bandwidth than text-based versions, this has to be accommodated, since the advertising is the actual 'payload' of the service and reductions of size or quality for this area are usually not acceptable.

Fig. 5. Advertisement reaching into Content Area

Different Screen Formats

Restricting the actual information to a specific area also allows to circumvent the problem of non-proportional scaling for different screen sizes. The information and advertising areas remain the same in both versions and only the navigational controls are moved for the wider screen format. Moving the controls to the free area at the side allows the information window to be enlarged proportionally, thus avoiding the need to author specific content for wide-screen television sets. This also allows the broadcasting of both versions with relatively little effect on bandwidth, since only the control data and the surrounding frame for wide-screen need to be added to the broadcast stream, while the actual information data, which accounts for the largest part of the broadcast data, is useable for both versions.

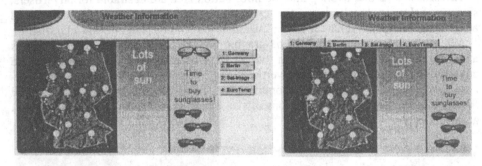

Fig. 6. Wide-screen and 4:3 layouts

Improvement over 'Classic' Teletext

To be accepted by the end user, new versions of Teletext need to provide an obvious additional value over the classic text- and block-graphic based systems. The best method to lure the user into accepting new services is to provide content that is graphic-based.

Providing a pleasing layout and as many pages as possible with information that is difficult or impossible to present with text based media is the easiest way to gain acceptance of new services. It also helps to add 'depth' to the layout, even if it is just achieved by using bitmaps that give the illusion of being three-dimensional to provide a visual 'added value' compared to traditional flat-looking Teletext applications.

Bandwith Considerations

While the surrounding frame and the navigational elements seem to be a waste of bandwidth, the actual impact on application size is minimal. Since the same layout is used for all pages within the application, the graphical layout elements need to be loaded only once for the whole application and generate only negligible overhead for the application size. The bandwidth required for the start-up page (which presents an application logo to the user, but is mostly there to ensure that all layout elements have been loaded from the broadcast stream before the application proceeds) and the graphical elements for both the 4:3 and the 16:9 version is roughly the same as for four content pages.

For the pages, most of the bandwidth is taken by the information data image, which, on the average accounts for 60% of the data, followed by the advertising, which takes another 30%, while the control information for the 4:3 and 16:9 displays require about 5% each. Replacing the information image by text usually reduces the data to the amount needed for control information.

Naming Scheme

To allow easy creation of pages by automatic tools and to reduce bandwidth, a simple naming scheme was used. All group index files are named <groupname>.<extension> and all pages within a group are named <groupname><page number>.<extension>, which allows easy automatic generation of the links for the green 'group index', yellow 'previous' and the blue 'next' button. The <groupname> is a short designation of the nature of the group, like 'news', 'sports' or 'misc'.

By keeping the extended title only in a specific location in the control file, it is not replicated with every link to the file, but still sufficiently accessible for automatic indexing, if the end system wants to provide it.

Authoring Techniques

The type of application presented above requires a new kind of application authoring technique. While it is feasible to develop a small number of pages manually, usually a Teletext service contains several hundreds of pages. Some of these pages are updated on a regular basis, sometimes automatically with data from an external source. Therefore, it is required to differentiate between direct authoring of an application and publishing, and continuous application production, which is the usual method for template-based applications.

Direct authoring is done manually by an application designer for each page separately. Here, an authoring tool is used, which offers facilities to create multimedia objects and arrange them on the page. Most tools also offer means to administrate the whole application, e.g. the transition between the different parts of an application. Those tools are, however, limited to a predefined set of functionality. An author can usually not perform special tasks with the tool, but has to program the application "by hand". This is not always possible, because the system may prevent direct access, and the author has to have specific knowledge about the system.

Another kind of authoring is to facilitate templates and building blocks, and to combine them in an authoring tool. The actual layout of an application is designed by an expert. Only predefined areas in the template can be filled with informational data for each page separately, and actions which are allowed by the template can be selected. An authoring tool supporting this approach can be fairly simple because it only has to support the application designer to fill in the gaps.

This authoring technique has the advantage of leading to stable and predictable applications, because an application author has no access to the underlying mechanisms of the applications. He provides the content data, defines the navigation, and can change only some of the content attributes like, for example, the colour of a text.

An enhancement to this technique is to provide the same means for automated authoring. Here again, the framework is supplied by an expert who designs the templates. The places where the content data should be inserted later (such as the information and advertisement areas described earlier) are marked by special elements. During the production process, the elements are replaced with the actual data by a filter program which can produce the application.

This way it is easily possible to build applications which have a mixture of static parts (such as an introduction screen) and dynamically generated pages.

Authoring with Scripting Support

The input for such an application generation system consists of a script interpreter and a number of scripts which contain procedures to assign content data to the predefined areas of the pages. The layout of the pages is defined by a sequence of procedures.

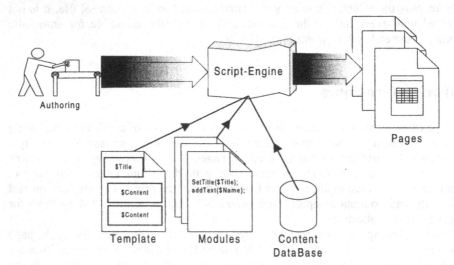

Fig. 7. Authoring of an application using various input sources

This method is well suited for applications in which the input data can be directly used without much calculation. In addition, it can be enhanced by using a scripting language. A template may define the general structure and layout of the application, but the content data for the areas is processed or created by scripts. This way, it is easy to perform conversions on the input data before inserting it into the template. Using only the general template method, the input data is most often pushed to the template process. With scripts, the input data can not only be processed, but additional input can be retrieved from different sources like, for example, different content databases. The following figure presents an example for a simple script to create a page for the presented Teletext application:

```
createPage($ID); % Create page and assign identifier
setTitle($TOPIC); % Assign the topic for the page
setMainImage($MAINIMAGE);    % Reference to main image
setAdvertisment($ADIMAGE);  % Reference to add image
if (numberOf($LABEL) > 0)   % Check for Label-defs
          createLabels($LABEL);
if (numberOf($KEYS) == 1)   % Check for Key-defs
          createKeyHandlers($KEYS);
```

Fig. 8. Script example

The marked keywords in the box above are examples of such procedures. The procedures assign data to a layout which is prepared by the expert. The *createPage*() procedure, for example, creates the general structure of a Teletext page and assigns an identifier to it. How this is done and what the appearance of the page actually is, is independent from the script. This way, an abstract authoring method can be used to describe the changing contents of the pages. How the other elements (such as the information data, the advertisement data, or the labels) are presented and the navigation is performed depends on the script.

```
PAGE>
<NAME>weather2</NAME>
          <TOPIC>Weather Information</TOPIC>
          <MAINIMAGE>weather2.gif</MAINIMAGE>
          <ADIMAGE>ad_weather2.gif</ADIMAGE>
          <LABEL TOP="false">
             <NAME>Germany</NAME>
             <TARGET>weather1</TARGET>
          </LABEL>
          <LABEL TOP="true">        <!-- current label -->
             <NAME>Berlin</NAME>
             <TARGET>weather2</TARGET>
          </LABEL>
          <LABEL TOP="false">
             <NAME>Sat-Image</NAME>
             <TARGET>weather3</TARGET>
          </LABEL>
          <LABEL TOP="false">
             <NAME>EuroTemp</NAME>
             <TARGET>weather4</TARGET>
          </LABEL>
          <KEYS>
             <RED>index</RED>
             <GREEN>weather</GREEN>
          </KEYS>
  </PAGE>
```

Fig. 9. Sample Input Data

Procedures can also perform more complex tasks. The *createLabels()* procedure in figure 8 is intended to provide some means to display all the defined labels and to create the necessary navigation functionality (for example Link-objects in MHEG-5) to move to a new page.

The data provided for the script is retrieved by the script interpreter. The pseudo-variables *$ID*, *$TOPIC* etc. refer to data record fields from an external data source, provided in XML for example. Figure 9 shows an excerpt from an XML input file for a Teletext application.

By providing very little additional information, like for example the TOP attribute to indicate the "current" label, the colour key navigation for the application can be determined automatically by the script.

The input data can be retrieved from a third party provider, or collected from various data sources and then combined into a single document. A stock market information page with business news from various news providers is an example for this approach.

The advantage of this scripting approach is that it is highly flexible and configurable. By including other, predefined procedures, the demand for building blocks can be satisfied. This way it is possible to include procedures for forms, tables, or even simple games. This method, however, has also some disadvantages. The whole system must be managed very carefully, and a very high standard for quality assurance must be set. Because of the scripting facility, programming and design errors can be harder to detect in advance.

Conclusion

This paper presents design consideration for Teletext applications for digital television under consideration of specific constraints of the broadcast environment, and discusses possible authoring technologies to create applications by scripts utilising a meta-language independent from the encoding. The approach allows to create applications for various environments and to re-use the application in the Internet.

Most of the work described in this paper was carried out within the framework of the European ACTS IMMP project. One of the goals of IMMP was to provide groundwork for future competitive authoring processes which exploit applications for digital television environments and broadcast networks. It is considered to enhance the experimental authoring system to a full product.

References

This section comprises a list of references to important documents. A list of resources in the WWW with further information with relevance for the digital television concludes this paper.

References

ISO/IEC IS 13522-5, Information technology — Coding of Multimedia and Hypermedia information — Part 5: MHEG Subset for Base Level Implementation.

ISO/IEC IS 13818-6, Information technology — Generic coding of moving pictures and associated audio information — Part 6: Extensions for Digital Storage Media Command and Control

Digital Terrestrial Television MHEG-5 Specification, version 1.04

EN 50221 Common Interface Specification for Conditional Access and other Digital Video Broadcasting Decoder Applications

ETS 300 706, Enhanced Teletext Specification

Klaus Hofrichter: MHEG 5 - Standardized Presentation Objects for the Set Top Unit Environment, in Proceedings of the European Workshop on Interactive Distributed Multimedia Systems and Services, Berlin, Germany, 4-6 March 1996, Springer Lecture Notes in Computer Science, 1996

Klaus Hofrichter, Andreas Kraft: An Approach for Script-based Broadcast Application Production, in Multimedia applications, services and techniques: 4th European conference; Proceedings, ECMAST'99, Madrid, Spain, May 26-18 1999, Springer Lecture Notes in Computer Science, 1999

Resources

DVB, Digital Video Broadcast: http://www.dvb.org

ATSC: Advanced Television Standards Committee: http://www.atsc.org

AICI: Advanced Interactive Content Initiative:
http://toocan.philabs.research.philips.com/misc/aici

ATVEF: Advanced Television Enhancement Forum: http://www.atvef.com

DAVIC: Digital Audio-Visual Council: http://www.davic.org

W3C: World Wide Web consortium: http://www.w3.org/MarkUp/Activity.html

IMMP: Integrated Multimedia Project: http://www-nrc.nokia.com/immp/

IMMP: GMD FOKUS IMMP page: http://www.fokus.gmd.de/magic/projects/immp

MHEG-5 UG: MHEG-5 users Group: http://www.fokus.gmd.de/ovma/mug

The MHEG Centre: The MHEG Centre: http://www.fokus.gmd.de/magic/projects/ovma

GMD FOKUS MAGIC: GMD FOKUS Competence Centre MAGIC
http://www.fokus.gmd.de/magic

XHTML: http://www.xhtml.org

XML: http://www.xml.org

Middleware Support for Multimedia Collaborative Applications over the Web: A Case Study

M. Pinto, M. Amor, L. Fuentes, and J.M. Troya

Depto. de Lenguajes y Ciencias de la Computación,
Universidad de Málaga. ETSI Informática.
Campus de Teatinos, s/n. 29071 Málaga, Spain.
email:{pinto,pinilla,lff,troya}@lcc.uma.es

Abstract. Software composition and frameworks are currently important technologies for multimedia services development, especially on the Web. These technologies can be successfully applied to multimedia collaborative services achieving the derivation of new services with shorter development time. In this paper we present the collaborative aspects of MultiTEL, a compositional framework for developing distributed multimedia applications over the Web, implemented using Java/RMI. MultiTEL is based on a new compositional model, which separate computation and communication aspects in different entities, components and connectors. MultiTEL also includes a middleware platform that supports local resource management, allowing the plug&play of multimedia device components. Currently, our goal is to extend MultiTEL with collaborative issues allowing the development of cooperative services, like application sharing, blackboard, etc. We will address the main collaborative application features of MultiTEL through a case study.

1 Introduction

The rapid evolution of our data communications infrastructure is making distributed projects increasingly viable [1]. The advances in communication technology and the impact of distributed multimedia systems make the computer support evolves from individual to group work and enhance the computer supported collaborative work (CSCW) application development.

CSCW comprises a large number of different applications. Collaborative interactions do not require that users must be active simultaneously. A CSCW application that specifically supports simultaneous activity is called synchronous or is known as a real-time service; otherwise, it is asynchronous or a non-real-time service. Due to technology limitations, for many years, commercial applications have been limited to asynchronous, text-based collaborations.

Recently, many factors are driving the appearance of synchronous CSCW applications. Concretely, two of them are especially relevant and interesting for us: the Internet and the compositional frameworks [2].

The increasing availability of the Internet and the WWW enables companies to develop cost-effective collaborative solutions [1]. The rapid adoption of Internet standards is producing an avalanche of new products and also the emergence of Web-based information networks called *Intranets*. However, the Web processing is tied to Web servers and it is not adequate for distributed applications running in client machines. The Java language, the WWW and CORBA [3]may be considered as complementary software technologies, which used together provide a powerful set of tools for developing and deploying multi-user distributed applications [4]. However, they do not provide a complete environment for the design and development of multimedia collaborative applications.

On the other hand, component-oriented architectures increase the reuse of standard components in open systems and help designers to cope with the complexity in multimedia telecommunication services development. In addition, frameworks are powerful because they define a generic architecture that provides a reusable context for implementing components. By combining compositional software with the power of frameworks, designers can try to short the development time of open multimedia services.

This paper addresses the implementation issues of groupware collaboration in MultiTEL, a distributed, compositional framework for the development of distributed multimedia applications over the Web [5]. MultiTEL is based on a new compositional model [6], which separates data processing from coordination patterns. The goal of this new model is to help service designers to modify a distributed application to obtain another one with the same patterns of communication and synchronization. Indeed, the service designer can reuse the difficult part (the concurrency management of components) while freely modifying the internal processing part of the application components.

Now, our goal is to extend MultiTEL for supporting CSCW services features like the management of latecomers, development of shared graphical user interfaces(GUIs), handle user event and maintain coherency between collaborators.

2 Overview of the MultiTEL Platform

An ideal reuse technology provides components with standard interfaces that can be easily connected to make a new system. Software developers do not have to know the internal implementation of a component, and it is easy for them to learn how to use it [7]. However, components are not isolated entities, they usually interact with each other to accomplish a certain task. Therefore, we distinguish between how a component achieves its computation and how that computation is combined with others in the overall system [8].

The MultiTEL compositional framework has been developed in order to integrate multimedia services on the Web. In this section, we present an overview of MultiTEL, that is a distributed component framework with a platform that uses Java/RMI and the Web for component communication [9]. MultiTEL implementation in Java implies the compositional model, which is the basis of the multimedia reference architecture. MultiTEL also offers a middleware platform that provides common services for accessing local resources, like local components, connectors, services and multimedia devices, like a camera or a microphone [10].

2.1 Compositional Model

MultiTEL defines a new compositional model which separates computation from coordination in two different units, *components* and *connectors* respectively, increasing the reusing of components and collaboration patterns. Components model real entities of the system and encapsulate computation. Connectors, which implement communication protocols, encapsulate the interaction and synchronization protocols between these entities.

Components Components are passive computational entities that encapsulate data and computation. Their interfaces defines the list of messages that can be received from connectors. The reception of a message causes changes in the internal state of components which are reported to the environment sending at least one event (Fig.1). Components have an identity independent of the different interactions in which they could engage. They do not know neither how the reception of a message and its computation influence the rest of the system, nor the connectors that will receive these events. This characteristic contributes to have a component market where components can be reused in new services, knowing only the interfaces of messages that offer.

Connectors Connectors are abstractions of coordination patterns. They implement communication protocols among two or more components and are implemented following the *State Design Pattern* [11]. They encapsulate the state transition diagram (STD) of a protocol and handle the events propagated by components (Fig.1). A connector interface defines the list of input events that triggers the STD. Connectors communicate with components by sending messages, therefore they have information about how the message exchange affects the behaviour of the system. Whereas components propagate events to unknown connectors, connectors have to specify the types of components they connect.

Dynamic Binding of Components and Connectors The *User Service Part* (USP) component is a persistent representation of a user in a service. The USP object contains information about the architecture of the application, that is, how components have to be plugged into connectors. This object constitutes the main part of MultiTEL platform, performing the dynamic composition of components and connectors at runtime according to the service architecture. This characteristic eases the design of open and dynamic systems, where the architecture of the system may change dynamically, adapting connections between components and connectors to the execution environment. Dynamic binding also enforces the construction of tailorable applications. The designer may define several connectors that might be candidates to catch the same event. Later the system will decide, for a certain customer, which connector matches their profile.

Figure 1 shows a simple CSCW service, with a *WhiteBoard* component that encapsulates a shared GUI and an *EventDispatcher* connector that encapsulates the *event broadcast protocol* and broadcasts the GUI's events to all collaborators.

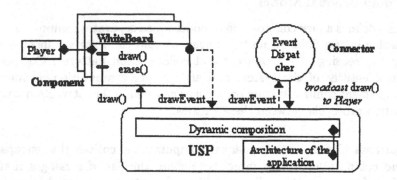

Fig. 1. Components and connectors model. Dynamic composition performed by the USP component

Users that interact with their GUI encapsulated in the *WhiteBoard* component provoke the sending of *drawEvent* events. The USP object provides information about the connectors that can handle the *drawEvent* event as it is sent from the WhiteBoard component. The USP looks at runtime which candidates' connectors are instantiated and finally sends the event to them. In the example, the USP will send the *drawEvent* event to the *EventDispatcher* connector. In this way if two implementations can receive the same set of events, it is possible to have different behaviours for different users in a system. By changing only the *EventDispatcher* connector implementation we could make collaborative all the events of the GUI or only part of them.

Message passing from connectors to components is also resolved inside the USP. When the *EventDispatcher* connector receives the *drawEvent* event, it broadcasts the *draw()* message to all collaborators through the USP. For broadcasting events we use the *broadcast event mechanism* implemented in the USP object. The USP allows to broadcast events by role type, so a connector may send events to all the components of the same type. In this example, we can broadcast the *draw()* message to all the components of type *Player*, where player is the role of the *WhiteBoard* component.

2.2 MultiTEL Platform

MultiTEL compositional applications will run on a distributed platform that uses Java/RMI and Web services. MultiTEL platform is structured in two levels: the compositional application components and a middleware platform.

Application Components (AC) MultiTEL framework encapsulates a multimedia telecommunication service (MTS) architecture which is implemented following the compositional model. MultiTEL helps designers to cope with the complexity of MTS dividing the architecture into three subsystems, each subsystem with a well defined scope. MultiTEL applications are instantiations of this architecture, including the CSCW applications that we will present in this paper. Figure 2 shows the application level composed by the following subsystems:

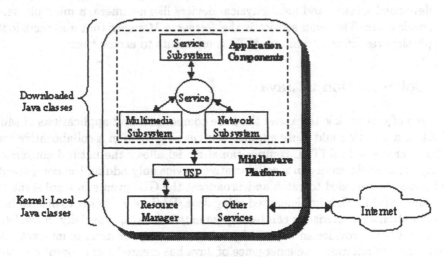

Fig. 2. MultiTEL platform

1. *Service subsystem*: Defines the service logic which is controlled by specific access and scheduling connectors. Participants can join a service playing one of these three roles:
   - *Organiser*: This user configures a generic service by giving values to component or connector parameters, and mapping network and multimedia resources to their local addresses.
   - *Manager*: The manager is in charge of starting an organised service. In multi-party services, the manager controls the service scheduling.
   - *Participant*: A simple participant can send/receive multimedia data from the service provider or other participants if the service logic allows it.
2. *Multimedia subsystem*: Defines a set of components and connectors for the allocation and control of multimedia devices.
3. *Network subsystem*: Defines a set of components and connectors for the management and reservation of broadband multimedia channels.

Middleware Platform The *Middleware Platform* is divided into two parts:

- *User Service Part* (USP): The USP is a Java applet that encapsulates the application architecture, that is, the components that define the service, and the current application context which comprises the components that are currently instantiated. Users join a multimedia service by running the corresponding USP applet. MultiTEL uses Java/RMI for communication of USPs among themselves and with the *Resource Manager* (Fig.2). The USP can modify its internal structures in order to add external components, such as vendor plug&play components, like multimedia devices or GUIs [10].

- *Kernel:* MultiTEL's kernel offers a set of services for allowing the control of software and hardware resources like components, connectors or previously developed services and local physical devices like a camera, a microphone, a speaker, etc. The main service is the *Resource Manager* that abstracts local physical resources and allows MTS components to access them.

3 Collaboration in Java

The aim of this work is to extend the base components and applications of MultiTEL in a way we could reuse applications running them in a collaborative way without changes. MultiTEL compositional model allows the natural conversion of any multimedia service in a collaborative service only adding the components and connectors needed to catch and broadcast the GUI events in a coherent order. Since MultiTEL is implemented using Java/RMI we are going to discuss how collaboration characteristics can be implemented in Java, especially in applets.

The WWW provides an unprecedented opportunity for users to interact with each other. In addition, the emergence of Java has created a consistent computing environment for supporting synchronous collaboration [2]. However, Java presents yet some problems that have to be resolved.

In this section we are going to discuss problems founded in Java and how these problems can be solved in MultiTEL.

3.1 A Few Problems

One of the main features of a collaborative application is its *shared environment.* The collaborators need to be aware of the actions of the others. One of the main problems in Java has been to maintain the knowledge of interactions between collaborators, like having different cursors for each one or seeing the graphical feedback of user interactions although this is been resolved in the new versions.

Another problem in Java is the management of events of the new Java event model. Each type of event handled for a GUI applet will have an *event listener* associated, which attends the events and executes the code associated with it. When a user interacts with a graphical component inside an applet, Java generates an event with a reference to this component. In a CSCW service, this event must be broadcasted to all the participants, but in the remote applets the component's reference is not valid and the event is lost and does not reach the remote component.

Another problem with events is to distinguish between *local* and *remote* events. Each component can have one or more *listener* but all receive the same events produced by the user interaction or posted by code. However, the reaction of the GUI component to local or remote events is different. When a user interacts with the GUI producing a *local* event, the listener must broadcast the event to all collaborators, but when the listener receives a *remote* event from the code, this event only must be sent to the local component. In the next section, we propose some solutions to these problems.

3.2 Proposed Solutions in MultiTEL

We propose a model for developing standard GUIs that help the management of local and remote events and the identification of remote events in its target. We will see how this prototype supports high reusing of GUIs in different services. We will use this prototype for developing MultiTEL GUIs, although it can be used for developing any Java application.

The issues of our proposal are:

1. Separate the GUI components from the *events listeners* associated with them in different Java classes. This follows the separation of concerns of the MultiTEL compositional model and allows us to obtain different behaviours of the same GUI applets by changing only the implementation of the *listeners*.
2. Identify the elements in a GUI component by string identification transporting it with the events during broadcast. This allows the identification of the element that must receive the events in the target.
3. Display the necessary information in the component in order to allow the reuse of a third-party GUI applet, without having the source code. Components should contain the necessary information (metadata), that allows designers to know by introspection how to reuse them in a new application.
4. Distinguish between local and remote events by extending the Java different event classes with a flag indicating if they are local or remote. The listeners' classes check whether the event is local — we broadcast it to all collaborators — or remote — we only send it to the local component —.

In order to ease the reuse of the GUI applets developed following this rules, we propose the implementation of a Java application or applet similar to the Beans Development Kit (BDK) [13] used for the development of JavaBeans. The input of this application builder will be a MultiTEL compound applet, where this application will show the applet tagged with the string identification of each component and the information needed about the component.

4 CSCW in MultiTEL: A Case Study

Typical multimedia services like Videoconference or Remote Education can be considered collaborative applications since they allow multiple users to interact remotely and collaborate to do the same task. However, there are other kind of collaborative applications which have a more restrictive definition. In this section, we are going to study applications that support groups of users engaged in a *common task* and that provide an interface to a *shared environment* [14].

MultiTEL is based on a replicated or *event broadcasting* architecture [12], where each collaborator maintains their own copy of the Java applet downloading it from the service home, but shares a *virtual instance* of it. The user's input events are broadcasted to each applet copy. In replicated architectures the primary issue is performance because they require less network bandwidth than a centralized one, since only state-changing information must be transmitted between the collaborators. In contrast, is more difficult maintaining consistency among the participants because the event control is distributed between the different instances of the application [12].

The key factors in CSCW applications are to maintain consistency between collaborators, manage latecomers, determine synchronous versus asynchronous interaction and What-You-See-Is-What-I-See issues. There are also other issues for groupware applications that are common to general multimedia services, like communication protocols or access control.

Now, we are going to see how all these factors can be modelled in MultiTEL, developing a CSCW application, and showing how the compositional architecture of a generic videoconference service can be extended with the components and connectors that model the specific characteristics of CSCW applications. Furthermore, we need to model the sending of events generated by users through GUI components. The example we present is similar to *Pictionary*© game, that we call *CWPictionary*.

The *CWPictionary* is a game board and its main elements are: the gameboard, a dice, a clock for controlling the turn timing, some cards and the paper for drawing. The participants are distributed in groups and the basic rules are: i) only a group have the turn each time; ii) inside the group with the turn, a player throw the dice, move inside the gameboard, take a card and draw the card's item; the rest of players in the group have to guess the drawing before timeout; iii) if the square have the "*all play*" icon inside, everybody in the game can guess the drawing; iv) if the square have the "*double time*" icon inside, the group have double time to guess. Figure 3 shows the game's collaborative elements.

Following the compositional MTS architecture of MultiTEL, the *CWPictionary* application design is divided in three subsystems: service, multimedia and network subsystem. For modelling group collaboration we are going to extend the service subsystem with the components and connectors needed.

4.1 Service Subsystem

Here we will discuss the components and connectors that determine the access and schedule of the service. First, we have decided which kind of access protocol for restricting user admission will be used [6]. In *CWPictionary* the participants play in groups, so the access protocol must keep this in mind to distribute the participants in groups. MultiTEL offers several possibilities like *manager access protocol*, in which the manager decides whether a user can join the service, or *database access protocol*, where there is an access control database with the users subscribed to a service.

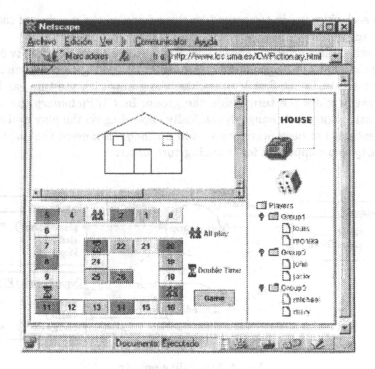

Fig. 3. CWPictionary gameBoard and the games collaborative elements

In *CWPictionary*, *latecomers* can only play in a existing group. When a *latecomer* joins the service they must obtain the actual instance of the *shared environment*, by using the *component and connector migration* in runtime offered by the USP object using *Java's Object Serialization*. In this case, once the *latecomers* has joined to a group in the service, the group's manager has to migrate the component that encapsulates the *shared environment* in the actual state.

Second, we have to choose the scheduling protocol. In *CWPictionary* we decide to use a rigid floor control with token passing, that is, only a group take the turn at a time. Inside a group only a member of it have the turn of throwing the dice, moving in the game board and drawing, and the rest of the group members have the turn to guess the drawing, except in special cases. In this control scheme, we distinguish between two different kind of tokens:

- *Draw Token*: There will be only one of it in the service and only the participant who has it can draw.
- *Guess Token*: There will be several in the service and the participants who have it can guess the drawing.

In *CWPictionary*, the scheduling protocol has two levels, because there will be *parallel conferences*, one between the leaders' groups and another one between the members of each group. In the first level is scheduled which group will have the turn each time and in the second level is scheduled the member of the group with the turn to draw.

The scheduling protocol showed in Fig.4 is the same for the two levels, but with different *EventAgenda* components. The EventAgent component stores scheduled events, in the first level decides the group with the turn, and in the second level decides the turn inside the group. In *CWPictionary* the order of group participation will change dynamically depending on the play evolution. In consequence, it has been necessary to add a *changeTurn* event that will be sent to *EventAgenda* component for changing turn order.

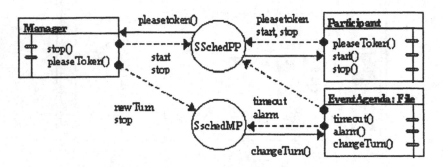

Fig. 4. Scheduling protocol

For CSCW application, the next step is to extend the service subsystem to manage collaborative features. These new components and connectors will receive the events generated by the GUI and broadcast them to all collaborators. The relation between them is showed in Fig.5. In MultiTEL the service is running in a distributed platform which implement inside the USP object the broadcast of an event to multiple participants, so it is not necessary to implement an additional event broadcaster. When a USP has to broadcast an event, it sends the event to the USPs of each receptor participant reaching the appropriate connector. For a particular service these components and connectors will depend on the floor control implemented. If the service uses a rigid floor control according to token passing control scheme, only the user that currently has the token can change the state of the CSCW environment and the coherency is enforced. In this case only the *EventBroadcaster* connector is needed. However, if the service needs multiple users to have the control over the state of the CSCW environment it is necessary to make an absolute ordering of user actions and requires the use of a *conflict resolution protocol* to maintain consistency. There are two possible approximations: *centralized control* versus *distributed control*.

– *Centralized solution*: In this case there would be a unique *EventMerged* connector, a *MergedInput* component and an *EventBroadcaster* connector in the service. The *EventMerged* connector receives the events from all participants, and sends them to the *MergedInput* component. The *MergedInput* component

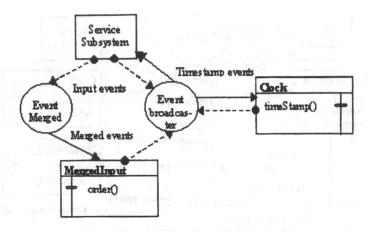

Fig. 5. Collaboration subsystem

sorts the events and sends them to the *EventBroadcaster* connector which broadcasts the events to all participants.

- *Distributed solution*: In this case there would be a *Clock* component and an *EventBroadcaster* connector for each participant in the service. When a participant sends an event, the *EventBroadcaster* receives it, adds a timestamp and broadcasts it to all participants. In this case, it is more important to assure a general coherency between all collaborators than to assure that the events are received in the same order that they were generated.

The centralized solution requires more computation since the events must be ordered and have the disadvantage of all centralized solutions, but maintaining the coherency between participant is easier.

MultiTEL can support different interaction modes. Changing the *Event-Merged* and/or *EventBroadcaster* implementation we can change between manage all input events in a collaborative way or only some of them. For example, a help button that explains the game's rules is not a collaborative component and if a user push this button the rest of collaborators do not need to be notified.

Multimedia Subsystem The multimedia subsystem is a collection of components and connectors that model multimedia devices and control data read/write operations (Fig.6). Our compositional model helps to insulate application developers from the difficulties of multimedia programming. Connectors characterise the dynamic relationships among *MultimediaDevice* components.

The *MultimediaSAP* component encapsulates the *logical connection graph* (LCG) which determines the devices to capture. In the *CWPictionary* service, a participant will join the service if the *AllocCont* connector can capture a microphone and a speaker or a chat device. If any of the participants do not have a microphone and a speaker, then everybody will use the chat device for communication. In addition, the use of a camera and a display device can be defined as optional. All these requirements will be specified in the LCG. The *AllocCont*

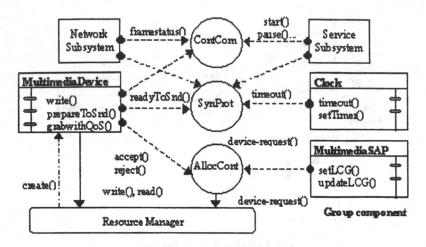

Fig. 6. Multimedia subsystem

connector will communicate with the *Resource Manager* to capture the devices. For each device, there will be a *ContCom* connector to control its behaviour. All interactions with the *MultimediaDevices* components will be through these connectors. If we want to synchronise different type of data, for instance, audio and video, we will use the *SynProt* connector.

Network Subsystem Network subsystem makes the management and reservation of broadband streams transparent to the rest of the system (Fig.7). The *ConnectC* connector initiates connections between local multimedia devices and the network according to the LCG. The architecture distributes an instance of a *VirtualConnection* component to each participant machine. This component encapsulates multicast transport connections of a single participant and is in charge of reading and writing multimedia data. We have implemented different versions of this component, where each one use a different communication protocol such as TCP, UDP, RTP and multicast. The *NetworkSAP* component abstracts the set of connections and network resources for a single service.

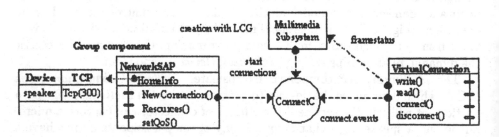

Fig. 7. Network subsystem

This component encapsulates transport information (protocol, port) about the current producers of the global conference sessions in the *HomeInfo* structure. However sometimes, collaborative applications demand configurations in which subsets of participants would maintain their own side-band conferencing sessions.

In MultiTEL, by assigning a *NetworkSAP* component representing transport connections of a group of participants, we can define parallel conferences without modifying the participant components.

5 Resource Management

MultiTEL platform has to manage network and multimedia resources. Consequently, the reservation of bandwidth channels for the transmission of data between distributed participants and the management of physical devices according to real-time features might be considered.

These resources are local to the client machine. They are not common to all participants as the rest of components and connectors, but they must be shared between all the applications running in the same machine. Furthermore, resource components encapsulate native code that does not allow the transportation of them. These features make the implementation of components that model physical devices different from other application components.

The *Resource Manager* (RM) is part of MultiTEL's kernel and its role is the abstraction of operating system resources. The RM offers a common representation through a common interface which defines the allocation, management and deallocation of resources and models multimedia devices like any other component in the system.

Since the RM creates resource components according to a common interface, a generic application is able to dynamically adapt its components to each participant local resources, attending to general constraints. RM creates software and hardware devices components with a requested data format where the hardware ones are associated with a hardware device. It's necessary to offer a common interface because device components must interact with the rest of components and connectors of the system.

5.1 Resource Manager Implementation

The RM is implemented as a Java remote object accessed by RMI. Communication between the components of the application level and RM is performed through the USP object, which knows its remote reference. The RM is the only part in MultiTEL that is machine dependent, however, only the native code of resource components needs to be changed to have a new version of RM for other operating systems. The rules for implementing and using the resource manager are the following:

1. The devices are globally identified in the system by a unique name, like "camera", "microphone", etc, established by MultiTEL. The service designer

only needs to know this identification and the interface the device implements to compose the service.

2. Device component must offer the same interface independently of the hardware device associated with it. This interface is also defined by MultiTEL.

3. RM maintains the rest of the information needed to create and manage the devices in an internal device database called *ResourceInfo* (Fig.8): i) *Device* unique identification; ii) *Device* type: software or hardware; iii) *Interface*: Interface implemented by the components; iv) *Implementation*: Java class that implements the component.

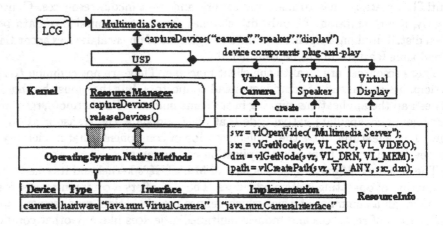

Fig. 8. Resource manager implementation

Figure 8 shows how resource manager operates. When the Multimedia subsystem wants to capture a list of devices invokes the *captureDevices()* method of the RM through the USP indicating the global identifier of the devices (e.g. "camera"). The RM finds the *Implementation* for this device looking at the *ResourceInfo* table (e.g. java.mm.CameraInterface) and then creates the component and add it to the USP application context, that is, RM performs device component plug&play. *VirtualDisplay*, *VirtualSpeaker* or *VirtualCamera* components have nodes which encapsulate the native methods (like *vlOpenVideo()* in Fig.8), but externally are considered as common components.

6   Conclusions

MultiTEL can be easily extended for developing a new kind of multimedia services, CSCW services, reusing or extending the framework architecture and only adding the connectors and components needed for broadcasting the graphical user interface events. In this paper, we have shown how compositional technologies facilitates the building of multimedia cooperative applications and how the new compositional model of MultiTEL increases the reutilization of components shorting the development time.

MultiTEL framework has been implemented in Java language JDK 1.1.4 and the native code in Digital Media Library. The complete framework and platform stands at roughly 60.000 lines of Java code, including multimedia application code but excluding C native code. We have derived different versions of multi-player games, VoD and videoconference services, including Lectures, Conferences, and Business Meetings, which are currently the most used especially in Intranet and Extranet areas. In *http://www.lcc.uma.es/lff/MultiTEL* can be found more information about MultiTEL.

We are now improving the *CWPictionary* prototype presented in this paper and also we are planning to develop new cooperative games with MultiTEL. Our future goal is to study collaborative features in electronic commerce applications.

References

1. Eric Ly. "Distributed Java Applets for Project Management on the Web". *IEEE Internet Computing*, pp. 21–26, May-June 1997
2. L. Fuentes, J.M. Troya. "Towards and Open Multimedia Service Framework". *Symposium on Object-Oriented Application Frameworks, ACM Computing Surveys*, 6 pages, Dec 1998
3. OMG. "Common Object Request Broker: Architectura and Specification". *OMG Document*, 1995
4. E. Evans, D. Rogers. "Using Java Applets and Corba for multi-user distributed applications". *IEEE Internet Computing*, pp. 43–55, May-June 1997
5. L. Fuentes, J.M. Troya. "MultiTEL: A Component-Oriented Framework in the Domain of Multimedia Telecommunication Services". *A Chapter of the book Object-Oriented Applications Frameworks*. Wiley&Sons. 24 pages, July 1999.
6. L. Fuentes and J.M. Troya. "A Component-Oriented Architecture to Design Multimedia Services on a Distributed Platform". In Takashi Masuda, Yoshifumi Masunaga and Michiharu Tsukamoto, editors, *Worlwide Computing and Its Applications* - WWCA'97. LNCS n 1274, pp. 90-105, Springer Verlag
7. R.E. Johnson. "Framework = (Components + Pattern)". *Communication of the ACM*, 40(10), pp. 39-42, October 1997.
8. O. Nierstrasz, J.G. Schneider, M. Lumpe. "Formalizing Composable Software Systems-A Research Agenda". *Proc. Of the 1st IFIP Workshop on Formal Methods for Open Object-based Distributed Systems* (FMOODS'96), France, March 1996
9. M. Amor, M. Pinto, L.Fuentes and J.M. Troya. "Multimedia Resource Management in a Framework for Internet Cooperative Services". *IDEAS'99*, April 1999
10. L. Fuentes and J.M. Troya. "A Java Framework for Web-based multimedia and collaborative applications". *IEEE Internet Computing*, 3(2), March-April 1999.
11. E.Gamma et al. "Design Patterns". *Addison Wesley*, Reading, Mass, 1995
12. J. Begole, C.A. Struble and C.A. Shaffer. "Leveraging Java Applets: Toward Collaboration Transparency in Java". *IEEE Internet Computing*, 1(2), pp. 57-64, March-April 1997.
13. Sun Microsystems. "Beans Development Kit", JDK Documentation, http://java.sun.com/beans/index.html, 1998.
14. C.A. Ellis, SL.J. Gibbs, G.L.Rein. "Groupware: Some Issues and Experiences". *Communications of the ACM*, 34(1), pp. 38-58, January 1991.

The "Virtual Interactive Presenter": A Conversational Interface for Interactive Television

Marc Cavazza[1], Walter Perotto[2], and Neil Cashman[2]

[1] Electronic Imaging and Media Communications Department, University of Bradford,
UK-BD7 1DP, Bradford, United Kingdom
M.Cavazza@Bradford.ac.uk
[2] Sony Digital Network Solutions Europe,
UK- RG22 4SB, Jays Close, Viables, Basingstoke, Hants, United Kingdom
{Walter.Perotto, Neil.Cashman}@Sonydnse.com

Abstract. With the advent of multi-channel digital TV, accessing Electronic Programme Guides (EPGs) will become increasingly complex. This paper describes ongoing work in the development of a conversational interface for Interactive TV. We claim that a conversational interface can support incremental refinement of user selections thus assisting user choice without requiring knowledge of editorial categories. We describe the system architecture, which integrates various components dedicated to language processing, EPG access and animation of the conversational character. Finally, we discuss the control strategy for the whole system from the perspective of human-computer dialogue.

1 Introduction: Information Access in Electronic Programme Guides

With the advent of 500-channels Digital TV, the selection of programmes will become an increasingly difficult task. It is thus necessary to develop new interfaces to the Electronic Programme Guides (EPG) that will enable the users to take full advantage of the diversity of channels offered. In this paper, we give an overview of the VIP (Virtual Interactive Presenter) project[1], which aims developing an intelligent interface based on a conversational character communicating with the user through speech recognition and speech synthesis.

An EPG is a structure containing schedule data, organised according to a description of categories and genres that should be meaningful both to the content provider and the user. The programme descriptors include generic information, such as genre and categories. However non-hierarchical information, such as starting time, parental rating, pricing information, or cast, is also highly relevant to the user. In this context,

---

[1] Partners of the VIP project include the University of Bradford, Cambridge University, Sony DNSE, the BBC and Advance Digital Communications plc. The project is funded by the DTI under the LINK DTI/EPSRC Broadcast Technology initiative.

the user should be enabled to make use of both kinds of descriptors without having to memorise editorial categories. Further, there is a need to accommodate idiosyncratic attributes, such as "entertaining", "boring", "funny", etc. Though there is some common ground knowledge about these features (political debates would rarely appear as entertaining), their detailed interpretation depends on specific user preferences. A related problem is the use of categories that do not appear as such in the set of editorial categories supporting the EPG, one example being, e.g. "soap".

The development of a conversational interface to the EPG is based on the assumption that human-computer dialogue is appropriate to express user preferences in terms of programme descriptors. In other words, interactive refinement of programme descriptions can be ensured through human-computer dialogue. It should be easier for users to discuss relevant criteria for the selection of their programmes in the course of dialogue, each step being dedicated to the refinement of a category or the choice between alternative subcategories, rather than to try to formulate complex queries. This can be illustrated by the sample dialogue of figure 1.

(1) System: What would you like to see?
(2) User: Something entertaining
(3) System: What about a movie?
(4) User: Yes, a comedy please
(5) System: There are two comedies tonight: "Sleepless in Seattle" and "Kindergarten Cop"
(6) User: I don't like Schwarzenegger
(7) System: What about "Sleepless in Seattle"
(8) User: Anything else?
(9) System: Still a comedy?
(10) User: No

Figure 1. Refining programme selection through dialogue.

Figure 1 illustrates how human-computer dialogue can guide the process of programme selection, through interactive refinement of programme description After an opening question from the system (1), the user requests a rather vague category (2), that the system has to interpret in terms of actual EPG categories. In return the system suggests a specific category, *movies* (3). Prompted by the system, the user refines this category into a subcategory, the *comedy* genre (4). The system can then access the database for instances corresponding to the user selection. It then outputs the corresponding instances, or a selection of these, presented as a list of alternative choices (5). The user implicitly discards one of these instances by rejecting one of its associated features, the cast (6). The system then seeks confirmation on the alternative selection (7), but the user asks for more movies (8). The system has to backtrack from the most specific point, i.e. from the last active subcategory (9).

In the field of Human-Computer Interaction, several projects have addressed conversational assistants [1] [2] [3]. These are generally animated characters with which the user interacts through spoken dialogue. The adoption of an agent-led dialogue influences the nature of linguistic exchanges between the system and the user. Having the user replying to system prompts simplifies both speech recognition and parsing. [4] [5] [6] [7]. Besides their user-friendly nature, conversational characters can also provide useful visual feedback in the course of the dialogue process (see below).

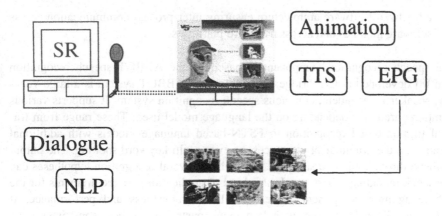

Figure 2: System Architecture.

2 System Architecture

The necessity to accommodate different software techniques within the same system requires appropriate architectural decisions. In this section, we introduce the overall software architecture and the individual components of the system. Control strategies, which are inspired by the dialogue process, will be discussed in the next section.

The VIP prototype is organised around the following software components:

- a speaker-independent speech recognition system that outputs the recognised ASCII string to the NLP module
- a NLP module which analyses the utterance in terms of semantic content to be used by the dialogue module
- a dialogue module that manages the linguistic exchanges between the user and the conversational character. As the interaction is organised around the dialogue itself, the dialogue module controls the interface display as well as access to the EPG database
- a user interface representing the TV set screen, displaying the Virtual Presenter, plus audiovisual information releted to the current programme selection. This includes a facial animation controller to animate the talking head in real time

- a database management system for the EPG. This database should be structured according to records and fields defined for the EPG, and includes both descriptor fields, free text summaries, still images and video trailers. Part of the database is mirrored as a semantic network in connection with the dialogue module. This is required in order to compute appropriate semantic distances [6].
- a hardware/software architecture ensuring inter-process communication across software modules and across hardware platforms

The speech recognition component is based on the ABBOT speech recognition system [8] developed at Cambridge University. The ABBOT system is a large vocabulary, speaker independent, continous speech recognition system. It supports various recognition paradigms depending on the language model used. These range from traditional trigram-based recognition to FSTN-based language models with additional constraints on the grammar of utterances, or even multi keyword spotting. The output is available through a lattice structure from which several recognition hypotheses can be extracted, including n-best, etc. In this kind of application, the requirements for the speech recognition component are essentially its robustness and performance. It should be ideally able to cope with largely unconstrained speech, sometimes even including hesitations. As proper names, which are not part of the initial vocabulary are an important part of the application, it should be able to generate a pronunciation form and recognise them accordingly. While in general interactive systems, the response time should be kept under one second, in a dialogue system which includes an information search component, the user can accept larger response times. It is however desirable for the overall response time to be less than 2-3 seconds This gives an upper bound for the speech recognition response time, depending on the performance of the other components.

The recognised sentence is then parsed by the NLP module. Speech understanding constitutes the first step of the dialogue process. The analysis of the user utterance should determine user input in terms of dialogue act (*enquiry*, *selection*, *refinement*) and relevant programme descriptors, in order to issue an appropriate query to the EPG database. Traditional parsing techniques can assemble semantic structures, which contribute both to the recognition of speech acts and the descriptors to be used as parameters for the EPG search. We have prototyped this approach with a TFG (Tree-Furcating Grammar) parser [9] which is a fully implemented parser integrating syntax and semantics. Fig. 3 represents the parsing of a user question, "is there a movie with James Woods?". Elementary trees associated to each word are combined using tree fusion operations (*S*ubstitution and *F*urcation) that are part of the syntactic formalism. Semantic features attached to these words are assembled synchronously to these operations and this is the basis for content processing. This processing technique is in our experience appropriate for sentences up to 15 words in length. The processing of a typical 10-12 word sentence is carried in approximately 100ms on a 150 Mhz R10000 processor. It should be noted from the example parse of figure 3, that a successful parse can be completed even in the absence, e.g. of determiners. However, it is not guaranteed that a system seeking to produce complete parses of the sentences will be robust enough to cope with the ungrammatical nature of spontaneous speech and with any remaining recognition errors from the speech recognition system. This is why we are currently modifying the parser towards a dependency-based system [10]. This is a natural evolution for the TFG formalism, especially from the standpoint of syntax-semantics integration. Relaxing syntactic constraints makes easier to process partially recognised utterances but more importantly, partial parses of a dependency system retain essential information and can be interpreted as such.

The target vocabulary size for the final the prototype is between 500 and 1000 words. Dialogue systems have been described with vocabularies as small as 40 words [2]. In our case, even for a proof-of-concept demonstrator, there is a need to reach a critical size on some aspects to validate the information access approach.

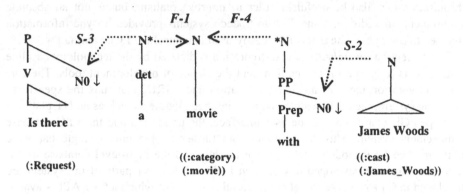

Figure 3: Parsing a user utterance with the TFG formalism.

Dialogue processing is based on the recognition and generation of specific dialogue acts that characterise the user intention and the system response. For instance, the user can express a request for a given programme category or can refine a previous category by specifying a subcategory of it. For instance, in reply to the suggestion "would you like a movie?" whenever the user replies "yes (acceptance of category), a thriller please (specifying a subcategory)" or "do you have any with Sean Connery?" (acceptance of the category, specifying the *cast* feature). Our central hypothesis is that the forms of query refinement can be described through a well-identified set of *speech acts* [11] [12]. The process of agent-led dialogue is equivalent to the definition of a specialised set of speech acts that are constrained by the nature of operations that differentiate between programmes. In the case of information access dialogues, several authors have suggested a specialisation of speech acts [5] [13] [14]. Examples of such speech acts are *Refine*, which refines the current selection by specifying an additional subcategory or feature.

The main goal of the user interface module is to provide visual feedback to the user during the dialogue process. There is wide agreement that visual feedback is an important aspect of dialogue-based interfaces [7] [2] [3]. In the VIP system, visual feedback is also required due to the flow of dialogue information and the limited persistence of spoken information. This can be easily dealt with with visual feedback, either through text or images from the selected movies. The other kind of visual feedback, which is specific to conversational interfaces is constituted by facial expressions as non-verbal information transmission from the artificial character. Previous work has demonstrated that within conversational interfaces, visual feedback provided by facial expressions of an animated character created a useful additional communication channel [1] [2]. In the specific case of information access dialogues, where the system has to react to the global status of the dialogue process, it is useful to reflect the system's difficulties through such a natural channel as facial expressions.

The animation of the virtual presenter can be based on any real-time animation package. For instance, some preliminary experiments have been carried with the SimpleFace system [15]. The facial animation system should be able to encode a set of parameters which can generate a subset of well-defined facial expressions. Lip synchronisation can also be useful in order to improve realism, but is not an absolute requirement. In addition, some Text-to-Speech systems provides lipsync information attached to each phoneme that can be easily used to control the facial animation.

The current hardware/software configuration is dictated by the technology baseline requirements (e.g., speech recognition) and the choice of development tools. The current configuration includes an SGI workstation under IRIX that runs the speech recognition system, the NLP module and the dialogue module as well as an NT platform with open GL board running the user interface, the animation and the EPG database management system. This configuration can further evolve into a single hardware platform. The NLP module, the dialogue module and the high-level database management control are developed in Common LISP, while other parts of the system are developed in C++ or make use of commercial systems for which a C++ API is available. The communication between the various modules takes place through sockets. This provides a unified mechanism for communication across programming languages and computer platforms. We have experimented various configurations for socket-based communication across platforms, one process running a C++ programme and the other process running a Common LISP programme. In order to validate this architecture, we have implemented both a LISP client and a LISP server, using the socket library part of the Allegro™ Common LISP implementation. The various modules communicate by exchanging command strings via these sockets. For instance, the dialogue module can issue commands to the interface module, specifying the trailers to be displayed and the facial expression of the virtual interactive presenter. In our first prototype, the EPG contents are represented as a set of feature structures within the dialogue/control module, while schedule data (video trailers, images, free text descriptions) can be stored in a standard database. This allows to search the EPG using pattern-matching procedures, as part of the dialogue process, without having to exchange SQL queries between the dialogue module and the database. Schedule data corresponding to a given programme selection can be retrieved from the databased by using a standard ID number, which is the only information that needs to be exchanged.

3 Dialogue as a Control Strategy for Information Access

As we have seen, the main design decision is that the conversational interface should control the global search strategy for the EPG. However, the dialogue process itself has to be supervised and this is a difficult technical problem in the general case [16]. In particular, recognition of speech acts at the user side is not sufficient to control the whole dialogue process. It can be shown that there is a need for controlling dialogue both locally, i.e., in terms of replying to the current user request and globally, in terms of dialogue history. Global dialogue control is in charge of monitoring dialogue progress, formulating alternative choices for the user and computing alternative categories when the user is not satisfied by the current selection, as illustrated by the sample dialogue of figure 1. Local control mainly interprets user speech acts and applies constraints on the speech acts that can be generated in response. One such example is the contextual interpretation of an utterance like "are there any movies with James Woods?". At early stages of the dialogue, this would have to be interpreted as a request, while at later stages, for which several categories are already active, it should be interpreted as refining the current selection. The local control is also in charge of formulating queries to the database from the current set of active descriptors. As the above example would illustrate, the common denominator between local control and global control lies in the current status of programme selection. A useful representation is constituted by the logical conjunction of features assembled from the sequence of speech acts produced by the user (search filter). The system should thus update a description of the active programme descriptors as they have emerged through the dialogue process.

The dialogue module also controls the visual feedback presented through the user interface. Visual feedback comprises two different aspects. The first one is that still photographs in the background of the talking face should be related to the programme categories under consideration at a given dialogue stage. The other well-known aspect stands in the display of facial expressions for the animated character that should reflect the dialogue status. This kind of feedback can depend on local data, for instance when the user input is not processed successfully. But it can also depend on global status, such as the fact that the user suddenly backtracks from the current selection.

From the above description, it does not appear possible to specify a fixed, procedural strategy for dialogue control in the general case. The reason is that many dialogue control rules are highly contextual and rely on previous dialogue acts, current status of the search, etc. It is possible however, to devise a set of control rules which should operate on the various situations. These rules would constitute a declarative control strategy and could be handled by a simple production system. This also allows flexibility and exploratory programming for the dialogue control module, hence for the system as a whole.

One key aspect of dialogue control is the strategy for suggesting possible choices to the user. The fact that the dialogue is led by the agent plays a key role in managing the dialogue. Firstly, it ensures that the selection offered to the user is always of manageable size (unlike traditional information retrieval techniques) and the set of alternatives is relevant. For instance, if the current selection contains 12 movies in three categories, it can be more appropriate to pick up one possible choice in each of those categories rather than the three first random occurrences. The latter aspect is also related to the contents of the EPG and the computation, whether static or dynamic, of "semantic distances" between the various categories and subcategories. This distance corresponds to information, which is not part of the EPG. It is specifically required as a global control mechanism to ensure the relevance of new suggestions produced in the case of local failure, rejection or backtracking. It should ensure that when the user has been rejecting a category and it is up to the system to come with new suggestions, the suggested categories provide a proper alternative. For instance, if the user has rejected *educational programmes* and *talk shows*, it could be more appropriate to suggest a *movie* than a *documentary*. In this way, the dialogue strategy can be related to the structure and contents of the EPG.

4 Conclusions

Conversational dialogue offers a novel approach to the problem of information access for a wide user population. By guiding the user in the selection process, it is likely to decrease the effort required from the user to browse the EPG. There appears to be a promising analogy between incremental refinement through dialogue and the actual way in which information is structured. Dialogue is one of the most advanced forms of interactivity. As a control mechanism, it can provide user-friendly access to complex information repositories regardless of their structure and contents. The conversational assistant, by mapping the user categories to the editorial categories provided by the EPG, ensures that the whole process is transparent to the user and can accommodate different proprietary formats for the EPG, or different standards. In its current development stage (speech recognition, NLP and basic dialogue functions have been implemented), the prototype only demonstrates the general validity of the concept, as it validates the fusion between specialised speech acts and the EPG search process. Full implementation and scaling up in vocabulary will be required to explore user-related aspects such as acceptance, effects on spontaneous speech and non-verbal communication. Another preliminary conclusion is that recourse to a conversational character influences the kind of NLP techniques to be implemented, and in that sense agent-based interaction differs from other speech-based information access systems, as also suggested by previous research in conversational agents.

Acknowledgements

Schedule data and images have been provided by the BBC (Adam Hume and David Kirby). The user interface layout has been designed by Daryl Fish.

References

1. Kurlander, D., Ling, D.T., 1995. Planning-Based Control of Interface Animation. *Proceedings of CHI'95 Conference*, ACM Press, New York, pp. 472-479.

2. Nagao, K. and Takeuchi, A. (1994). Speech Dialogue with Facial Displays: Multimodal Human-Computer Conversation. In: *Proceedings of the 32nd Annual Meeting of the Association for Computational Linguistics* (ACL'94), pp. 102-109.

3. Beskow, J., & McGlashan, S. (1997): Olga - A Conversational Agent with Gestures. In: *Proceedings of the IJCAI'97 workshop on Animated Interface Agents - Making them Intelligent*, Nagoya, Japan, August 1997.

4. MGuide (1998). *Guidelines for Designing Character Interaction*. Microsoft Corporation. Available on-line at http://www.microsoft.com./workshop/imedia/agent/guidelines.asp

5. Busemann, S. Declerck, T., Diagne, A., Dini, L., Klein, J. and Schmeier, S. (1997). Natural Language Dialogue Service for Appointment Scheduling Agents. In: *Proceedings of ANLP'97*, Washington DC.

6. Sadek, D. (1996). Le dialogue homme-machine: de l'ergonomie des interfaces à l'agent dialoguant intelligent. In: J. Caelen (Ed.), *Nouvelles Interfaces Homme-Machine*, OFTA, Paris: Tec & Doc (in French).

7. Yankelovich, N., Levow, G.-A. and Marx, M. (1995). Designing Speech Acts: Issues in Speech User Interfaces. *Procedings of CHI'95*, Denver.

8. Robinson, T., Hochberg, M. and Renals, S. (1996). The use of recurrent neural networks in continuous speech recognition. In: C. H. Lee, K. K. Paliwal and F. K. Soong (Eds.), Automatic Speech and Speaker Recognition – Advanced Topics, Kluwer.

9. Cavazza, M. (1998). An Integrated TFG Parser with Explicit Tree Typing. In: *Proceedings of the fourth TAG+ workshop*, IRCS, University of Pennsylvania.

10. Nasr, A., 1995, A Formalism and a Parser for Lexicalized Dependency Grammars, *Proceedings of the Fourth International Workshop on Parsing Technologies*, Prague.

11. Searle, J. (1959). *Speech Acts*. Cambridge: Cambridge University Press.

12. Cohen, P.R. and Perrault, C.R.. (1979). Elements of a plan-based theory of speech acts. *Cognitive Science*, 3(3):177-212.

13. Traum, D. and Hinkelmann, E.A. (1992). Conversation Acts in Task-Oriented Spoken Dialogue. *Computational Intelligence*, vol. 8, n. 3.

14. Allen, J. and Core, M. (1997). DAMSL: Dialogue act markup in several layers. Draft contribution for the Discourse Resource Initiative.

15. Parke, F. and Waters, K. (1996). *Computer Facial Animation*, A.K. Peters.

16. Bunt, H.C. (1989). Information dialogues as communicative action in relation to information processing and partner modelling. In: Taylor, M.M., Néel, F. and Bouwhuis, D.G. (Eds.), *The Structure of Multimodal Dialogue*, Amsterdam, North-Holland.

A Video Compression Algorithm for ATM Networks with ABR Service, Using Visual Criteria

S. Felici[1] and J.Martinez[2]

[1] Institut de Robotica, Universistat de Valencia, Spain
santiago.felici@uv.es
[2] Dept. Comunicaciones Universidad Politec. Valencia, Spain

Abstract. In this paper [3] is presented an adaptive video compression algorithm designed for the video transmission over *best-effort* network with services that can adapt to changing network conditions, specifically Available Bit Rate (ABR) services in Asynchronous Transfer Mode (ATM) networks.

The proposed algorithm tries to minimise the impact that cell losses could have on the perceived quality, using perceptual bit allocation procedures and Rate-Distortion minimisation techniques. Furthermore the algorithm adaptively estimates the compression ratio by means of a forecast mechanism based on the value of the feedback signal sent by the network. It should be noticed this algorithm is based on a 3D subband coding that uses wavelet filter banks but it does not use motion estimation with block matching procedures to prevent from error propagation effects.

1 Introduction

A new generation of systems called Networked Multimedia Systems are currently being the subject of intense study. Different network technologies capable of supporting efficiently such systems are being analysed to determine which one offers better performance. The Broadband Integrated Service Digital Network (B-ISDN) based on the Asynchronous Transfer Mode (ATM) is one of these candidates. ATM networks provide different service categories [1] to support multimedia services. The applications select a service category based on their QoS requirements. The ABR class of service was initially conceived to support data traffic. Its service model is based on the *best-effort* paradigm but enhanced by some specific characteristics: *fair sharing of the available resources among the contending ABR connections* and *a closed-loop feedback mechanism*.

Video-based applications that are rate adaptive, can obtain substantial benefits by using ABR connections. These benefits can be summarised in the following three aspects. First, these applications typically require some guarantee on bandwidth, for example a minimum encoding rate for a video stream, but can take

---

[3] This work has been supported in part by the Spanish Science and Technology Commission (CICYT) under grant TIC96-0680

advantage of spare bandwidth. This can be supported by an ABR connection by defining a Minimum Cell Rate at connection set up. Second, when explicit rate feedback is used and the ABR connections supporting these applications are multiplexed on a dedicated queue at the switches, the cell transfer delay is more predictable because the congestion control mechanism keeps the queues almost empty. And third, the feedback mechanism keeps each source informed of the available bandwidth it has at their disposal. This information can be used by the quantizier to adapt quickly to new network conditions [2]. But, when a video signal that has been processed by a typical video compression standard algorithm, is transmitted over an ABR connection, it is difficult to obtain the best visual quality, as perceived by a human viewer, as we will see in section 2. We take a new approach to solve this problem by designing an adaptive video compression algorithm specific for ABR connections, which incorporates a visual criteria to improve its efficiency.

The usage of visual criteria is still an open question and different studies can be found in the literature. For example [3] introduces a 3D subband coding (with Haar and 16-QMF filters) which uses a Just Noticeable Distortion criteria for 64 Kbps wireless channels. This approach differs from the one described in this paper because, first it uses constant bit rates rather than ABR services and second, it uses QMF filters which are difficult to implement if linear phase response, similar to the HVS, is required. [4] proposes an optimization technique for adaptive quantization of image and video under delay constraints for ATM VBR connections. In this case, the study focuses on the Rate-Control as an optimal Trellis-based buffered compression, where the Distortion is computed using Mean Square Error measures, which is not related to the HVS.

The rest of the paper is structured as follow. Section 2 describes an experiment in which a H.263 video sequence is transmitted over an ABR connection subjected to different kinds of perturbations. Section 3, proposes a new compression technique using subband decomposition, which is based on the Wavelet Transform(WT) instead of on the DCT(Discrete Cosine Transform) as in H.261, H.263, MPEG-1, MEPG-2, etc. After the decomposition, a bit allocation is performed based on two limiting factors: HVS perception requirements and available bandwidth on the network. Section 4 describes the operation of the coder. Section 5 evaluates the performance of the proposed compression technique and provides numerical results. Finally, section 6 presents the conclusions and ideas for future work.

2 Performance Evaluation of a H.263 Video Transmission System over an ABR Connection

This section evaluates by simulation how an H.263 video signal degrades when it is transmitted over a connection offering a *best-effort* service. The degradation of the video signal will be measured by computing its Signal to Noise Ratio (SNR) and by simple visual perception of the received frames. This experiment highlights the fact that a quantitative performance parameter like the SNR is not always linearly related to the perceived human quality.

2.1 Reference Network Configuration

The network configuration used in the experiment can be observed in figure 1. As can be seen, it is a simple two node configuration in which all the connections have their bottleneck at the same link (D11). This link is the backbone link. Source A has just one connection set up with destination A and this connection is supporting the H.263 video signal. The other connections are ABR connections supporting greedy sources, that is, sources that will use as much bandwidth as it available for them.

The capacity of the access links, as well as the capacity of the backbone link, is 10 Mbps. The length of the access links is 0.2 Km, while the length of the backbone link is 2 Km. The ABR switch algorithm used is an improved version of the CAPC algorithm as described in [2]. The values used for the source parameters and for the switch parameters have also been taken from [2].

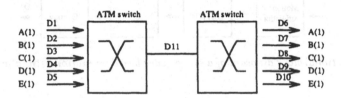

Fig. 1. *The network configuration with 2 ATM switches*

As the objective of the experiment is to stress as much as possible the video signal, two major disturbances have been introduced in the experiment. The first one takes place at 400 ms and emulates the set up of a new connection that consumes a bandwidth of 5 Mbps. The second one takes place at 900 ms and emulates the set up of new connection that consumes an additional portion of 3 Mbps. These events could emulate, for example, the set up of two CBR connections with different Peak Cell Rates. Also, multiplexing greedy sources on the same bottleneck link used by the video connection ensures that the video connection will have to contend for bandwidth, and it will never get more than its fair share.

Given that the capacity of the backbone link is 10 Mbps ($23.58\frac{cells}{ms}$) and that there are 5 ABR connections sharing this capacity, each connection has $\frac{23.58}{5} = 4.71\frac{cells}{ms}$ available during the first 400 ms. The video coder uses a Source Input Signal (SIF) format of 352x288 pixels and, therefore it requires a compression ratio of 11.55:1 during this time interval. The compression ratio required during the 400 to 900 ms interval is 23.10:1 and 57.77:1 after 900 ms. The Minimum Cell Rate for the video source has been set to 360 Kbps which ensures that the compression rate will never be lower than 64:1.

2.2 Results of the Experiment

The most common compression technique used by video standards is motion estimation with DCT. This technique is also used by the ITU-T H.263 recommendation. Figure 2 describes the basic functional blocks diagram of a video coder: motion estimation, DCT, quantization and entropy coding. As can be observed, the quantization step can be changed by a external signal. Basically, this external signal controls the rate of the video signal at the output of the coder. During the experiment we have made use of this control signal to make sure that the video coder output rate is always consistent with the bandwidth that the connection has available.

Due to motion estimation, a H.263 video coder generates a hierarchical data structure. This structure introduces frame dependencies as can be seen in figure 3, where *I intra-frames* are independent ones, *P predictive-frames* are dependent on the I frame and *B bidirectional-frames* are dependent on the I and P frames.

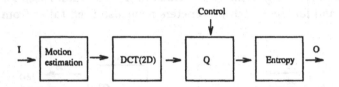

Fig. 2. *Functional block diagram of a video coder based on motion estimation and DCT*

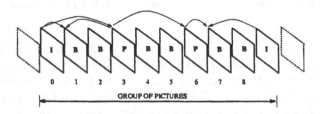

Fig. 3. *Group of pictures and the dependences among I intra-frames, P predictive-frames and B bidirectional-frames*

When using this type of video coder in a scenario where the available bandwidth is not constant and cell losses can occur, different kind of problems can appear. First, the hierarchical frame structure makes possible the error propagation. Second, the DCT does not operate in the same way as the HVS, under very low bit rates [5] and then, the video perception that a viewer has is unpredictable. And third, the mosquito noise due to the rigid 8x8 pixel block size [5] appears after quantization.

Figure 4 shows three frames of the H.263 sequence under evaluation. The selection of these three frames have been done as follows. The first one corresponds to the frame that was being transmitted by the source when the first bandwidth reduction took place at 400 ms. The second one corresponds to the frame that was being transmitted by the source when the second bandwidth reduction took place at 900 ms. And the third is a frame selected at random when the source was using the smaller bandwidth portion, that is, after 900 ms. The SNR (db) for each of the three frame is: 36.82, 15.2 and 32.82 (from left to right). As it can be observed, the quality perceived from the second frame cannot be anticipated by its SNR value.

Finally, it should be pointed out that the H.263 coder used does not implement any type of progressive transmission method to improve the quality of the received signal[6], because as previously discussed, under low bit rates, it still works in a bad way.

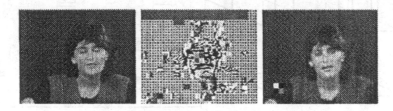

Fig. 4. *The three selected frames from the Miss America sequence as perceived at destination*

3 The Proposed Adaptive Video Compression Algorithm

Video images have a lot of spatial, frequential and temporal redundancy, besides the subjective redundancy. To take advantage of this redundancy different techniques have been proposed to reduce the bandwidth required to transmit a video signal. Over networks that offer *best-effort* services the best choice, as it will be shown, is to mimic the behaviour of the HVS. In these network scenarios subband coding methods have a better response when the signal has to be transmitted.

This paper proposes to mimic the HVS in two ways: first by performing a multiresolution decomposition of the video signal using a 3D WT[7] and second, by introducing a visual criteria to help the coder to determine which subbands need a prioritized treatment. As it will shown, a subband with higher priority will get more bits allocated and will be transmitted with less delay. The perceptual criteria used for the experiments described in this paper is the one presented in [8]. Figure 5 shows a functional block diagram of the proposed video coder. Notice, that in figure 5, the quantization step can also be modified by the control input.

The right side of the figure 6 shows a fast separable 3D WT decomposition using recursive biorthogonal filter banks, where the high-pass is called *wavelet*

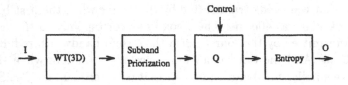

Fig. 5. *Functional block diagram of a the proposed video coder based on the Wavelet Transform*

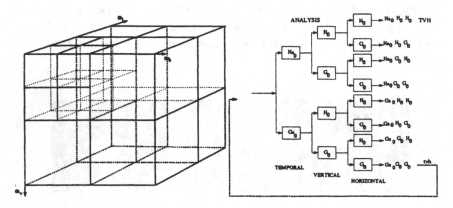

Fig. 6. *Left: Frequency division for the horizontal, vertical and temporal domains into two resolution levels using a 3D dyadic separable WT. Right: Its filter bank implementation.*

function or filter G, and the low-pass is called *scaling function*, or filter H. As it can be observed, multiresolution is obtained by new iterations, where the absolute low pass subband is feedbacked as a new input to the filter bank. This decomposition is done by splitting each frequency domain (horizontal, vertical and temporal) into two parts, as can be seen on the left of figure 6. Initially by this process we get 2x2x2 frequency shares (called subbands) that compose the first resolution level. If the absolute low pass subband (from the output of every low-pass filter of each domain), is processed again, we get 2x2x2 subbands more, that compose the second resolution level. The output of each filter is undersampled by 2 (dyadic sampling) to keep constant the amount of information[9], without adding any redundancy. In this way we also achieve a frequency response that is approximately constant logarithmic, similar to the HVS[7].

The notation that will be used to identify each subband is: t and T for the temporal low and high responses, h and H for the horizontal low and high responses and v and V for the vertical low and high responses.

To perform the inverse transform the same filter structure is used in reverse mode but changing the analysis filters by synthesis filters. Both the analysis and synthesis filters have to meet the following requirements: perfect reconstruction, null aliasing and null distortion[5][9]. These biorthogonal filters have been designed to be linear, using Daubechies' method, avoiding in this way damaging the image after the quantization process [5]. It should also be noticed that a 2 coefficient filter (Haar filter) have been used in the temporal axis to reduce the number of frames that need to be stored by the coder. These filters are labelled with subindex a_0 in figure 6.

The transfer functions of the analysis (with subindex $_0$) and synthesis (with subindex $_1$) filters, both for the horizontal and vertical domains, are:

$$
\begin{aligned}
H_0(z) &= \tfrac{1}{4}(1 + 2z^{-1} + z^{-2}) & H_1(z) &= \tfrac{1}{4}(-1 + 2z^{-1} + 6z^{-2} + 2z^{-3} - z^{-4}) \\
G_0(z) &= \tfrac{1}{4}(1 + 2z^{-1} - 6z^{-2} + 2z^{-3} + z^{-4}) & G_1(z) &= \tfrac{1}{4}(1 - 2z^{-1} + z^{-2})
\end{aligned}
$$

$$(1)$$

4 Operation of System

When a system, as the one described in the previous section, is fed with a video sequence of 25 $\frac{frame}{sec}$, then a set of 4 frames (4x40=160 ms) are needed to perform a complete 3D decomposition, with two resolution levels . This number of frames represent a trade off between the decorrelation ratio and the number of frames that need to stored at the coder. The process can be observed in figure 7.

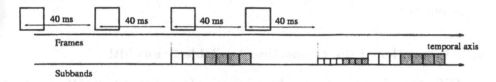

Fig. 7. *Subband generation using 3D Wavelet Transform with two resolution levels. Different frames are processed every 40 ms*

For example, lets assume that our system is going to perform the decomposition of 4 frames, that we label as frames 1, 2, 3 and 4. The system uses the pair of frames 1-2 and the pair 3-4 to obtain the first resolution level. This process generates 8 subbands from each pair of frames. Then we use the pair of subbands *tvh* from each of the original pair of frames (1-2 and 3-4) to generate the second resolution level. By this process we obtain 8 additional subbands. Therefore, at the end of the decomposition process we obtain 7+7+8 subbands[4]. The absolute low pass subband (or *tvh*) of the second resolution level is coded using Differential Pulse Code Modulation (DPCM) because it has a uniform pixel distribution.

The subbands in figure 7, with white boxes represent subbands with low temporal frequencies while with dark boxes represent subbands with high temporal frequencies. For an example of the 8 subbands from the second resolution level of Miss America, see figure 8.

---

[4] Note, that the 2 low pass ones of the first resolution level (denoted *tvh*) have been used for the second resolution and will not be transmitted.

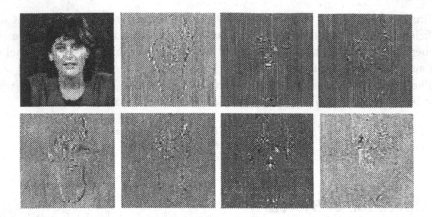

Fig. 8. *Subbands from the second resolution level of Miss America. From top to bottom and from left to right: tvh, tvH, tVh, tVH, Tvh, TvH, TVh and tVH*

After this decomposition, the Bit Allocation Algorithm estimates the number of bits per coefficient (or the total number of bits for each subband), as will be seen in next subsections. The Bit Allocation Algorithm requires two inputs: the bandwidth that the connection has available at any point in time and the priority of the subband.

4.1   Estimation of the Connection Available Bandwidth

An ABR connection receives periodic information from the network of the bandwidth it has available. This information is conveyed by a special type of cells called Resource Management (RM) cells. The RM cells are inserted by the source in the normal data cell flow sent toward the destination terminal. Once the RM cells get to the destination terminal then they are sent back to the source, collecting on their way congestion state information supplied by the switches. The transmission rate of an ABR source is computed by taking into account both the information conveyed by the RM cells and a set of source behaviour rules [10]. The rate at which an ABR source can transmit at any given time is called *Allowed Cell Rate* (ACR). The proposed video coder tracks the values taken by the ACR to estimate the bandwidth that a connection has available.

The value of the ACR changes in a very small time scale (in the order of hundreds of μs) and cannot be used directly by the coder. Bear in mind that the coder works at the frame time scale [5], therefore it only requires to estimate the connection available bandwidth at this time scale. This can be achieved by filtering the values taken by the ACR. One technique that is easy to implement is to perform an exponential weighted averaging of the ACR samples, that is $MACR = MACR + \alpha(ACR - MACR)$, where MACR (Mean ACR) is the

---

[5] In fact, the proposed coder works at a time scale equivalent to four frames (160 ms) according to the decomposition method descried.

estimated available bandwidth and α determines the cut-off frequency of the low pass filter. The value of α can be related to the sampling frequency by

$$\cos w_c = \frac{\alpha^2 + 2\alpha - 2}{2(\alpha - 1)} \tag{2}$$

One of the problems to determine the value of α is that the sampling frequency is not constant. It depends on the inter-arrival times of the RM cells, which itself is a function of the available bandwidth that changes with time. A common trade-off value is $\alpha = \frac{1}{16}$

The video coder uses the value of MACR as a forecast of the bandwidth that the connection will have available during the next 160 ms. Figure 9 on the left shows how the MACR estimator (continuous line)performs, during an experiment carried out in the same conditions of the experiment described in section 2, but using the proposed video coder instead of a H.263 coder. The same figure shows how the ACR (on the left of figure 9, dotted line) of the ABR video source adapts to the changing network conditions and the backbone link utilisation factor, see figure 9 on the right.

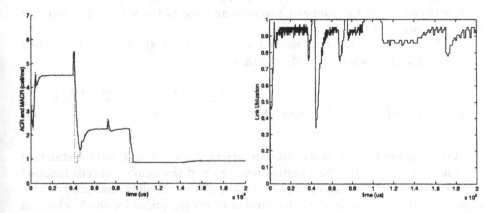

Fig. 9. *Left: Evolution of the ACR (dotted line) and of the MACR (continuous line) as a function of time for the H.263 video transmission experiment. Right: backbone link utilisation for the same experiment.*

4.2 Bit Allocation Algorithm

The bit allocation procedure is done applying the Rate-Distortion theory. This requires to know the probability density function (pdf) of the coefficients in every subband in order to estimate the distortion noise introduced by every quantization process. These pdfs are well characterised by generalised Gaussian distributions with zero mean [5]. In general, subbands with lower energy should have fewer bits, but the subbands with more perceptual weight (like the low pass ones) get more bits per pixel.

The Rate-Distortion theory (equations 3 and 4) is based on the computation of the Mean Square Error (MSE) which it is not an HVS mechanism [5]. The proposed video coder introduces weighted perceptual factors [8] to achieve a better perceptual bit allocation. The bit allocation is estimated by minimising the next equations:

$$Distortion: \quad D(b) = \sum_{k=1}^{M} \alpha_k \omega_K c_K 2^{-2b_k} \sigma_k^2 \tag{3}$$

$$Rate: \quad R(b) = \sum_{k=1}^{M} \alpha_k b_k \leq 4 * \frac{(MACR) * 48 * 8}{N * f_{rate}} = R \tag{4}$$

where M is the number of processed subbands (in our case 7+7+8), N is the total number of pixels available (in our case the number of pixels within 4 frames), $MACR$ is the forecasted available bandwidth at the time the bit allocation procedure is executed, $f_{rate} = 25 \frac{frame}{sec}$, c_k is a parameter which depends on the pdf of coefficient of each subband, α_k is the relative subband size respect to the original frame size, $b = (b_1, b_2, ..., b_M)$ represents the bit per coefficient for each subband, σ_k^2 is the subband variance and ω_K is the weighted perceptual factor per subband.

By the Lagrange theorem we can minimise D(b) given R(b), by finding $min(D(b) + \lambda R(b))$ where $\lambda' = \frac{\lambda}{c_k}$ and then

$$b_k = 0.5 * \log_2(\frac{2\ln(2)\omega_k \sigma_k^2}{\lambda'}) \quad \text{with} \quad \lambda' = 2^{\frac{\left[\sum^{22} \alpha_k (\log_2(2\ln 2) + \log_2(\omega_k \sigma_k^2)) - 2*R\right]}{\sum^{22} \alpha_k}} \tag{5}$$

Assuming the maximum absolute coefficient value as four times the standard deviation [5], we can get the quantization step and the number of cells required for every subband. An example of operation of the bit allocation procedure is shown in figure 10, which plots the number of cells required for the 8 subbands of the second resolution level, as a function of MACR. If a subband has no bit allocated to it, an infinite quantization step is assumed and no information will be transmitted for it.

4.3 Adaptation to the Service Provided by the Network

Once the information generated by each subband is available, the coder creates an information unit per subband. Each information unit contains: the frame number, subband number, and the quantization step. The information unit just described forms, what in ATM jargon is called the *Service Specific Convergence Sublayer* Packet Data Unit (SSCS-PDU). This PDU is then transferred to the *Common Part Convergency Sublayer* (CPCS), where it is encapsulated in a CPCS- PDU and then segmented by the Segmentation and Reassemble

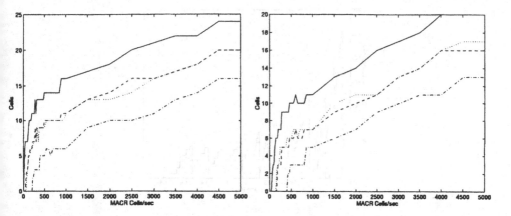

Fig. 10. *Bit allocation algorithm results: Number of cells for each subband of the second resolution level. Subband identification: i) Left: tvh (solid), tvH (dashed), tVh (dotted) and tVH (dash-dot). ii) Right: Tvh (solid), TvH (dashed), TVh (dotted) and TVH (dash-dot)*

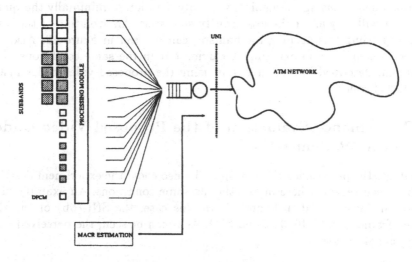

Fig. 11. *Functional block diagram of the transmission subsystem: Processing Module, Transmission Buffer and MACR Estimation.*

(SAR) sublayer. In this study we have used the ATM Adaptation Layer 5 which CPCS-PDU format is defined in [11].

The order in which the different SSCS-PDUs get transferred to the CPCS will define the order in which the different subbands will be transmitted. This order is determined by their perceptual priority, which is related to their energy and to the HVS criteria selected. Once the subbands have been ordered, the highest priority subband is transmitted first. This ordering mechanism is implemented in the transmission subsystem, shown in figure 11.

The available bandwidth forecast computed by the video coder can be some-

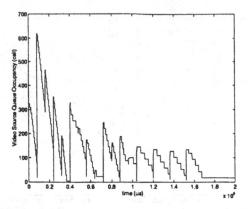

Fig. 12. *Queue occupancy at the video transmission system*

times too optimistic. In these cases, when a new set of subbands arrive to the ATM layer ready for transmission, any stale information must be flushed. In general, this queue flushing mechanism will only deteriorate minimally the quality because it will only affect the low priority subbands. An evolution of the queue occupancy using this flushing mechanism, can be seen in figure 12. Also, this has been taken from an experiment carried out in the same conditions of the experiment described in section 2, but using the proposed video coder instead of an H.263 coder.

5 Performance Evaluation of the Proposed Video Coder over an ABR Connection

To evaluate the performace of the proposed video coder the experiment described in 2 has been repeated here in exactly the same conditions. An example of the results can be observed in figure 13. In this case, the SNR(db) of the three selected frames are: 37.10, 35.7 and 31.4. As it can be seen, the perceived visual quality is better now.

Fig. 13. *Three different frames from Miss America sequence using the adaptive compression system. The first three are at next simulation transitions: switch from 10 to 5 Mbps, switch from 5 to 2 Mbps and finally at 2 Mbps*

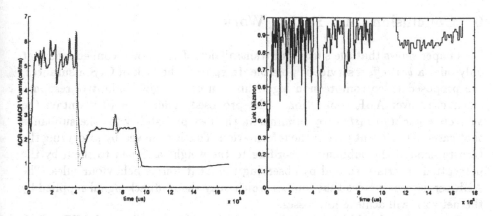

Fig. 14. *Left: ACR (dotted line) and MACR (continuous line). Right: Backbone link utilisation.*

Furthermore, a new set of experiments have been carried out to generate a statistical study or to evaluate the percentage of transmitted frames that are received successfully at destination. By successfully received frames we mean both the frames that have arrived with no errors and the frames that have arrived on time. A frame arrives on time when its end-to-end transfer delay is less than 21 ms, which has been chosen by observing the peak queue occupancy at the switch output port attached to the bottleneck link. Each one of these experiments has been performed in exactly the same scenario described for the experiment of the section 2, but this time the ABR data connections have a bursty traffic profile instead of being greedy sources. For simplicity, both the burst lengths and the inter-burst intervals have been generated randomly, but with such probability to keep every bursty source with a mean cell rate of $6.28\frac{cell}{ms}$, with burst lengths of 2, 16 and 32 cells.

For the 8 subbands of the second resolution level, the results are shown in Table 1. The values represent the successful arrival probability conditioned to the fact that the subband under consideration has been transmitted. As can be observed, the successful arrival probability of the most important subbands is quite high, which insures a good reception quality even under very demanding network scenarios. Figure 14 shows how the ACR, the MACR and the backbone link utilization factor change with time during one of the experiments.

subband	$\hat{P}(C/T)$	subband	$\hat{P}(C/T)$
tvh	0.9900	Tvh	0.8142
tvH	0.9495	TvH	0.7347
tVh	0.9011	TVh	0.7426
tVH	0.8542	TVH	0.7663

Table 1. *$P(C/T)$ for the 8 subbands of second resolution level*

6 Conclusions and Future Work

This paper shows that the successful transmission of video over connections that only offer a *best-effort* service is possible. In spite of the lack of QoS assurances, the proposed video compression algorithm can offer a good adaptive response when used over ABR connections. The proposed coder tries to maintain the *subjective quality constant* by minimizing the loss probability for the subbands that carry the relevant part of the information. This is achieved by prioryzing the transmission of the subbands according to the weight assigned to them by the perceptual criteria used and by observing the ABR source behaviour rules. The ABR service guarantees that sources which obey the feedback signal supplied by the network will achieve low losses.

Future work should be oriented towards designing an enhanced ABR switch algorithm. This algorithm should be able to guarantee that the important subbands will never be lost.

References

1. Rec. I371 Traffic Control and Congestion Control in B-ISDN ATM. *ITU-T*, May 1996.
2. J.R. Vidal J. Martinez and L. Guijarro. A Low Complexity Congestion Control Algorithm for the ABR Class of Service. In *Internation Distributed Multimedia Systems IDMS98, Oslo (Norway)*, pages 219–230, September 1998.
3. C Chou and C Chen. A perceptually optimized 3d subband codec for video communication over wireless channels. *IEEE Trans. on Circ. and Syst. Video Tech.*, 1996.
4. A. Ortega Chi-Yuan Hsu and A. Reibman. Joint Selection of Source and Channel Rate for VBR Video Transmission under ATM Policing Constraints. *IEEE Journal on Selected Areas in Commun.*, 1997.
5. Gilbert Strang and T. Nguyen. *Wavelets and Filter Banks*. Wellesley-Cambridge Press, USA, 1996.
6. K. R. Rao and J. J. Hwang. *Techniques and Standards for Image , Video and Audio Coding*. Signal processing series. Prentice Hall, New Jersey, 1996.
7. A. B. Watson. Efficiency of a model human image code. *Journal of the Opt. Soc. of Am.*, 1987.
8. J. Malo A. Pons A. Felipe J. Artigas. Characterization of the human visual system threshold performance by a weighting function in the gabor domain. *Journal of Modern Optics*, 44(1):127–148, 1997.
9. P. P. Vaidyanathan. *Multirate systems and filter banks*. Prentice Hall, 1993.
10. S.Fahmy R. Goyal R. Jain, S. Kalyanaraman and S. Kim. Source Behavior for ATM ABR Traffic Management: An Explanation. *IEEE Communications Magazine*, 34:50–57, November 1996.
11. Rec. I363 B-ISDN ATM Adaptation Layer Spec. *ITU-T*, 1993.

Content-Fragile Watermarking Based on Content-Based Digital Signatures

Jana Dittmann, Arnd Steinmetz, Ralf Steinmetz

Jana.Dittmann@darmstadt.gmd.de, Arnd.Steinmetz@darmstadt.gmd.de,
Ralf.Steinmetz@KOM.tu-darmstadt.de
GMD - German National Research Center for Information
Technology, Institute (IPSI)
Dolivostraße 15,
D-64293 Darmstadt, Germany
+49-6151-869-845

Abstract.

The development of new multimedia services and environments requires new concepts both to support the new working process on distributed computers and to protect the multimedia data during the production and distribution. This article addresses image/video authentication and copyright protection as major security demands in digital marketplaces. First we present a content-based signature technique for image and video authenticity and integrity. Based on this technique, we introduce a tool for interactive video authentication and propose content-fragile watermarking, a concept which combines watermarking and content-based digital signatures to ensure copyright protection and detection of integrity violation.

1 Motivation

The expansion of digital networks all over the world allows extensive access to, and reuse of visual material. Problems include unauthorized copying, reading, manipulating or removing of data, which might lead to financial loss or legal problems. Modifications range from scaling, rotating, stretching to direct changes of the image content. For example persons who never were at the same time at the same place appear in a film together, like in"Forest Gump": J.F. Kennedy in a scene with Tom Hanks. These kind of modifications can be easily done with digital image data and can be performed without specific knowledge about image processing today. However, the advantages decrease the credibility of digital data and justify the increasing demand for technologies to prove the integrity and to ensure the authenticity of the images.

Thus, designers, producers and publishers of video or multimedia material are seeking technical solutions to address these problems associated with copyright protection and authentication of origin. The goal is to ensure trustworthiness. This paper presents in a very dense way a new digital signature technique for image authentication, which also enables digital watermarking to search for manipulated yet copyrighted material.

First the paper explains the need and the problem of content-based digital signatures, which are not equivalent in the security comparing with general cryptographic signatures, but helps to find image manipulations depending on a expected security level. We present an approach for content-based digital signatures based on an edge detection algorithm. To give the user interactive support for visual video authentication, we present the VideoCubeAuthenticator.

Based on this discussion, we describe a watermarking technique which enables the copyright holder to search for manipulated images using the proposed signature technique. With current technology, the copyright holder can for example search over the internet for copyright violations with the private watermarking information, the copyright holder can detect his images but is not able to detect which images have been manipulated. It is not possible to find attackers which have copied and also manipulated the image illegally. Furthermore for public watermarks, where the user is informed about existing copyrights, it is necessary to verify if the material has the original impression and content. Otherwise, the copyright holder could be suspected for probative facts. Based on this discussion we present our new approach which combines content-based signatures and digital watermarking to ensure copyrights and integrity of the data. We call it the content-fragile watermarking approach, CFW. It enables the copyright holder to search for manipulated images and enhances the public watermarking schemes.

2 Content-Based Digital Signatures

The success of the Internet depends on the contents offered. Increasingly the Internet users seek for information with integrity and authenticity. Authentication schemes fulfil an increasing need for trustworthy digital data in commerce, industry, defense, *etc*. The concept of digital signatures is to ensure authentication, [1]. The importance and the need of trustworthiness is reflected by the German Digital Signature Act (enacted on 1st August 1997) and some Digital Signature Acts of several States of US. These laws are forerunners at the level of the state regulation.

Since especially for images and video the editing or modifying of the content can be done efficiently and seamlessly, the credibility of digital data is compromised. The concept of digital signatures for image/video authentication could use the known public key algorithm and public key infrastructures to ensure trustworthiness. However, the direct application of digital signatures to digital image data is vulnerable to image processing techniques like conversions, compression or scaling. The image material is changed irreversible without content modifications. Although the content of the image has not been changed and the viewers still have the same image impression, the signatures verification would fail. Consequently, a direct comparison of image data cannot lead to a secure evidence of the image's correctness and authenticity. So we have to differ from content-preserving and content-changing manipulations, and cannot apply the digital signature algorithm directly to the image binary data. A draft classification can be found in the tables 1 and 2.

Content-preserving manipulations	Content-changing manipulations
• Transmission errors- • Noise • Data storage errors • Compression and quantization • Brightness reduction • Resolution reduction • Scaling • Color convertions • γ-distortion • Changes of hue and saturation	• Removing image objects (persons, objects, etc.) • Moving of image elements, changing their positions • Adding new objects • Changes of image characteristics: color, textures, structure, impression, etc. • Changes of the image background (change of the day time or location (forest, ocean)) • Changes of light conditions (shadow manipulations etc.)

Table 1. Content-preserving and content-changing manipulations

Content-preserving image effects	Content-changing image effects
• Loss of details and depth of focus • Loss of color resolution, color shifting • Whole image effected (except of transmission error rates)	• Mostly no loss of details and depth of focus • Changes influences usually only image parts • All changes manipulate the image content

Table 2. Effects of content-preserving and content-changing manipulations

The main concept for image authentication is to extract the image characteristics of human perception, called content. Content-based digital signature techniques are needed for verifying the originality of video content and the "intactness" of images/videos, [2], [7-9], [12-15]. Authentication signatures are expected to survive only acceptable transcoding or compression and reject other manipulations. Very important is that content-based signatures cannot prevent forgery but could be used to determine wether an image/video is authentic.

Content-based signatures can be classified into the content, the image attributes, which is used as the input for the digital signature algorithm. The content extraction is called feature extraction. First approaches can be found in [2] based on histogram techniques. A feature vector is based on image characteristics that carry the image's meaning, it should not be influenced by the preserving image alterations. These feature vectors are used to create the content based signature, see the next figure.

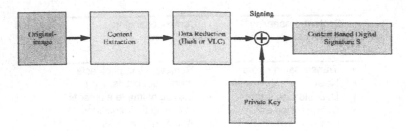

Fig. 1. Creation of the content based signature

In order to judge the usability of feature codes, their manipulation weakness and robustness against most critical content-preserving alterations has to be examined. The most content-preserving alterations are scaling and quantization as they have rather powerful effects on the data and they are used rather frequently.

Today there are the following main approaches for the feature code extraction:

- the comparison of the DCT-coefficient characteristics introduced by [12], [13]
- the approach using the sign of the DC-DCT-coefficients of succeeding blocks introduced by [17], and
- the use of intensity/color/luminance histograms or textures information introduced by [2], [15], [18].

All three approaches have several problems with quantization, partly with resolution reduction and mainly with scaling as content preserving manipulations. The extracted feature code is fragile to the powerful effects of scaling and high quantization or resolution reduction. Highly compressed and scaled/re-scaled images are mostly detected as manipulated. Detailed results can be found in [10]. Only with very large tolerant levels the feature code is invariant against scaling, resolution reduction and high MPEG compression. But the use of high tolerance levels includes less sensitivity against content changing manipulations like changes of image characteristics or light models. Based on our experience with the proposed feature extraction schemes, [2], [12], [13], [15], [17], [18], [10], we designed a new approach based on the edge characteristics of the image data.

3 A New Approach: Edge-Based Digital Signatures

The idea is to determine the edge characteristics of the image or single video frame and transform them into a feature code for the content-based digital signatures. The edge characteristics of an image give a very good reflection of the image content, because they allow the identification of object structures and homogeneity of the image. We are using the canny edge detector described as the most efficient edge separator in [6]. In the following section we describe the algorithm for generating and verification of the edge based feature code for the digital signature. We show our test results and suggest improvements to the feature code generation to be robust against compression and scaling.

3.1 The edge detection

Canny has defined three criteria to derive the equation of an optimal filter for step edge detection: (1) good detection, (2) good localization, and (3) low-responses multiplicity. The edge detection process serves to simplify the analysis of images by drastically reducing the amount of data to be processed, while at the same time preserving useful structural information about object boundaries. In his report [11] Canny developed mathematical formula for a one-dimensional edge profile in order to facilitate the analysis. The result of the edge detection process is a new image consisting of gray and black values for the edges and white for the background, called edge characteristic C. The algorithm does not give contours as result. Our task is now to transform this image into a binary contour characteristic: the so called binary edge pattern EP. The edge image is processed in the following way:

1. Processing the image edges: The Canny algorithm produces edges with different gray values, background is white. Values close to the color black represent more intensive edges and are replaced by black. Edges close to the value white are replaced by white. With the parameter *detail [0...1]* we can influence the recognized edge details, default is 0.5. As the result we get a new black/white edge image with the most relevant edges represented by black.
2. The black/white edge image is now scanned:. If the pixel value is black then we store a '1' else the value is white and we store a '0'. Finally we get the binary edge pattern EP of the image which will be used in the following algorithm for content based signatures. This post processing is intended to provide a conversion from the edge detection result image to almost binary representation of edges. The ideal result should be a sequence of points along with topological connections indicating the path of edges. Here we provide only a binary scheme for edge location.

3.2 The signature algorithm

The edge-based signature is calculated in the following way: First we extract the edge characteristics C_I of the image I or video frame with the Canny edge detector E (feature extraction). Second we transform it into a binary edge pattern EP_{CI} introduced in the section above. Third we make a VLC(Variable Length Code) for data reduction to produce the feature code instead of a hash value and finally we calculate the image signature $SigI$ (initialized with a private key of the originator) of the feature code. The signature generation *sign* calculates the hash over the VLC code and signs the hash value, instead of using directly the VLC code (here we would have had to handle the block size):

1. *Feature extraction:* $C_I = E(I)$
2. *Binary edge pattern:* $EP_{CI} = f(C_I)$
3. *Feature code:* $VLC(EP_{CI})$
4. *Sign feature code:* $SigI = sign(Hash(VLC(EP_{CI})))_{PrivatKey}$

The signed VLC code is now added to the image or video data for example into the private user section of images or MPEG system layer: $VLC(EP_{CI})) + SigI$.

The verfication process is performed in the following way: If the user gets the image or video T he first calculates the actual image edge characteristic C_T and the binary edge pattern EP_{CT} based on the original image size (check image size). Second he extracts and verifies the signed feature code, the original edge characteristic, with the appropriate public key. Third he compares both characteristics. Verfication steps:

1. *Feature extraction:* $C_T = E(T)$, $EP_{CT} = f(C_T)$
2. *Extract original feature code:* $EP_{CI} = VLC(EP_{CI})$, *verify* $Hash(VLC(EP_{CI})) = Decrypt(SigI)_{PublicKey}$
3. *Check* $EP_{CI} = EP_{CT}$

We have decided to embed the original VLC in addition to the signed VLC hash, because hash codes do not offer the possibility to verify which parts of the image were manipulated.

3.3 Test results and improvements

Our tests are based on 20 different images. We have performed 5 content manipulations: single pixel manipulations, adding/removing edges, adding/removing objects in different sizes, manipulations of the textures and manipulations of the image background. As major content-preserving manipulations we tested intensity enhancements, compression, quantization, scaling, low-pass filtering, edge enhancement, lossy and non lossy format conversions. Our first approach recognizes all image manipulations except single pixel modifications, which do not effect edges. The approach is invariant to main content-preserving manipulations, because we address the main impression of an image: the edge characteristics. Beside these advantages we have still some problems with high compression, quantization and scaling, because the edges are slightly moved. The authenticator has problems with intensive low-pass filtering and edge enhancement. However, if the changes of pixel values are not too great, we can still consider them as some kind of noise and use larger tolerance values, see improvements. The following figures show the results of the verification processes of a manipulated bus frame (the horse was cut out, the original bus frame can be seen in the quantization example), a quantized and a scaled bus frame without content manipulations (single frame of our bus video).

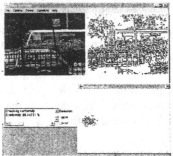

Fig. 2. Verification process results of a manipulated bus frame, detection of a manipulated region in the difference image (the horse edges are missing) without accessing the original data

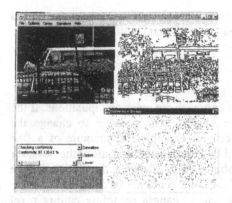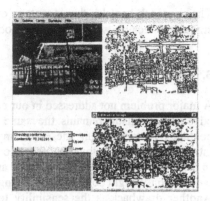

Fig. 3. Left the results of the verification process after 50% compression rate, the edges are wider and we get errors although there are no content changes, **Fig. 4.** Right the results of the verification process after image scaling. The edges are slightly moved and we get high error rates although no content change was performed

3.4 Improvements

Our second step was to integrate a tolerance threshold which can improve the recognition of content-preserving changes like quantization and scaling without loosing the sensibility of the manipulation recognition. We have developed two approaches for the implementation of the threshold:

Pattern search: Based on the edge characteristic we perform a local qualified search: the EP_{ci} is overlayed to the EP_{cr} to find conformity in the surrounding areas. It can be seen as a search for orginal edge pattern EP_{ci} in the actual pattern EP_{cr}.

Block approach: Based on the edge characteristic we build a binary block edge pattern over 8x8 blocks of the image, see next figure. If the 8x8 block is crossed by an edge, then the binary block edge pattern contains a "1" else a "0". So we get another abstraction level based on 8x8 blocks. The block size is variable to determine the acceptance rate of content-preserving and content-changing modifications.

Fig. 5. Block structure for the binary block edge pattern

The results of the first approach are satisfying, manipulations are recognized and the algorithm is tolerant against quantization and scaling up to 70 %. The problem is poor runtime performance.

The second approach offers beside the recognition of content-changes a high tolerance level against content-preserving manipulations. A manipulation recognition can be seen in figure 6. The problem of the block approach is the choice of the block size. A block size higher than 16x16 is not recommendable to avoid non recognized content manipulations. Depended on the block-size the security level can be chosen.

The approach has also the advantage of the reduction of the data rate for the digital signature, because we observe only blocks instead of all pixel values.

3.5 Open problems

A major problem not addressed in our current approach are color manipulations. If the edge characteristic remains the same (same object), it is possible to change the content by manipulating the color impression, for example the structure of a flag (stripes) could be the same but the color could be manipulated to confuse the observer in a news channel. Furthermore there are problems with images without edges where we have only slight color effects. Changes of the color could not be detected.

Another drawback is the sensibility to very small changes regarding cutting pixel lines or edges. The problem is to find a threshold to recognize real content-changes when the edges or pixels are manipulated slightly, like for example in the block oriented approach 2. First results show, that a 8x8 block allows very low modification, but they can be evaluated as not significant. A complex test scenario with different kinds of attacks is necessary and will be performed.

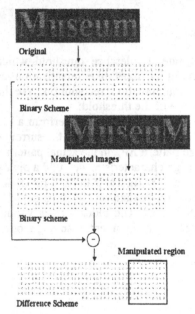

Fig. 6. Binary block edge pattern detection and manipulation recognition

4 New Interactive Tool for Video Authentication: VideoCubeAuthenticator

The recognition of video authentication is more complex than for still images, because the user has to prove the integrity for every frame and the surrounding frames, [8]. To solve the problem of succeeding frames we sign the whole video including all frames hashes. It is also possible to divide the video in essential video sequences and sign these parts to protect the scenes separately. First we build for every frame the signed feature code for single frame authentication, described in the section before. Then all hash values of the frames, using the already calculated VLC hash value of the edge characteristic, are used for the global video hash. One hash over the whole video allows the check of succeeding frame manipulations. The global hash value is signed and also stored in private user sections of the video. To give the user a navigation tool to verify the integrity and authenticity of the video, we use the VideoCube introduced in [4]. We use a video visualization in the form of a 3D-cube, [16]. The frames of a selected video are displayed as a floating 3D block, where the current image is represented as full image, while the next images form the lateral surface of the cube. Beside the normal video view the user can switch to the edge presentation of the video frames and perform the integrity check and verify the authentication of the video. The correctly detected edges are displayed transparent and the incorrect edges are black in order to track the manipulations visually through the whole video. The user can switch from the edge view to a transparent video view, where the correct content is also transparent and the manipulated region can be seen. The tool performs an automated tracking of manipulated frames and shows the changes in the video. The next figure illustrates the recognized manipulations: the horse was cut out and a 3D presentation visualizes the manipulations.

Fig. 7. Manipulation view of the bus video, the transparent edge view , **Fig. 8.** View to the manipulated areas in the check video, manipulated regions are displayed opaque

The advantages of the edge characteristic is the possibility to find the same edge characteristic in following frames to build the manipulation 3D-view. The tool provides the view inside the video sequence. Thus, the VideoCube represents content aspects of the video sequence. Features, such as manipulation tracking and video authentication are clearly visible in the pattern appearing in the transparent display of the cube. The user can stroke across the cube with the cursor to access any part of the video, even a single frame, like thumbing through pages of a flip book or switch to manipulation view.

5 A New Approach: Content-Fragile Watermarks

Content-based signatures are usually added to the private or user sections of the images or videos. But it is also possible to combine digital signatures and digital watermarking techniques[1], which usually embed copyright information invisible into multimedia data. The first advantage of using signatures as watermarking information is, that the signature can not be removed by an attacker, because the signature is directly involved with the original, Second advantage the copyright holder could also search for copyrighted and manipulated images. Furthermore we could solve the problem with public watermarking [2], where the user gets the information about the copyright holder. The public watermarking problem with high robustness watermarking mechanisms of the new generation is that the watermarking information remains present though the image content could be changed. The following figures show examples, where the content is changed and the watermarking information is still present.

Fig. 9. Left: watermarked original frame, (bus and president&wife), right: manipulated watermarked frame with still correct watermark retrieval

The consequence is, that if we use public watermarks, for example DigiMarc in Adobe PhotoShop, the copyright hint gives the user the impression that the image is untouched. Beside this confusion the user could hold also the copyright holder responsible for the manipulated image content. Obviously there is a misleading correlation between robust watermarking technology and content changes.

The problem to embed the content based signature as a watermark is that watermarking techniques usually cannot embed more than 10 to 100 bytes. It is therefore impossible to embed the content-based signature with a data rate higher than 1 kByte. Thus we have designed a solution called content-fragile watermarking which combines our watermarking technique, see [5] and the edge-based characteristic for image integrity detection. Our watermarking method which is working in the spatial domain is based on overlaying a pattern with its power concentrated mostly in low frequencies. The pattern is created using a pseudo random number generator and a cellular automaton with voting rules. We add a 8x8 pattern over every 8x8 Block of the frame. To embed binary code words, we defined additional modification rules of the overlaying 8x8 pattern. The robustness of the method has been shown in [5].

---

[1] Digital watermarking is the enabling technology for proof of ownership on copyrighted material, detect the originator of illegally made copies, monitor the usage of the copyrighted multimedia data and analyze the spread spectrum of the data over networks and servers.

[2] Public watermarking describes watermarking information which can be retrieved by everyone, whereas private watermarks can only be retrieved by the copyright holder with a private user key. A secure public watermarking is today not known, because today's watermarking algorithm are symmetric. [3]

5.1 Embedding

The idea is to generate two watermarking patterns. The first pattern $M_{Private}$ is for private copyright information using the private user key and may consists of textual information, for example the copyright holders name. The second pattern $M_{Content}$ is for content proof. The private copyright information is then embedded with $M_{Private}$ pattern. The content pattern generation during the watermarking process is initialized with the feature code (VLC(EP_{CI}) based on the edge characteristic C_I of the image I, see next figure. Watermark generation steps:

1. *Feature code: VLC(EP_{CI})*
2. *Generation: $M_{Private(privateKey)},$ $M_{Content(VLC(EPCI)}$*
3. *Embedding: $I_w = I + M_{Private} + M_{Content}$*

| Image I | Binary edge pattern | VLC compression | Watermarking pattern, $M_{Content}$ |

Fig. 10. Generating the public watermarking pattern

The image feature code as initializing value ensures the integrity of the image data and performs the construction of the content-fragile watermark $M_{Content}$.

5.2 Retrieval

The retrieval is performed in the following way:
First it is possible to search for the private watermarking information to track the usage or find illegal customer copies T. With the increasing robustness, the copyright information remains present although manipulations of the image/video content are done.

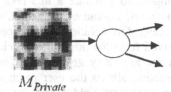

$M_{Private}$

Fig. 11. Private watermarking search with pattern $M_{Private}$

Second if the copyright holder wants to know if there is a image content violation, the copyright holder can search for the $M_{Content}$ in the found image T. In the use case of public watermarking the user loads an image T (for example over the internet), he could also verify the public content-fragile watermark. For the verification process we need the actual edge characteristic of the image T. We get the actual $VLC(EP_{CT})$ and generate the appropriate watermarking pattern, illustrated in the next figure. Afterwards the watermarking search for the generated pattern can start in image T. If the pattern will be found then the image is unchanged, the image content is the same. Otherwise the image is manipulated.

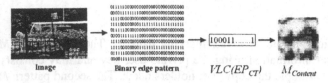

Fig. 12. Generation of public watermarking pattern for the integrity check

Verification Process:
1. Pattern generation: $M_{Content(VLC(EPCt))}$
2. Pattern search with the watermarking algorithm
3. Integrity decision

Using content-fragile watermarks, we cannot show the manipulated areas like introduced in section 4. However, it gives a new opportunity to evaluate wether the image is unchanged using digital watermarking techniques combined with the edge detection algorithm. The approach produces both a robust and a content-fragile watermark.

5.3 Enhancements for secure private or public content-fragile watermark

The problem with the proposed content-fragile watermark is, that an attacker could add a new $M_{Content(VLC(EPCt))}$ after manipulating the image because our watermarking technique is public. The robust copyright watermark would be still present and when the copyright holder or the user in the case of public watermarking tries to check the content watermark, he would generate the actual (manipulated) edge characteristic EP_{CT}, generate the $M_{Content(VLC(EPCt))}$ and search for this faked pattern. We propose the following solution:
1. For private watermarking we use a private user key, like for the $M_{Private}$, as additional input parameter for the $M_{Content}$ generation: $M_{Content(privateKey + VLC(EPCt))}$. It is now impossible for the attacker to add a new valid $M_{Content}$ without the knowledge of the private user key.
2. For public watermarking it is more complex to generate a non-fakable content watermark, because we cannot rely on a secret. Instead of the secret, we have to use an additional publicly verifiable parameter. Our solution is based on a publicly verifiable key, *ContentKey*, which is derived from the original edge characteristic and signed by the copyright holders private key. Key generation: *ContentKey* = $sign(Hash(VLC(EP_{Ct})))_{PrivatKey}$. This construction allows the user to check first the *ContentKey* with the public key of the copyright holder and the actual edge characteristic $VLC(EP_{CT})$. If the verification is successful he can then use the *ContentKey* for the pattern generation. If we use this solution it is not necessary to use the actual edge characteristic again for the pattern generation, because it is already a parameter for the *ContentKey*. An attacker is not able to generate a new *ContentKey* or a new content watermarking pattern, because he does not know the private key of the copyright holder. The additional problem with this approach is the transmission of the *ContentKey* and the identification of the copyright holder and his public key. For both information we could use public watermarking or embed the data into user data of image or video. Altogether the attacker could only replace the information by using his own or a wrong private key. The advantage of

this solution is that a successful verification also proves the identity of the copyright holder and ensures authenticity beside integrity.

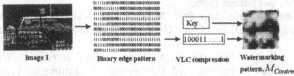

Fig. 13. Secure content-fragile watermark by using additional key information

5.4 Problems

The use of the content-fragile watermark as additional private watermark for manipulation detection is very usefully for copyright holders to track and trace copyright violations. The use as public information is possible but more complex and is subject to practical evaluation.

The detection problems of content manipulations are in the field of color manipulations, because the edges are currently not sensitive to color changes without edge changes. There are also problems with high compression and scaling transformations or color modifications as content-preserving manipulations, see section 5. However, the advantage of this first approach is that the first watermark is very robust against attacks and the second watermark is content-fragile for manipulation recognition. Signed originator *ContentKeys* can be useful for originator authentication.

6 Summary and Further Work

Our current research is focused on the definition of improved methods for content-based signatures and the combination with watermarking to get content-fragile watermarks. The future work will concentrate on sophisticated attribute extraction describing the image content and their combination with watermarking techniques. The edge detection approach offers first interesting results regarding the image manipulation detection. The development of attribute extraction will also be combined with object recognition. Further work will address the object characteristics of images and video frames to be tolerant against high quantization and high scaling as allowed image manipulations, which are not recognized as image content changes. Another topic is the integration of color characteristics and the design of an intelligent cellular automaton for generatation of watermarking patterns, which are robust against quantization and scaling in addition to robust feature codes.

References

[1] Bearman, D., and Trant, J. Authenticity of Digital Resources: Towards a Statement of Requirements in the Research Process. D-Lib Magazine, June 1998.

[2] Chang, Shih-Fu und Schneider, Marc: A Robust Content Based Digital Signature for Image Authentication, Proceedings of the International Conference on Image Processing, Lausanne, Switzerland, September 1996.

[3] Cox, Ingemar J, and Linnartz, Jean-Paul M.G.: Public watermarks and resitence to tampering, Proceedings of IEEE Int. Conf. O Image Processing, 1997, available only on CD-ROM

[4] Dittmann, Jana, Nack, Frank, Steinmetz, Arnd, Steinmetz, Ralf: Interactive Watermarking Environments, Proceedings of International Conference on Multimedia Computing and Systems, Austin, Texas, USA, 1998, pp. 286-294

[5] Dittmann,, Jana, Stabenau, Mark, Steinmetz, Ralf: Robust MEG Video Watermarking Technologies, Proceedings of ACM Multimedia'98, The 6th ACM International Multimedia Conference, Bristol, England, pp. 71-80

[6] Fischer, Stephan: Indikatorenkombination zur Inhaltsanalyse digitaler Filme, D 180 (Diss. Universität Mannheim), 1997, Shaker Verlag Aachen.

[7] Friedman, G.L. *The Trustworthy Digital Camera: Restoring Credibility to the Photographic Image*. IEEE Trans. on Consumer Electronics, Vol.39, No.4, pp.905-910, Nov. 1993.

[8] Gennaro, R., and Rohatgi, P. *How to Sign Digital Streams*. CRYPTO '97, Santa Barbara, CA, USA, August 1997, pp.180-197.

[9] Gennaro, R., Krawczyk, H. and Rabin, T. *RSA-based Undeniable Signatures*. CRYPTO '97, Santa Barbara, CA, USA, August 1997

[10] Haberhauer, Maike und Dittmann, Jana: "Das Bild lügt nicht?" – Untersuchung und Entwicklung von Lösungsansätzen zur Sicherstellung von Authentizität und Integrität von digitalen Bilddaten, Technical Paper, GMD, September 1998

[11] J. F. Canny, "A computational approach to edge detection", IEEE Trans. on Pattern analysis and Machine intelligence. vol. PAMI-8, Nov. 1986.

[12] Lin, C.-Y. and Chang, S.-F. *A Robust Image Authentication Method Surviving JPEG Lossy Compression*. SPIE Storage and Retrieval for Image and Video Databases, San Jose, CA, USA, Jan. 1998.

[13] Lin, C.-Y. and Chang, S.-F. *Issues and Solutions for Authenticating MPEG Video*. SPIE Storage and Retrieval for Image and Video Databases, San Jose, CA, USA, Jan. 1999.

[14] Quisquater, J.-J., Macq, B., Joye, M., Degand, N. and Bernard, A. *Practical Solution to Authentication of Images with a Secure Camera*. SPIE Storage and Retrieval for Image and Video Databases, San Jose, CA, USA, Feb. 1997.

[15] Schneider, Marc and Chang, Shih-Fu: Digital Watermarking and Image Authentication, 1996, http://www.ctr.columbia.edu/~mars/papers/reports/water/doc.html.

[16] Steinmetz, A.: DiVidEd A Distributed Video Production System, work in Progress, Proceedings of Visual96 Information Systems, February 1996

[17] Storck, D.: A New Approach to Integrity of Digital Images, Proceedings of IFIP World Conference- Mobile Communication, Canberra, Australia, 1996.

[18] Zhong, D. und Chang, Shih-Fu: Video Object Model and Segmentation for Content-Based Video Indexing, IEEE International Conference on Circuits and Systems, Hong Kong, June 1997.

New Structures for the Next Generation of IDMS

Dick Bulterman

Oratrix Development BV Amsterdam

Abstract. Interactive multimedia is not new. As early as 1991, sophisticated interactive multimedia systems were available that allowed
CDROM-based presentations to be created and distributed to mass audi-
ences. Typical applications for CDROM-based multimedia were art and
entertainment, product catalogues, computer-based training and information kiosks.

The nature of the CDROM itself meant that the presentation structure
remained static even as the contents appeared to be dynamic. Given the
long production cycles of CDROM presentation, and the conventional nature of CDROM distribution channels, CDROM-based multimedia provided only an incremental improvement over conventional documents.

The mass acceptance of the World-Wide Web has significantly changed
the landscape of presenting distributed information. Users have come to
expect that the Web is a generic source for all of their information needs,
including information that is multimedia in nature.

This talk will focus on the challenges that a Web-like environment presents
to the design and support for the next generation of interactive distributed multimedia systems. These challenges include: - adapting content to the needs to the user - defining time relationships in an uncertain
timing environment - problems with scalability and interoperability of
applications - problems with protecting the rights of IDMS content own-
ers and users.

The talk will present recent development in the standardization of languages and architectures for IDMS support. In particular, the facilities
provided by SMIL and HTML+TIME will be contrasted in their support
for general IDMS applications. We will provide examples of new developments in each language and show developments in new user interfaces
that make IDMS's more accessible to a wide range of users.

Multi-drop VPs for Multiparty Videoconferencing

on SONET/ATM Rings

-- Architectural Design and Bandwidth Demand Analysis

Gang Feng   Tak-Shing Peter Yum

Department of Information Engineering

The Chinese University of Hong Kong

N.T. Hong Kong

e-mail: gfeng5@ie.cuhk.edu.hk

Abstract. This paper proposes a scheme for implementing multiparty videoconferencing service on SONET/ATM rings. Different multicasting methods on SONET/ATM rings are discussed and compared. A new multicast VP called " Multi-drop VP" which is particularly suitable for SONET/ATM rings is proposed. An Add-Drop Multiplexer(ADM) structure for rings capable of multi-dropping is also presented. Several VP assignment schemes are proposed and their bandwidth utilizations are compared.

1. Introduction

Videoconferencing is expected to be one of the most important services in broadband networks. Different kinds of videoconferences have different configurations, user-interactions, quality of service (QoS) requirements, and network resource requirements[1-5]. Among the various videoconferencing methods, "Speaker-video" conference, for which only the video and voice of the current speaker are broadcast to all other conferees, demands the least amount of equipment and bandwidth as compared to that of selectable media, common media [3], and virtual space conferences[2].

Multiparty videoconferences can be implemented on various kinds of networks. The ITU H.320 [15] is a recently adopted standard for videoconferencing on narrowband ISDN. Other implementations include those on Ethernet ring, Token Ring, IP networks and ATM-based BISDN [16-18]. In this paper, we focus on the multicasting aspects and the virtual path (VP) assignment schemes for implementing multiparty videoconferencing on SONET/ATM type rings.

The increasingly popular SONET ring has the advantages of standard signal interfaces, economic adding and dropping of traffic streams, self-healing capability and the support of operation, administration and maintenance [19]. Present SONET rings use synchronous transfer mode (STM) for signal multiplexing and switching and support non-switched DS1 and DS3 services. Recently, Virtual Path (VP)-based Asynchronous Transfer Mode (ATM) technology is being introduced into SONET rings as a means to reduce cost and to provide flexible bandwidth demand arrangement [8,12]. This cost reduction is achieved by the use of non-hierarchical path multiplexing.

The ATM VP-based SONET ring architecture [6,7] is essentially a combination of the SONET/STM architecture and ATM virtual channel (VC)-based network architecture. As a result, it keeps the simplicity of the SONET/STM network while retaining the flexibility of the ATM technology. In [8], a SONET/ATM ring using point-to-point virtual path (SARPVP) was proposed. It was assumed there that the ATM ADM (Add-Drop Multiplexer) for the VP-based ATM rings can be built from the SONET ADM by replacing the STS-3 termination cards by the ATM STS-3c line cards. Other ATM ring architectures proposed can be found in [8,10-12] and topics such as bandwidth allocation[10], self-healing mechanism and VPI assignment[11] were studied.

In this paper, we study the multicasting and VP assignment problems for multiparty videoconferencing on SONET/ATM rings. In Section 2, we introduce the network configuration, the service characterization and the SONET/ATM ring architecture. In Section 3, we discuss the use of various multicasting methods on

276

SONET/ATM rings and a new method called Multi-drop VP multicasting is proposed. An ADM structure suitable for Multi-drop VP is presented in Section 4. In Section 5, we discuss the conference management issues on SONET/ATM rings. In Section 6, we propose several multi-drop VP assignment schemes. After evaluating and comparing the bandwidth utilization of these schemes in Section 7 and Section 8 , we conclude this paper in Section 9.

2. Videoconferencing on SONET/ATM Rings

SONET self-healing ring can be unidirectional or bidirectional at normal working state. In this paper, we assume the ADMs can support bi-directional transmissions. To carry conferencing traffic over SONET/ATM rings, a two-layer subscriber network architecture similar to that in [13] can be used. This architecture consists of a transfer network and an access network as shown in Fig.1. The transfer network is a SONET/ATM ring with Add-Drop Multiplexers (ADMs). The access network links up business and residential customers to the ADMs.

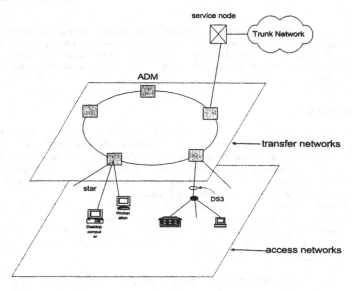

Fig.1. A two-layer subscriber network architecture

Video is the most troublesome traffic in videoconferencing. Fundamental issues regarding its transmission over ATM networks remain unresolved. For example, there is no general consensus on whether variable bit rate (VBR) schemes are better than constant bit rate schemes (CBR) for video services. ITU Recommendation H.320[15] is a collection of standards for videoconferencing and videotelephony systems. It is intended for systems with channel capacity up to T1 or E1 rates and includes recommendations for audio coding, video coding, multiplexing and system control. In this paper , we assume that the audio, video and control data can all be multiplexed onto a fixed rate channel such as DS1 on SONET rings or equivalently a CBR channel on SONET/ATM rings and the H.320 conferences belong to this type.

 ATM networks can be either VC or VP based. The VC-based ATM network , as depicted in Fig.2, consists of ATM VC switches and manages VC connections on a VC-by-VC basis. It is sometimes referred to as the *full* ATM switched network and is characterized by its very flexible and efficient bandwidth management capability. On the other hand, it is also complex and expensive compared to the ATM VP switches[8,19].

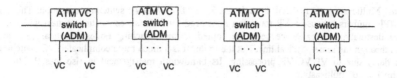

Fig.2. A section of VC-based SONET/ATM ring

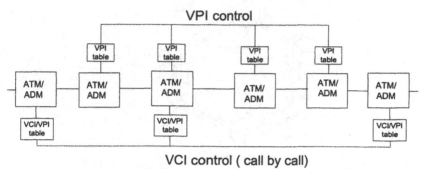

Fig.3. A section of VP-based SONET/ATM ring with multipoint connection

To reduce the signal transport complexity while preserving some flexibility of bandwidth management at intermediate nodes, the VP-based ring augmented with multicast function can be used (Fig.3). Here the intermediate nodes perform the functions of cell routing by VPI, while the end nodes perform the functions of call setup, call admission control, VP assignment, routing, VP capacity allocation and traffic control.

3. Multicasting on SONET/ATM Rings

Multicast connections can be set up in many kinds of networks. For example, according to Q.931[14] and Q.93B[14], the ITU recommended call setup protocol for ISDN and B-ISDN respectively, a multicast connection is set up by establishing multiple point-to-point connections. Efficient techniques for multicasting on ATM networks are still being intensively researched on. In this Section, we discuss the use of four multicasting methods on SONET/ATM rings. The first three are from [21] and the fourth called Multi-drop VP is new.

A. Multiple Point-to-point VCs Scheme

This scheme makes use of the existing point-to-point communication and control protocols to set up a point-to-point connection for each destination[20]. It is simple but is also bandwidth wasteful as it requires identical cells to flow through the same physical links.

B. VP Augmented VC Multicasting Scheme

This is a VC multicasting method augmented by point-to-point VPs. On a ring using point-to-point VPs like SARVP[8], the intermediate nodes for a multicast connection are also the junction nodes between the point-to-point VPs. At these switching nodes (in this case, VC switches), the cells are passed from the VP processing layer to the VC processing layer. After copying at the VC processing layer and translating the VPI numbers, the copied cells are returned to the VP processing layer for onward transmission in the next VP. Here multicasting is performed at the VC layer but the same VP route is used by different VC connections. Multicasting is realized through this kind of VP-VC-VP switching operation. This method can save VPI numbers because a VP route can be shared by different connection request with the same source and destination nodes. Since cells are copied at the switching nodes, identical cells are not transmitted through the same physical link. But the tradeoff is a much more complicated VC switching and the added delay due to VP-VC-VP processing. Its bandwidth management is also less flexible. Fig.4 illustrate this kind of multicasting.

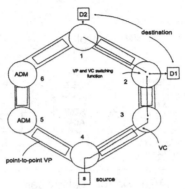

Fig.4. VP augmented VC multicasting on an SONET/ATM ring

C. VP Multicast Scheme

This method requires all possible point-to-multipoint VPs be established. A point-to-multipoint connection can then be established by reserving bandwidth on the corresponding point-to-multipoint VP for multicasting. Here, cells from the source node are copied at the intermediate ADMs on the SONET/ATM ring based on the VPI's. This method has the advantages of efficient bandwidth utilization, simple transport processing and flexible bandwidth management. But it has the problem of requiring a large number of VPI for all possible multicasting routes.

In the example showing in Fig.5, VP 1 is for the multicast connection from node 4 to nodes 2 and 1. VP 2 is for the multicast connection from node 4 to nodes 3 and 1. Note that setting up and tearing down VPs "on the fly" are possible. But this is confusing with VC and contradicts the basic VP design philosophy of aggregating multiple VCs for better reliability and performance.

In VP multicasting, many VPs need to be set up for all combinations of source and destination groups. All together, $(i+1)\binom{N}{i+1}$ VPs are required for connections with i destinations on an N-node ring. Since the number of destination nodes can range from 1 to $N-1$, the total number of VPs needed is $\sum_{i=1}^{N-1}(i+1)\binom{N}{i+1}$.

Even if we limit the maximum number of destination nodes allowed in a multicast to be M, $(M < N)$, the total number of VPs needed is still very large. Fig.6 shows the number of VPs need vs. network size N under several M values. It can be seen that the multicast VP scheme is impractical for the huge number of VPs needed.

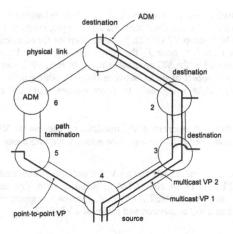

Fig.5. VP multicast on a SONET/ATM ring

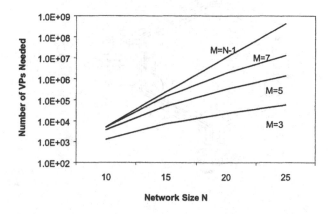

Fig.6. Number of VPs needed for multicast VP scheme.

D. Multi-drop VP

Multicasting on SONET/ATM rings requires the dropping and forwarding functions at the nodal transceivers. Based on these requirements, we design the "Multi-drop VP" for multicasting on SONET/ATM rings. It has all the advantages of VP multicast without requiring VPs to be established for all multicast combinations. The traditional point-to-point VP[9] has only one exit point. A multi-drop VP allows multiple drops or exits. To distinguish these two types of VPs, we use one bit in the GFC (Generic Flow Control) field of the cell header as the "Multi-drop VP Indicator"(MDI). Specifically , *MDI=1* means that the cell concerned has to be copied (or tapped out) at all intermediate ADMs and *MDI=0* means that the cell is on a point-to-point VP. Note that the GFC field is used for traffic control data on a multi-access network. But on SONET/ATM rings, a VC connection corresponds to a DS1 channel and these is no statistical multiplexing among the VCs. Therefore, cell-level flow control is not needed at the user-network interface(UNI) and the GFC field can be used to indicate the multi-drop nature of the VP.

All tapped-out cells are passed to the VC processing layer. Those belonging to the local destinations are passed there and the remaining ones are discarded. As an example, consider Fig.7 where two multicast VCs are carried on a single Multi-drop VP (VP 2). VC 1 is set up for the connection from node 4 to nodes 2 and 1. At the intermediate node (i.e., node 3), the cells belonging to VC 1 are tapped out from the VP processing layer but get discarded at the VC processing layer. In fact, VP 2 can carry all multicast traffic from node 4 to node 1 and any subset of nodes between them. In other words , multi-drop VP keeps the advantages of multicast VP while being able to accommodate multicast connections with various destination combinations.

Multi-drop VP is simpler than VP augmented VC multicasting because no VC switching function and therefore no VPI translations are needed at intermediate nodes. It requires a much smaller number of VPIs than VP multicast.

The VPI values are assigned on a global basis and no VPI translation at intermediate nodes is required. As there are 12 bits for the VPI field in the NNI ATM cells, up to 4096 VPs can be defined. If the DS1 channels via circuit emulation on the SONET/ATM ring are used to support videoconferencing service, then each DS1 can be treated as a VC connection and assigned a VPI/VCI.

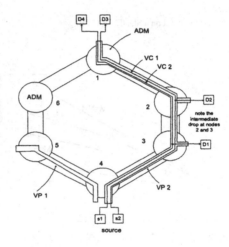

Fig..7. Multi-drop VP in the SONET/ATM ring

4. Add-Drop Multiplexer(ADM) for SONET/ATM Ring

ADMs for SONET/ATM rings can be implemented in different ways depending on the actual SONET STS-Nc terminations. A kind of ADM architecture for point-to-point SONET/ATM rings was introduced in [8]. In this section, we modify the hardware architecture in [8] to accommodate multi-drop VPs. The most commonly proposed ATM STS-Nc terminations are STS-3C, STS-12c, and STS-48c. Fig.8 shows the ADM hardware architecture for STS-3c terminations. It consists of the SONET layer, ATM layer and the service mapping layer. As the SONET layer is identical to that in [8], we focus only on the latter two.

The ATM layer performs the following functions:

1. ATM/SONET interface --- convert the STS-3c payload to ATM cell stream and vice versa;
2. Cell type classifying --- check the MDI of individual cells and copy out those with MDI=1.
3. Cell addressing --- For cells with MDI=0, check their VPIs to determine if they should be dropped (for local termination) or forwarded (for termination in the down-stream nodes);
4. Idle cell identifying --- Identify the idle cell locations for cell stream insertion via a sequential access protocol .

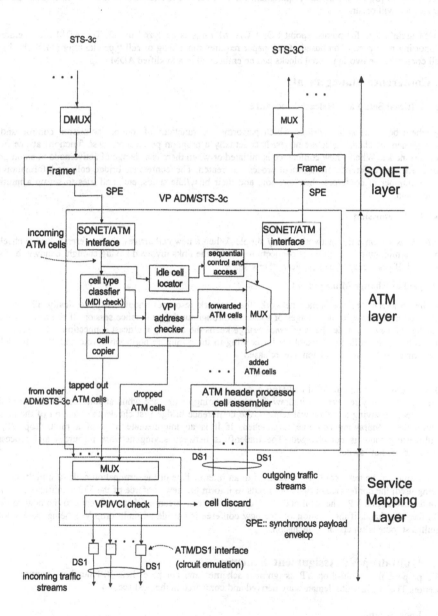

STS-3c

STS-3C

DMUX

MUX

Framer

Framer

SPE

SPE

SONET
layer

VP ADM/STS-3c

incoming
ATM cells

SONET/ATM
interface

SONET/ATM
interface

cell type
classfier
(MDI check)

idle cell
locator

sequential
control and
access

VPI
address
checker

forwarded
ATM cells

MUX

ATM
layer

cell
copier

added
ATM cells

from other
ADM/STS-3c

tapped out
ATM cells

dropped
ATM cells

ATM header processor
cell assembler

DS1

DS1

MUX

outgoing traffic
streams

Service
Mapping
Layer

VPI/VCI check

cell discard

SPE:: synchronous payload
envelop

ATM/DS1 interface
(circuit emulation)

DS1

DS1

incoming traffic
streams

Fig.8. A simple ADM hardware architecture suitable for Multi-drop VP

The service mapping layer maps the input cells to their corresponding DS1 cards based on their VPI/VCI values. Cells from different STS-3c payloads are first multiplexed into a single cell stream. Their VPI/VCI are checked. Those correspond to the local terminations are passed there while the rest are discarded. According to [8], the bandwidth requirement for DS1 service is allocated on the peak rate basis and so no congestion will occur.

ADM architecture for point-to-point SONET/ATM rings is analyzed in [8]. The ADM architecture for supporting multicasting proposed in this paper requires the adding of cell type classifier (MDI check) and cell copier. These two functional blocks can be embedded in a modified ADM chip.

5. Conference Management

A. Multicast Setup and Release Procedure

Let there be a *conference bridge* which performs the functions of routing, admission control, and the management of changing active nodes. It is actually a program performing these functions at one of the network nodes. When a new conference is initiated or when there is a change of active node in an on-going conference, a conference management process is created. The conference bridge collects information such as the number of conferees, their location, and their busy/idle status, etc., and tries to set up a multicast connection.

B. Call Admission

Call admission on a ring network is very simple. When a new call arrives, the conference bridge checks if there is a minimum hop multicast connection with all the links involved having enough bandwidth for the new call. If yes, accept the call, reject otherwise.

C. Speaker Change Management

In the speaker-video conference network, the network resources should be dynamically allocated and retrieved in response to the changes of speakers throughout the conference session. If the next speaker is attached to the same node, the *conference bridge* keeps the existing multicast connection. Otherwise, a new connection is identified and established according to the minimum hop routing rule and the channels in the former multicast connection are released.

D. Conferee Joining and Withdrawing

One conferee may request to withdraw from an ongoing conference while another conferee may wish to join. Upon receiving a withdrawal request, the conference bridge first checks the location of the node to which the withdrawing conferee is attached. If it is an intermediate node of a multi-drop VP, the multicasting route is not changed. The tradeoff is between saving network resources and processing overhead of "hot" switching.

For joining, if the new conferee is attached to an intermediate or the termination node of a multi-drop VP being used, the conference bridge only needs to inform the new conferee of the VC identifier used by the conference in that VP. The local node then outputs the cell stream of that conference to the new conferee. On the other hand, if the location of the new conferee is outside all multi-drop VPs being used, a longer multicast route is set up for its inclusion.

6. Multi-drop VP Assignment Schemes

We propose five multi-drop VP assignment schemes and compare their VPI numbers required in this section. Their bandwidth demands are derived and compared in the next section.

A. Loop Scheme

In this scheme, each source node sets up a loop multi-drop VP that passes through all other nodes, as shown in Fig.9. The total number of VPs required is therefore N. To balance the traffic on the clockwise and the counter-clockwise directions, the VPs for source nodes $1,3,5, ...$ can be assigned on one direction and the VPs for source nodes $2,4,6, ...$ on the other direction.

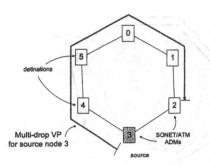

Fig.9. Loop Multi-drop VP Assignment scheme

B. Double Half-Loop Scheme

In the Double Half-loop scheme, two multi-drop VPs on the two sides of the source node are set up for embracing the rest of the nodes. Fig.10 shows such an assignment for node 3 being the source node. The number of multi-drop VPs required for encircling assignment is $2N$ and each VP has length of approximately $N / 2$ hops.

Under this scheme, when all destination nodes are on one side of the source node, only one VP is needed. This results in a higher bandwidth efficiency than the Loop Assignment scheme.

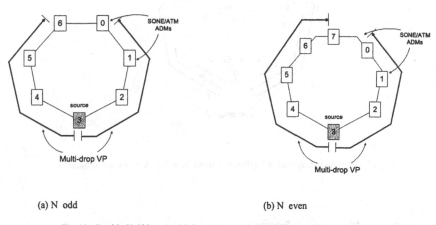

(a) N odd   (b) N even

Fig.10. Double Half-loop Multi-drop VP Assignment

C. Single Segmental Minimum-hop Scheme

For each source node, we set up multi-drop VPs to all other nodes on one direction, as shown in Fig.11. Under this scheme, a minimum-hop route *within one segment* can be found for any multicast connection. Again, to balance the traffic on the two directions on a bi-directional ring, the VPs for nodes *1,3,5, ...* can be assigned on one direction and the VPs for nodes *2,4,6, ...* on the other direction. Obviously, the total number of VPs required is *N(N-1)*.

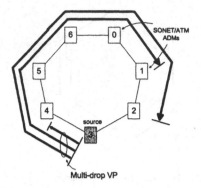

Fig.11. Segmental Minimum-hop Multi-drop VP Assignment

D. Minimum-Hop Within Half-Loop Scheme

In the Double Half-loop scheme, if we add VPs for all the sub-segments of the half-loop VPs as shown in Fig.12, the bandwidth utilization can be increased. We call this the *Minimum-hop within Half-loop Scheme*. Obviously, its bandwidth efficiency is higher than the first three schemes. The minimum number of VPs required is $N(N-1)$.

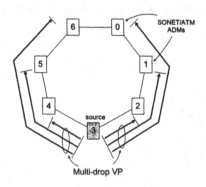

Fig.12. Minimum-hop within Half-loop Multi-drop VP Assignment scheme

E. Unconstrained Minimum-hop Scheme

A *minimum hop route* does not waste any bandwidth resources. Such route is always available if multicast VPs are set up for all combination of source and destinations. As discussed in Section 3(c), the total number of VPs needed is huge. On the other hand, with the use of multi-drop VPs, the same can be achieved when multi-drop VPs are set up for all combinations of "source and farthest destination" pairs. To do so, for each node as source node, we assign *N-1* multi-drop VPs on the clockwise direction around the ring to each of the other nodes and another *N-1* multi-drop VPs on the counter-clockwise direction to each of the other nodes as well (Fig.13). The total number of VPs required is *2N(N-1)*. When all these VPs are available and used, we call this the *Unconstrained Minimum-hop* scheme.

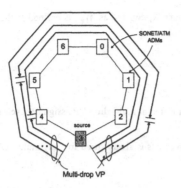

Fig.13. Minimum Hop Assignment scheme

Table 1 compares the VP numbers required for the five multi-drop VP assignment schemes to that of the VP multicast scheme.

Table 1. Comparison of VP number required on an N-node ring

Loop Scheme	Double Half-Loop Scheme	Single Segment Minimum-hop Scheme	Minimum-hop within Half-loop Scheme	Unconstrained Minimum-hop Scheme	VP Multicast Scheme
N	2N	N(N-1)	N(N-1)	2N(N-1)	$\sum_{i=1}^{N-1}(i+1)\binom{N}{i+1}$

7. Bandwidth Demand Analysis

In this section, we analyze the bandwidth demand of k party conferences for the five multi-drop VP assignment schemes assuming that the source and k-1 destinations are randomly located on an N-node ring. For convenience, we refer the five schemes as Scheme A, B, C, D and E according to their order of presentation in the last section. Without loss of generality, we can let node 0 be the source and let $\mathbf{A} = (A_1, A_2, ..., A_{N-1})$ be a random vector with $A_i = 1$ indicating that node i is a destination and $A_i = 0$ otherwise. In addition, let $\mathbf{a} = (a_1, a_2, ..., a_{N-1})$ be a binary vector and

$$\Omega_k = \left\{ \mathbf{a} \mid \sum_{i=1}^{N-1} a_i = k-1 \right\} \tag{1}$$

Due to the symmetry, the destination distribution takes any pattern in $\mathbf{a}$ with the same probability. Given N and k, the total number of patterns is simply $\binom{N-1}{k-1} = \frac{(N-1)!}{(k-1)!(N-k)!}$. The probability that $\mathbf{A}$ will take on any specific pattern $\mathbf{a}$ is just one over that total number, specifically,

$$\Pr ob[\mathbf{A} = \mathbf{a}] = \begin{cases} \dfrac{(k-1)!(N-k)!}{(N-1)!} & \text{for } \mathbf{a} \in \Omega_k \\ 0 & \text{otherwise} \end{cases} \tag{2}$$

Let $h_X(\mathbf{a})$ be the number of links used by a specific connection request of size k with destination distribution $\mathbf{a}$ under multi-drop VP assignment scheme X. Averaging over all destination distributions $\mathbf{a}$ in Ω_k, we get the expected number of links required as

$$E[h_X(\mathbf{a})] = \sum_{\mathbf{a} \in \Omega_k} h_X(\mathbf{a}) \Pr ob(\mathbf{A} = \mathbf{a}) \tag{3}$$

The bandwidth demand factor under scheme X, denoted as η_X, is defined as the average number of links used normalized by the ring size N. In other words

$$\eta_X = \frac{E[h_X(\mathbf{a})]}{N} \qquad (4)$$

In the following, we derive $h_X(\mathbf{a})$ under different multi-drop VP assignment schemes.

A. Loop Scheme

Under this scheme, a route of $N-1$ hop counts is always used for any multicast connection request. Thus we have

$$h_A(\mathbf{a}) = N - 1 \qquad (5)$$

B. Double Half-loop Scheme

Under this scheme, all VPs have length about $N/2$. If all destinations are clustered within one of the two half-loops, one VP is enough. Otherwise, two VPs are required. Specifically, if N is odd, we have :

$$h_B(\mathbf{a}) = \begin{cases} (N-1)/2 & \text{if } \sum_{i=1}^{(N-1)/2} a_i = k-1 \text{ or } \sum_{i=(N+1)/2}^{N-1} a_i = k-1 \\ N-1 & \text{otherwise} \end{cases} \qquad (6)$$

If N is an even number, we have

$$h_B(\mathbf{a}) = \begin{cases} N/2 & \text{if } \sum_{i=1}^{N/2} a_i = k-1 \\ N/2-1 & \text{if } \sum_{i=N/2+1}^{(N-1)} a_i = k-1 \\ N-1 & \text{otherwise} \end{cases} \qquad (7)$$

C. Single Segmental Minimum-hop Scheme

Under this VP assignment scheme, the number of links used by the multi-drop VP is numerically equal to the VP hop count from the source to the farthest destination. Specifically,

$$h_C(\mathbf{a}) = \max\{j \mid a_j > 0, \forall j\} \qquad (8)$$

D. Minimum-hop within Half-loop Scheme

Let u and v be the node numbers of the farthest destinations from the source within the two half loops. For a specific $\mathbf{a}$, they are given by,

$$u = \begin{cases} \max\{j \mid a_j > 0, j \le (N-1)/2\} & \text{for } N \text{ odd} \\ \max\{j \mid a_j > 0, j \le N/2\} & \text{for } N \text{ even} \end{cases} \qquad (9)$$

$$v = \begin{cases} \min\{j \mid a_j > 0, j \ge (N+1)/2\} & \text{for } N \text{ odd} \\ \min\{j \mid a_j > 0, j > N/2\} & \text{for } N \text{ even} \end{cases} \qquad (10)$$

With that, the number of links used is simply

$$h_D(\mathbf{a}) = u + (N - v) \qquad (11)$$

E. Unconstrained Minimum-hop Scheme

Under this scheme, a minimum hop route can always be used for any multicast connection request. On a ring, the minimum hop route can use either one VP in the clockwise direction or one VP in the counter-clockwise direction or two VPs fanning out from the source in both directions. Let $w_1, w_2, \ldots w_{k-1}$ be the node numbers of the destinations in ascending order, i.e., $w_1 < w_2 < \cdots w_{k-1}$. After using the minimum hop route for the multicast connection, there will be an idle segment left on the ring. The length y of the idle segment is just the number of links between the two adjacent connection nodes that are farthest apart. Enumerating all such segment lengths and finding the largest one, y is obtain as

$$y = \max \left\{ (N - w_{k-1}), (w_{k-1} - w_{k-2}), \ldots, (w_2 - w_1), w_1 \right\} \qquad (12)$$

The number of links used under the unconstrained minimum hop scheme is therefore

$$h_E(\mathbf{a}) = N - y \qquad (13)$$

7. Numerical Results

Fig.14-17 show the bandwidth demand factors for some network sizes and call sizes under the various multi-drop VP assignment schemes. Fig. 14 is for the call size $k=2$, i.e., for point-to-point call. We can find that: (1) Bandwidth demand factor for Scheme E is significantly smaller than those for Schemes A, B, C and D; (2) Schemes C and D have the same η values; (3) η_C and η_D do not change with the call size.

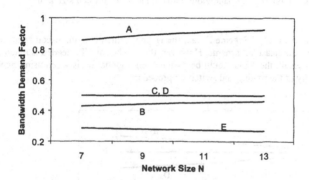

Fig.14. Comparison of bandwidth demand factor η for call size k=2

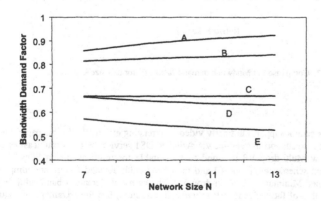

Fig.15. Comparison of bandwidth demand factor η for call size k=4

Fig. 15 shows the results for $k=4$. Here, η_A and η_B increase with the network size while η_D and η_E behave the opposite. Fig.16 and Fig.17 show the results for $k=6$ and 8 respectively. Here, we see that for $k \geq 6$, η_A and η_B become virtually indistinguishable. While η_C is always a constant, η_D and η_E both decrease slowly as N increases.

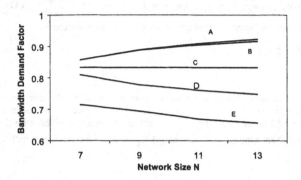

Fig.16. Comparison of bandwidth demand factor η for call size k=6

Note that the bandwidth demand of Scheme E, i.e., the Unconstrained Minimum-hop Multi-drop Scheme is the *same* as that of the multicast VP scheme. However, the number of VPI needed is much smaller than that of the former. As a result, the VPIs could be assigned on a global basis when using Scheme E. This can reduce the complexity of the routing and switching procedure.

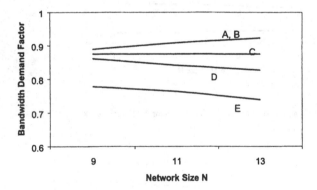

Fig.17. Comparison of bandwidth demand factor η for call size k=8

9. Summary

Current SONET rings cannot support multiparty videoconferencing efficiently. In this paper, we propose to use SONET/ATM rings to support this service via switched DS1 service. Various multicasting methods are discussed and the new Multi-drop VP is found to be suitable for multicasting on SONET/ATM rings. Several VP assignment schemes are proposed and their bandwidth demand factors are compared. Among them, the Unconstrained Minimum-hop Multi-drop VP scheme has the smallest bandwidth demand factor which is identical to that of the multicast VP scheme. It, therefore, has the advantage of requiring much much smaller number of VPs to be set up and is therefore the preferred VP assignment scheme for multiparty video conferencing service on SONET/ATM rings.

Reference

1. Sabri, S. and Prasada, B., "Video conferencing systems," Proc. of IEEE, vol.73, No.4, pp.671-688, Apr.1985.

2. Haruo Noma, Yasuichi Kitamura, et.al, "Multi-point virtual space teleconferencing system," IEICE Trans. Commun., Vol. E78-B, No.7, July 1995.

3. Y. W. Leung, Tak-shing Yum, "Connection optimization for two types of videoconferences," IEE Proc. Commun ., Vol.143, N0.3, June 1996.

4. T. S. Yum, M. S. Chen and Y. W. Leung, "Video bandwidth allocation for multimedia teleconferences," IEEE

   Trans. on Commun., pp. 457-465, Feb 1995.

5. Turletti, T. Huitema and C. Journal, "Videoconferencing on the Internet," IEEE/ACM Transactions on Networking, Jun., 1, 1996, v 4.

6. CCITT Study Group XVIII, Report R34, COM XVIII-R 34-E, June 1990.
7. K.Sato,S. Ohta, and I. Tokizawa, "Broadband ATM Network Architecture Based on Virtual Paths," IEEE Trans Commun., vol. 38, no.8, Aug. 1990.
8. T. H. Wu, et.al, "An Economic Feasibility Study for a Broadband Virtual Path SONET/ATM Self-Healing Ring Architecture," IEEE Journal on Selected Areas in Commun ., Vol.10, No.9, December 1992.
9. CCITT Study Group XVIII, " Draft recommendation I.311, B-ISDN general network aspects," SGXVIII, January, 1990.
10. Atsushi Horikawa, et.al, "A new bandwidth allocation algorithm for ATM ring networks", IEEE GLOBECOM'95, pp404-409.
11. Yoshio Kajiyama, et.al, " An ATM self-healing ring," IEEE J. of Selected Areas in Communications, vol.12, no.1 , pp 171-178,Jan. 1994.
12. Jay J. Lee and Kwi-yung Jung, " An algorithm for determining the feasibility of SONET/ATM rings in broadband networks," IEEE Fourth International Conference on Computer Communication and Networks, 1995, pp356-360.
13. N.Tokura, K. Kikuchi and K. Oguchi, "Fiber-optic subscriber networks and systems development,""Trans. IEIEC, vol.e-74, no.1,1991.
14. ITU-T draft Recommendation Q.931, Q.93B.
15. ITU-T Recommendation H.320, " Narrow-band ISDN Visual Telephone Systems and Terminal Equipment", 1996.
16. D. Saha, D. Kandlur, T. Barzilai, Z.Y. Shae, and M. Willebeek-LeMair, " A videoconferencing testbed in ATM: Design, implementation, and optimazations," in Proc. ICMCS, 1995.
17. Y. C. Chang, Z.Y. Shae, and M. Willebeek-LeMair, "Multiparty videoconferencing using IP multicast," in SPIE Proc.Networking Commun., San Jose, CA,Feb. 1996.
18. T. Turletti, "The INRIA videoconferencing system (IVS)," ConneXions, vol.8, no.10, pp.20-24 ,1994.
19. A. S. Tanenbaum, *Computer Networks*, Prentice-Hall International, Inc., 3rd edtion,1996. pp125-126.
20. Hideki Tode, et.al, "Multicast routing schemes in ATM," International Journal of Commun. Systems. Vol.9,185-196,1996.
21. Hideki Tode, *et.al,* "Multicast routing schemes in ATM", International Journal of Communication Systems, Vol. 9, 185-196 (1996).

A Generic Scheme for the Recording of Interactive Media Streams

Volker Hilt, Martin Mauve, Christoph Kuhmünch, Wolfgang Effelsberg

University of Mannheim, LS PI IV, L 15,16
68131 Mannheim, Germany
{hilt,mauve,kuhmuench,effelsberg}@informatik.uni-mannheim.de

Abstract. Interactive media streams with real-time characteristics, such as those produced by shared whiteboards, distributed Java applets or shared VRML viewers, are rapidly gaining importance. Current solutions to the recording of interactive media streams are limited to one specific application (e.g. one specific shared whiteboard). In this paper we present a generic recording service that enables the recording and playback of this new class of media. To facilitate the generic recording we have defined a profile for the Real-Time Transport Protocol (RTP) that covers common aspects of the interactive media class in analogy to the profile for audio and video. Based on this profile we introduce a generalized recording service that enables the recording and playback of arbitrary interactive media.

1 Introduction

The use of real-time applications in the Internet is increasing quickly. One of the key technologies enabling such transmissions is a transport protocol that meets real-time requirements. The *Real-Time Transport Protocol* (RTP) has been developed for this purpose [16]. The RTP protocol provides a framework covering common aspects of real-time transmission. Each encoding of a specific media type entails tailoring the RTP protocol. This is accomplished by an *RTP profile* which covers common aspects of a media class (e.g. the RTP profile for audio and video [14]) and an *RTP payload* specifying the transmission of a specific type of media encoding (e.g. H.261 video streams).

While the class of audio and video is the most important one and is quite well understood, interactive media streams are used by several applications which are gaining importance rapidly. Interactive applications include shared whiteboard applications [3], multi-user VRML models [9] and distributed Java animations [7]. Many existing protocols for interactive media are proprietary. This prevents interoperability and requires re-implementation of similar functionality for each protocol. For this reason, we have defined an RTP profile [10] that covers common aspects of the distribution of interactive media. It can be instantiated for a specific interactive media encoding.

The RTP profile for audio and video has enabled the development of generic recording services like those described in [4][15]. The RTP audio and video recorders operate independently of a specific video or audio encoding. Instead of decoding incoming RTP packets and storing video and audio content (e.g. in H.261 or MPEG format), they operate on entire RTP packets. This has the major advantage that the mechanisms implemented in the recorder (e.g. media storage or media synchronization during playback) are available for all video and audio formats.

Recent developments extend these RTP recorders to the proprietary protocols of specific applications. In general, interactive media streams require additional functionality in an RTP recorder since certain information about the semantic of an interactive media stream must be considered. In particular *random access* to an interactive media stream requires mechanisms to provide the receivers with the current media state. For example, if a recorded shared whiteboard stream is accessed at a random position, the contents of the page active at that time must be displayed to the user. Thus, a recorder must provide the receiving shared whiteboards with the page content before the actual playback is started.

Our RTP profile for interactive media provides a common framework that enables the development of a generic services like recording or late join for the class of interactive media. In this paper we discuss the principles of such a generic recording service. We present mechanisms that are required for the recording and playback of interactive media streams, and we show that random access to these media streams can be achieved by these mechanisms without having to interpret media-specific data.

The remainder of this paper is structured as follows: Section Two provides an overview over related work. Section Three introduces a classification of different media types. Section Four provides a short overview of our RTP profile for interactive media on which the presented recording scheme is based. Section Five describes the basic architecture of an RTP recording service. Section Six discusses fundamentals of random access to stored interactive media streams, and Section Seven describes two mechanisms that realize media independent random access to these media streams. Section Eight describes the current state of the implementation. Section Nine concludes the paper with a summary and an outlook.

2 Related Work

Much work has been done on the recording of media streams. The rtptools [15] are command-line tools for recording and playback of single RTP audio and video streams. The Interactive Multimedia Jukebox (IMJ) [1] utilizes these tools to set up a video-on-demand server. Clips from the IMJ can be requested via the Web.

The mMOD [13] system is a Java-based media-on-demand server capable of recording and playing back multiple RTP and UDP data streams. Besides RTP audio and video streams, the mMOD system is capable of handling media streams of applications like mWeb, Internet whiteboard wb [6], mDesk and NetworkTextEditor. While mMOD supports the recording and playback of UDP packets, it does not provide a generalized recording service with support for random access.

The MASH infrastructure [11] comprises an RTP recording service called the MASH Archive System [12]. This system is capable of recording RTP audio and video streams as well as media streams produced by the MediaBoard [18]. The MASH Archive System supports random access to the MediaBoard media stream but does not provide a recording service generalized for other interactive media streams.

A different approach is taken by the AOF tools [2]. The AOF recording system does not use RTP packets for storage but converts the recorded data into a special storage format. The AOF recorder grabs audio streams from a hardware device and records the interactive media streams produced by one of two applications AOFwb or the Internet whiteboard wb. Random access as well as fast visual scrolling through the recording are supported but the recordings can only be viewed from a local hard disk or CD. The recording of other interactive media streams is not possible.

In the Interactive Remote Instruction (IRI) system [8] a recorder was implemented that captures various media streams from different IRI applications. In all cases a media stream is recorded by means of a specialized version of the IRI application that is used for live transmission. This specific application performs regular protocol action towards the network but stores the received data instead of displaying it to the user. For example, a specialized version of the video transmission tool is used to record the video stream. Such a specialized recording version must be developed for each IRI tool that is to be recorded.

One of a number of commercial video-on-demand servers is the Real G2 server. The Real G2 server is capable of streaming video and audio data as well as SMIL presentations to RealPlayer G2 clients. A SMIL presentation may contain video and audio as well as other supported media types like RealText, RealPix and graphics. In contrast to the recording of interactive applications, a specialized authoring tool is used to author SMIL presentation, which consist of predefined media streams displayed according to a static schedule.

3 Interactive Media

3.1 Classification of Interactive Media

Before discussing the recording of interactive media streams, it is important to establish a common view on this media class. Basically, we separate media types by means of two criteria. The first criterion distinguishes whether a medium is discrete or continuous. The characteristic of a *discrete medium* is that its state is independent of the passage of time. Examples of discrete media are still images or digital whiteboard presentations. While discrete media may change their state, they do so only in response to external events, such as a user drawing on a digital whiteboard. The state of a *continuous medium*, however, depends on the passage of time and can change without the occurrence of external events. Video and animations belong to the class of continuous media.

The second criterion distinguishes between interactive and non-interactive media. *Non-interactive media* change their state only in response to the passage of time and do not accept external events. Typical representations of non-interactive media are video, audio and images. *Interactive media* are characterized by the fact that their state can be changed by external events such as user interactions. Whiteboard presentations and interactive animations represent interactive media. Figure 1 depicts how the criteria characterize different media types.

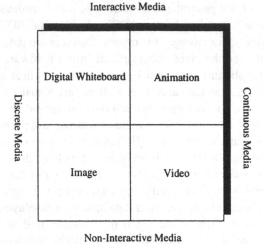

Fig. 1. Examples of Media Types

3.2 Model for Interactive Media

An *interactive medium* is a medium that is well defined by its current state at any point in time. For example, at a given point in time the medium "Java animation" is defined by the internal state of the Java program that is implementing the animation. The *state* of an interactive medium can change for two reasons, either by the passage of time or by *events*. The state of an interactive medium between two successive events is fully deterministic and depends only on the passage of time. Any state change that is not a fully deterministic function of time is caused by an event. A typical example of an event is the interaction of a user with the medium. An example of a state change caused by the passage of time might be the animation of an object moving across the screen.

In cases where a complex state of an interactive medium is transmitted frequently by an application, it is necessary to be able to send only those parts that have changed since the last state transmission. We call a state which contains only the state changes that have occurred since the last transmitted state a *delta state*. A delta state can only be interpreted if the preceding full state and interim delta states are also available. The main advantages of delta states are their smaller size and that they can be calculated faster than full states.

In order to provide for a flexible and scalable handling of state information, it is sometimes desirable to partition an interactive medium into several *sub-components*. In addition to breaking down a large media state into more manageable parts, such partitioning allows participants of a session to track only the states of those sub-components they are actually interested in. Examples of sub-components are VRML objects (a house, a car, a room), or the pages of a whiteboard presentation.

To display a non-interactive media stream like video or audio, a receiver needs to have an adequate *player* for a specific encoding of the medium. If such a player is present in a system, every media stream that employs this encoding can be processed. This is not true for interactive media streams. For example, to process the media stream that is produced by a shared VRML browser, it is not sufficient for a receiver to have a VRML browser. The receiver will also need the VRML world on which the sender acts; otherwise the media stream cannot be interpreted by the receiver. But even if the receiver has loaded the correct world into its browser, the VRML world may be in a state completely different from that of the sender. Therefore, the receiver must *synchronize* the state of the local representation of the interactive medium to the state of the sender before it will be able to interpret the VRML media stream correctly.

Generally speaking, it does not suffice to have a player for an interactive media type. Additionally, the player must be initialized with the *context* of a media stream before that stream can actually be played. The context is comprised of two components: (1) the environment of a medium and (2) the current state of the medium. The *environment* represents the static description of an interactive medium that must initially be loaded into the media player. Examples of environments are VRML worlds or the code of Java animations. The *state* is the dynamic part of the context. The environment within a player must be initialized with the current state of the interactive medium before the stream can be played. During transmission of the stream, both sender and receiver must stay synchronized since each event refers to a well-defined state of the medium and cannot be processed if the medium is in a different state.

4 RTP Profile for Interactive Media

In order to be able to develop generic services which base solely on our RTP profile for interactive media, common aspects of interactive media streams which are not already handled by RTP must be supported by the profile. These aspects can be separated into two groups: information and mechanisms. Information is needed, so that a generic service can analyze the semantics of the application level communication. The information provided by the RTP profile is: identification of application-layer packet content, identification of sub-components and sequence numbers. Mechanisms are needed by a generic service to take appropriate actions on the medium. The mechanisms provided within the RTP profile are: announcement of sub-components, requesting state transmissions and mapping of sub-component IDs to application-level names.

The remainder of this section discusses basic concepts of the RTP profile for interactive from the view of a generic recording service. A detailed description of the profile can be found in [10].

4.1 Structure of Data Packets

The model presented in Section 3.2 illustrates that states, delta states and events of an interactive medium must be transmitted in real-time. We define the structure of data packets containing theses basic elements of interactive media as depicted in Figure 2 within our RTP profile; for the general structure of RTP packets see [16]. The most important fields of these packets are type, sub-component ID and data. The type field is needed to distinguish the different packet types state, delta state and event defined in the profile. This is especially important for the recording service, which must be able to identify the type of content transported in an RTP packet without having to interpret the data part of the packet. In state and delta state packets the sub-component ID field holds the sub-component ID of the state included in the data part of the packet. In event packets this field identifies the sub-component containing the "target" of an event. The data field of the packet contains the definition of states, delta states or events specific to the payload type.

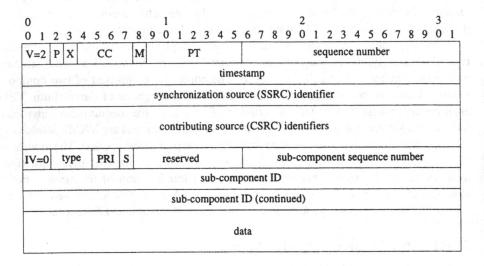

Fig. 2. RTP Packet Structure for States, Delta States and Events

Since setting the state of a sub-component can be costly and might not always be reasonable, state and delta state packets contain a priority (PRI) field. This priority can be used by the sender of the state to signal its importance. A packet with high priority should be examined and applied by all communication peers which are interested in the specific sub-component. Situations where high priority is recommended are resynchronization after errors or packet loss. Basically a state transmission with high priority forces every participant to discard its information about the sub-component and requires the adoption of the new state. A state transmitted with low priority can be ignored at will by any participant. This is useful if only a subset of communication partners is interested in the state. An example of this case is a recorder that periodically requests the media state in order to insert it into the recording.

4.2 Announcement of Sub-Components

For the implementation of an efficient recording service it is important that the sub-components present in a session are known. Furthermore it should be possible to distinguish those sub-components which are currently needed to display the medium. Those sub-components are called *active*. An example for active sub-components are the currently visible pages of a shared whiteboard. All remaining sub-components are *passive* (e.g. those shared whiteboard pages which are currently not visible for any user). Declaring a sub-component active does not grant permission to modify anything within that sub-component. However, a sub-component must be activated before a session participant is allowed to modify (send events into) the sub-component. The knowledge about active sub-components in a session allows a recording service to transmit only those sub-components during a playback that are actually visible in the receivers.

The profile provides a standardized way to announce the sub-components of any application participating in an interactive media session and allows to mark sub-components as active. Active and passive sub-components are announced by selected participants in regular intervals within RTCP reports.

4.3 Requesting State Transmissions

In many cases it is reasonable to let the receivers decide when the state of sub-components should be transmitted. Thus, a receiver must be able to request the state from other participants in the session.

As the computation of state information may be costly, the sender must be able to distinguish between different types of requests. Recovery after an error urgently requires information on the sub-component state since the requesting party cannot proceed without it. The state is needed by the receiver to resynchronize with the ongoing transmission. These requests will be relatively rare. In contrast, a recording service needs the media states to enable random access to the recorded media. It does not urgently need the state but will issue requests frequently. For this reason, the state request mechanism supports different priorities through the priority (PRI) field in the state query packet. Senders should satisfy requests with high priority (e.g. for late joiners) very quickly, even if this has a negative impact on the presentation quality for the local user. Requests with low priority can be delayed or even ignored, e.g. if the sender currently has no resources to satisfy them. The sender must be aware that the quality of the service offered by the requesting application will decrease if requests are ignored.

5 RTP Recording Service

An RTP recording service such as the MBone VCR on Demand (MVoD) [4] usually handles two network sessions (see Figure 3). In the first, the recorder participates in the multicast transmission of the RTP media data. Depending on its mode of operation (recording or playback), it acts as a receiver or sender towards the other participants of the session. A second network session can be used to control the recorder from a remote client, e.g. using the RTSP [17] protocol. During the recording of an RTP session, the recorder receives RTP data packets and writes them to a storage device. Packets from different media streams are stored separately. When playing back, the recorder successively loads RTP packets of each media stream and computes the time at which each packet must be sent using the time stamps of the RTP headers. The recorder sends the packets according to the computed schedule. A detailed description of the synchronization mechanism implemented in the MVoD can be found in [5].

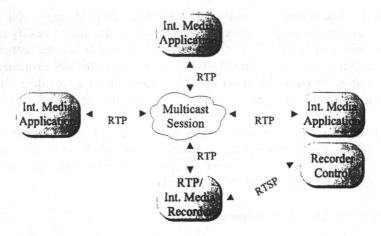

Fig. 3. Scenario for the Recording of an RTP Session

6 Random Access

In contrast to the traditional media types where random access to any position within a stream is possible, interactive media streams do not allow easy random access without *restoring the context* of the stream at the desired access position. For example, jumping directly to annotations on a whiteboard page only makes sense if the right page is shown on the screen. To restore the context of a recorded stream in a receiver, two operations have to be performed: First, the environment has to be loaded into the receiver. The environment can be provided by the recording service or by a third party, e.g. an HTTP server. Then the receiver must get the state of the interactive medium at the desired access position within the recorded stream. Let us come back to our whiteboard example. If we want to jump to minute 17 of a recorded teleconferencing session we must be able to show the contents of the page active at that time, together with the annotations made by the speaker. If we did not restore the state of the whiteboard, the page (which might have been loaded originally at minute 12) would not be visible.

6.1 Recovering the Media State

The state of an interactive medium can be recovered from a recorded media stream. Note that the generic recorder is not able to interpret the media-specific part of the RTP packets and thus cannot directly compute the media state and send it to the receivers. But the recorder may re-send existing RTP packets that are stored within the recorded media stream. Thus, it is our goal to compose a sequence of recorded RTP packets containing states and events that put a receiver into the desired state. The task a recorder has to accomplish before starting a playback is to determine the appropriate sequence of recorded packets.

In an interactive media application the current state is determined by an initial state and a sequence of events applied to that state. In a discrete interactive medium the event sequence is not bound to specific points in time. Thus, the application of an event sequence to an initial state of a discrete interactive medium will always result in the same media state, independent of the speed at which the sequence is applied. In contrast, the event sequence for a continuous interactive medium is bound to specific points in time. A sequence of events that is applied to the state of a continuous interactive medium will leave the system in the correct state only if each event is applied at a well-defined instant in time.

This main difference between discrete and continuous interactive media must be considered when computing the sequence of event and state packets to recover the media state. In the case of a discrete medium, such a sequence can be computed to recover the media state at any point in a recorded stream. In contrast, the media state of a continuous medium can only be recovered at points within a recording where a state is available; events cannot be used for state recovery because they must be played in real-time. Therefore, random access to an interactive continuous media stream will usually result in a position near the desired access point. The more often the state is stored within a stream, the finer is the granularity at which the stream of a continuous interactive medium can be accessed.

Interactive media applications usually send the media state only upon request by another application. Thus, the recorder must request the state at periodic intervals. The requests use a low priority because a delayed or missing response reduces the access granularity of the stream, which can be tolerated to some degree.

7 Mechanisms for Playback

The mechanisms presented in this section implement the recovery of the media state from recorded media streams. Both mechanisms can be implemented completely in the recorder. The receiving applications need not recognize the recorder as a specific sender, nor does the recorder need to interpret media-specific data. All applications that use a payload based on the RTP profile for interactive media can be recorded, and will be able to receive data from the recorder.

7.1 The Basic Mechanism

This simple mechanism is able to recover the media state from interactive media streams which do not utilize multiple sub-components. When starting playback of such a stream, the best case is if the state is contained in the recorded stream at exactly the position at which the playback is to start. Then playback can begin immediately. But in general, the playback will be requested at a position where no full state is directly available in the stream.

Let us consider, for example, a recorded media stream that consists of the sequence S_0 containing a state, three successive delta (Δ) states and several events (see Figure 4). If a user wants to start playback at position t_p from the recording, the state at t_p must be reconstructed by the recorder. A continuous interactive medium does not allow direct access to t_p because the recorder cannot determine the state at t_p since there is no state available at t_p in the recorded stream. However, access to position $t_{\Delta 3}$ within the stream is feasible, because $t_{\Delta3}$ is the location of a delta state. The complete media state at $t_{\Delta3}$ can be reconstructed from the state located at position t_s and the subsequent delta states until position $t_{\Delta3}$, which is the position of the last delta state before t_p. The events between t_s and $t_{\Delta3}$ can be ignored, because all modifications to the state at t_s are reflected in the delta states. The packets that contain states can be sent at the maximum speed at which the recorder is able to send packets. If required by the medium, the internal media clock is part of the media state. Thus, after applying a state, the media clock of a receiver will reflect the time contained in the state. When the recorder finally reaches $t_{\Delta3}$ (and has sent $\Delta3$), fast playback must be stopped, and playback at regular speed must be started. The start of the regular playback may not be delayed because events must be sent in real-time relative to the last state. This is important since for continuous interactive media the events are only valid for a specific state that may change with the passage of time. Altogether, the recorder will play back sequence S_1 shown in Figure 4.

For discrete interactive media, fast playback of events is possible. Therefore, random access to position t_p can be achieved by also sending the events between $t_{\Delta3}$ and t_p at full speed. The resulting sequence S_2 is also shown in Figure 4.

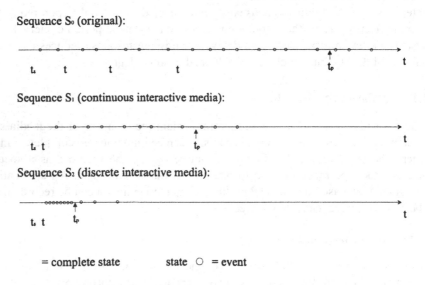

Sequence S_0 (original):

Sequence S_1 (continuous interactive media):

Sequence S_2 (discrete interactive media):

= complete state   state ○ = event

Fig. 4. Playback of a Recorded Sequence of States, Delta States and Events

7.2 Mechanism with Support for Sub-components

In a more sophisticated mechanism the existence of sub-components can be exploited to reduce the amount of data required for the recovery of the media state. Using sub-components, the state of an interactive medium can be recovered selectively by considering only those sub-components which are actually required to display the medium in its current state.

Let us take a closer look at the shared whiteboard example of Section 6 where we wanted to access minute 17 of the recording of a teleteaching session. Without the use of sub-components, the recorder would have to recover the complete media state valid at minute 17, which comprises all pages displayed so far. But if the shared whiteboard has divided its media state into several sub-components (e.g. a whiteboard page per sub-component) the recorder is able to determine the sub-components that are active at minute 17 and may recover them selectively.

In general, when a recorded stream is accessed, the set of active sub-components at the access position can be determined and their state can be recovered. This is suffi- cient to display an interactive medium at the access position. However, it must be assured, that a receiver is enabled to display all subsequent data in the recorded stream. If the subsequent data contains the re-activation of a passive sub-component (e.g. a jump to a previous page in the recording of a shared whiteboard session), a receiver would not hold the state for this sub-component as passive sub-components were not recovered initially. Consequently, the receivers would not be able to decode data refer- ring to that sub-component. Thus, the recorder must assure that the state of a sub-com- ponent is present in the recorded stream at any position where a passive sub- component is re-activated. This can be accomplished at the time of recording if the recorder requests a state transmission for each sub-component that gets activated and inserts the retrieved state into the recording. For discrete media streams this scheme can be optimized by not requesting the state of a sub-component if the recorder can reconstruct it from previous data within the recorded stream.

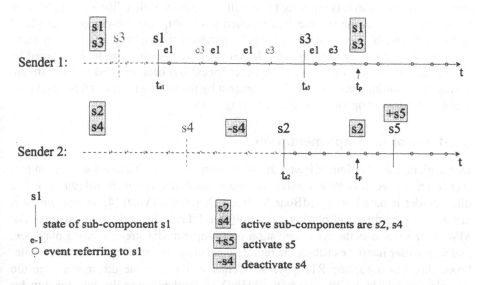

Fig. 5. Playback of a recording containing sub-components. Greyed states and events of the recorded streams are filtered during the recovery of the state at t_p

The example shown in Figure 5 depicts recorded streams of a continuous interac- tive medium with two senders. Sender 1 operated on sub-components 1 and 3, whereas the recorded stream of sender 2 contains packets for sub-component 2 and 4 and later for 2 and 5. If these recorded streams are accessed at position t_p, the recorder has to compute the list of sub-components which are active at t_p. In our case these are s1, s2 and s3. For each of these sub-components, the position of the most recent sub-component state before t_p must be located in the recorded stream. As a result, the recorder gets the positions of sub-component states t_{s1}, t_{s2}, t_{s3}. (For the sake of simplicity, we have only considered states; support for Δ states can be achieved similar to the basic mechanism.) s1 is the sub-component whose state is farthest from t_p (here $t_{s1} < t_{s2} < t_{s3}$). Thus the recorder has to start playback at position t_{s1} and recovers the state s1. The recorder must continue with the playback because events referring to s1 are located between t_{s1} and t_p. Notice that we are considering a continuous interactive medium where all events must be played in real time. During the playback of the stream between t_{s1} and t_p two problems may occur: At first, events may be located in the stream which refer to states that have not yet been sent. The sending of these events must be suppressed because a receiver can not interpret them correctly. In our example, events concerning s3 and s4 are filtered out. Secondly, there may be sub-component states in the stream that are not in the set of active sub-components at t_p (s4 in our example) and thus are not needed for playback at t_p. Therefore the state of s4 (and all events referring to s4) must also be filtered out.

Summing up, the recorder will start playback at position t_{s1}, sending the state of sub-component s1 and events referring to s1. All other states and events will be filtered out. The next required state is s2, which will be sent as soon as it shows up and, after that, all subsequent events referring to s2 will also pass the filter. The same holds true for s3. Finally, once the recorder has reached position t_p, the sub-components s1, s2 and s3 will have been recovered, and regular playback without any filtering may start. After the start of the regular playback, the set of active sub-components is enlarged by s5. As the state of a newly activated sub-component has been inserted into the stream during the recording, the state of s5 can be sent by the recorder. Thus, all receivers are enabled to interpret upcoming events referring to s5.

8 Status of the Implementation

Our work on the recording of interactive media streams initially started with the implementation of a recorder for a shared whiteboard, the digital lecture board (dlb) [3]. The dlb recorder is based on the MBone VCR on Demand (MVoD) [4] service which is capable of recording and playing back multiple RTP audio and video streams. The MVoD server assures the synchronization of multiple media streams during playback, and a Java user interface enables the remote control of the recorder. The shared whiteboard dlb uses a specific RTP payload format to transmit the dlb media. The dlb recorder module basically extends the MVoD by implementing the functionality for random access to recorded dlb media streams. To achieve random access, the dlb recorder uses a mechanism that is specific to the dlb media stream.

Based on the experiences from the implementation of the dlb recorder, we are currently implementing the recording service for interactive media described in this paper and we are now finishing a very early alpha version. Like the dlb recorder, the interactive media recorder is realized as an extension to the MVoD. Thus, the existing algorithms for the synchronization of multiple media streams as well as the recording facilities for audio and video streams can be reused. The implementation of the mechanisms for random access allows the presence of discrete and continuous interactive media streams as well as audio and video streams within the same recording.

9 Conclusion

We have presented a generic recording service for the class of interactive media with real-time characteristics. Examples of such media are shared whiteboards, multi-user VRML worlds and distributed Java applications. In analogy to RTP video and audio recorders, we have developed a generic recording service that is based on an RTP profile for the interactive media class. The profile covers the common aspects of this media class. We have presented the basic ideas of the RTP profile, pointing out the features that enable the recording and playback of interactive media regardless of a specific media encoding.

We have described the key concepts of the generic recording service. An important aspect of this recording service is that it enables random access to recorded streams. The media context of a recording is restored before playback is started. We have showed that the context of a medium can be recovered relying only on the RTP profile and we have presented two recovery mechanisms. We are currently finishing the implementation of a first prototype of the described interactive media recording service.

In future work we will implement the RTP payload-type specific functionality for distributed Java animations and multi-user VRML. Our recording service will then be tested and validated with those media types. Furthermore, we are working on a second generic service that will implement a late join algorithm. During the implementation and testing of the RTP profile, the payload types and the generic services, we expect to get enough feedback for a full specification of the profile and the payload types. We intend to publish those specifications as Internet drafts.

Acknowledgments. This work is partially supported by the BMBF (Bundesministerium für Forschung und Technologie) with the "V3D2 Digital Library Initiative" and by the Siemens Telecollaboration Center, Saarbrücken.

References

[1]   K. Almeroth, M. Ammar. *The Interactive Multimedia Jukebox (IMJ): A New Paradigm for the On-Demand Delivery of Audio/Video*. In: Proc. Seventh International World Wide Web Conference, Brisbane, Australia, April 1998.

[2]   C. Bacher, R. Müller, T. Ottmann, M. Will. *Authoring on the Fly. A new way of integrating telepresentation and courseware production*. In: Proc. ICCE '97, Kuching, Sarawak, Malaysia, 1997.

[3]   W. Geyer, W. Effelsberg. *The Digital Lecture Board - A Teaching and Learning Tool for Remote Instruction in Higher Education*. In: Proc. ED-MEDIA '98,

Freiburg, Germany, AACE, June 1998. Available on CD-ROM, contact: http://www.aace.org/pubs/.

[4]   W. Holfelder. *Interactive Remote Recording and Playback of Multicast Videoconferences*. In: Proc. IDMS '97, Darmstadt, pp. 450-463, LNCS 1309, Springer Verlag, Berlin, September 1997.

[5]   W. Holfelder. *Aufzeichnung und Wiedergabe von Internet-Videokonferenzen*. Ph.D. Thesis (in German), LS Praktische Informatik IV, University of Mannheim, Shaker-Verlag, Aachen, Germany, 1998.

[6]   V. Jacobson. *A Portable, Public Domain Network 'Whiteboard'*, Xerox PARC, Viewgraps, April, 1992.

[7]   C. Kuhmünch, T. Fuhrmann, and G. Schöppe. *Java Teachware - The Java Remote Control Tool and its Applications*. In: Proc. of ED-MEDIA '98, Freiburg, Germany, AACE, June 1998. Available on CD-ROM, contact: http://www.aace.org/pubs/.

[8]   K. Maly, C. M. Overstreet, A. González, M. Denbar, R. Cutaran, N. Karunaratne. *Automated Content Synthesis for Interactive Remote Instruction*, In: Proc. of ED-MEDIA '98, Freiburg, Germany, AACE, June 1998. Available on CD-ROM, contact: http://www.aace.org/pubs/.

[9]   M. Mauve. *Transparent Access to and Encoding of VRML State Information*. In: Proc. of VRML '99, Paderborn, Germany, pp. 29-38, 1999.

[10] M. Mauve, V. Hilt, C. Kuhmünch, W. Effelsberg. *A General Framework and Communication Protocol for the Transmission of Interactive Media with Real-Time Characteristics*, In: Proc. of IEEE ICMCS'99, Florence, Italy, 1999.

[11] S. McCanne, et. al. *Toward a Common Infrastructure for Multimedia-Networking Middleware*, In: Proc. of NOSSDAV '97, St. Louis, Missouri, 1997.

[12] S. McCanne, R. Katz, E. Brewer et. al. *MASH Archive System*. On-line: http://mash.CS.Berkeley.edu/mash/overview.html, 1998.

[13] P. Parnes, K. Synnes, D. Schefström. *mMOD: the multicast Media-on-Demand system*. 1997. On-line: http://mates.cdt.luth.se/software/mMOD/paper/mMOD.ps, 1997.

[14] H. Schulzrinne. *RTP Profile for Audio and Video Conferences with Minimal Control*, Internet Draft, Audio/Video Transport Working Group, IETF, draft-ietf-avt-profile-new-05.txt, March 1999.

[15] H. Schulzrinne. *RTP Tools*. Software available on-line, http://www.cs.columbia.edu/~hgs/rtp/rtptools/, 1996.

[16] H. Schulzrinne, S. Casner, R. Frederick, V. Jacobson. *RTP: A Transport Protocol for Real-Time Applications*. Internet Draft, Audio/Video Transport Working Group, IETF, draft-ietf-avt-rtp-new-03.txt, March, 1999.

[17] H. Schulzrinne, A. Rao, R. Lanphier. *Real Time Streaming Protocol (RTSP)*. Request for Comments 2326, Multiparty Multimedia Session Control Working Group, IETF, April 1998.

[18] T. Tung. *MediaBoard: A Shared Whiteboard Application for the MBone*. Master's Thesis, Computer Science Division (EECS), University of California, Berkeley, 1998. On-line: http://www-mash.cs.berkeley.edu/dist/mash/papers/tecklee-masters.ps.

A Framework for High Quality/Low Cost Conferencing Systems

Mirko Benz[1], Robert Hess[1], Tino Hutschenreuther[1], Sascha Kümmel[1], and Alexander Schill[1]

[1] Department of Computer Science, Dresden University of Technology,
D-01062 Dresden, Germany
{benz, hess, tino, kuemmel, schill}@ibdr.inf.tu-dresden.de

Abstract. This paper presents a framework for the development of advanced video conferencing systems with very high quality. The design and performance of the developed components like session management, real-time scheduler and a specific transport system are outlined. The core of the framework consists of a toolkit for the processing of audio and video data. It integrates capturing, compression, transmission and presentation of media streams into an object oriented class library. The specifics of audio/video hardware, preferred operation modes, compression algorithms and transport protocol properties are encapsulated. Due to the achieved abstraction level complex application software can easily be developed resulting in very short design cycles. Contrary to existing systems the developed application is regarded in its entirety. The focus on highest possible performance throughout all components is unique to our approach. Due to a number of optimisations, specific compression hardware is not required which leads to low overall system costs.

1   Introduction

Current conferencing systems based on pure software compression techniques for video streams often exhibit a lack of presentation quality that is not acceptable for business communication. The common approach to tackle these problems are dedicated hardware solutions. However there are a number of drawbacks. First usually only one compression algorithm is implemented. This often collides with the requirement of concurrent support for ISDN- (H.261) and IP-based (H.263) conferencing. Another aspect is that it is difficult to keep pace with the ongoing development as to be seen in the H.263+ area. Furthermore compression hardware contributes to additional, relatively high costs and has to be installed on each computer which makes this strategy scale poorly.

As a consequence, compression algorithms and methods were improved to make software compression feasible. However many research projects concentrate on the optimisation of this process alone. Furthermore, often required portability restricts the use of processor extensions which can lead to impressive gains. In contrast, our approach targets standard desktop hardware and operating systems within a corporate networking environment. The conferencing system is designed in an integrated manner. This means that optimisations are not restricted to several independent components like the compression or transmission of video streams. In contrast, we use layer overlapping techniques and integration of several processing steps. This leads to highly efficient cache usage and reduces memory accesses. Moreover, INTEL MMX processor extensions were used to speed up several computations.

The installation of modern high-speed communication systems opened the way to a novel distributed compression approach as introduced in [10]. In this scenario dedicated servers can be installed for the demanding compression effort while desktop systems perform only a light weight preparation of the video processing. The flexibility of the whole system is usually not influenced by exploiting compression servers since a modification of the compression process only requires a replacement of software components at the client side. Hence scalability and moderate costs can be achieved.

More important than improved presentation quality and low costs are security issues for the business application of conferencing systems. In this environment encryption of voice data is often required. Furthermore, the misuse of application sharing must be made impossible. Due to the achieved performance gains, this is now possible without reducing video quality. On the other hand conferencing systems should support more than two participants and integrate a quality of service management for real-time media streams as designed and implemented here.

All these goals were considered within the design and implementation of the framework. It consists of session management, application sharing and a specific transport system. Furthermore, it integrates a toolkit for the processing of audio and video data. Capturing, compression, transmission and presentation of media streams are supported via an object oriented class library. The specifics of audio/video hardware, preferred operation modes (e.g. colour models), compression algorithms and transport protocol properties are encapsulated. Resources for object-chains for real-time processing of media streams can be reserved by an integrated scheduler. As a result, high performance conferencing systems can be developed in a relatively short time without requiring thorough knowledge of implementation details like the used compression techniques or multimedia hardware specifics.

The next section outlines our current conferencing system architecture. It includes the following components: session management, application sharing, transport system, soft real-time scheduling and VTToolkit – an object-oriented class library for multimedia data handling. Chapter 3 presents more details concerning the core functionality of VTToolkit. Furthermore, it highlights the general architecture, object structures and interfaces. The components for the audio/video manipulation will be discussed as well. Afterwards, the applied optimisation methods and the achieved performance results will be discussed.

2 Conferencing System Architecture

In different research projects we developed special architectures, mechanisms and protocols that are integrated to form a framework for the development of high performance/low cost conferencing systems (figure 1). For example, compression algorithms were designed and optimised by our group as outlined in [9]. Furthermore, specific transport protocols as well as quality of service management [18] and scheduling mechanisms were developed to support real-time multimedia applications.

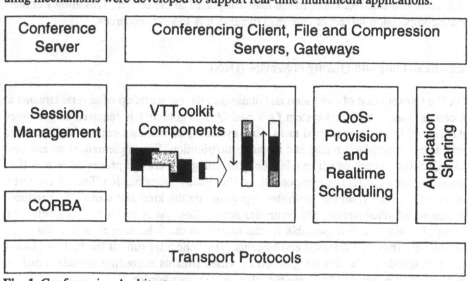

Fig. 1. Conferencing Architecture

Within all these areas the achievement of the highest possible performance or presentation quality is of key interest. Hence, the work is focused on high speed networks (ATM, Fast/Gigabit Ethernet, Myrinet), modern standard platforms (X86/MMX) and operating systems (Windows NT). The support of different hardware architectures and operating systems is not of primary concern.

The core of the architecture is VTToolkit, an object-oriented class library. It provides components that handle specific tasks within multimedia applications like capturing, compression and transmission of audio and video streams. At runtime the components are arranged to processing chains that are controlled by the real-time scheduler.

Due to the structuring into components, it is fairly easy to implement complex audio and video transmission applications without prior knowledge of the internal design, device specifics or transport protocols. Application level software like conferencing clients, compression servers, transport gateways and video on demand servers have been implemented based on these framework. Currently it is used as a high quality conferencing system in teleworking and teleteaching projects [4], [19].

The following sections outline the functionality and design of several components and provide performance details where appropriate.

2.1 Session Management

The main tasks of the Session Management component are the registration and administration of resources, participants and services like gateways or compression servers. Furthermore, the establishment of conferences and media streams is performed.

During start-up of a conference, the client registers system properties. If a session gets active the conference server connects the clients according to their capabilities and requirements. During the lifetime of a connection the current bandwidth and CPU utilisation is measured. Furthermore loss is reported to the sender. This allows to integrate compression servers or gateways dynamically to save resources.

2.2 Real-Time and Quality of Service (QoS)

For the transmission of live video and audio and the presentation of several streams at a certain quality both end system QoS and QoS negotiation is required to guarantee the execution of the requested tasks. End system QoS includes control algorithms to schedule competitive streams and resource distribution. We implemented an end system QoS framework which enables the scheduling of periodic processes in real time under Windows NT. The components of the Video Transmission Toolkit request a certain CPU-time from the controller depending on the kind and size of the produced or consumed video stream. The controller accumulates the requests and forwards them to the Negotiator. It is responsible for the request at the Scheduler as well as for negotiation with the other involved end systems. The Scheduler runs at the highest priority level. It schedules the threads grouped by video streams according to their actual requirements concerning CPU-time bounded by an upper border.

The scheduling is based on initial reservation by the thread and subsequent monitoring of actual resource consumption. The displacement of the other application processes is realised by temporarily increasing the thread priority. So this thread can only be disturbed by threads with the same priority or system activities like DMA or interrupts. As a results an accuracy of 1 ms can be achieved. The time slice given to the thread in this case is 2 ms. The thread can apply for a part of the whole CPU-time and the scheduler provides this as long as resources are available. It reserves always a user determined amount of time for system processes and conventional applications. If sufficient resources are not available for the requesting process the scheduler offers the remaining resources to this application thread and it decides whether to reduce its request or to refuse the connection setup at all.

2.3 Transport Protocol Development

Due to the nature of traffic sources within the conferencing system classic transport protocols like UDP and TCP are not fully adequate[1]. This stems from the variety of

---

[1] Approaches like RTP [16] and RSVP [17] rectify some deficiencies. However they provide additional layers and complexity that might degrade performance. For interoperability with existing systems however, support is currently integrated in the VTToolkit transport objects.

traffic characteristics of audio/video transmissions, application sharing or file exchange and the resulting performance and quality of service requirements. This includes request/response as well as stream-based communication scenarios with constant and variable bit rates respectively. Furthermore, security, real-time and multicast aspects have to be considered. Hence, the protocol mechanisms like multiplexing, segmentation, flow/congestion control or retransmission have to be configurable to match the application requirements and to exploit the network features in order to avoid duplicate and thus redundant processing.

In our research group the Configurable Connection-oriented Transport Protocol (CCTP) was developed that addresses the above mentioned aspects. The implementation is based on the connection oriented media (e.g. ATM) support within Windows NT 5 and provides a Windows sockets compatible interface. To achieve higher performance widely appreciated mechanisms like the reduction of copy operations, context changes and interrupt combination as well as architectural and implementation concepts like application layer framing (ALF) [6] and integrated layer processing (ILP) [5] were applied. Combined with each other these optimisations lead to higher network communication performance. However, more important than the pure performance advantage is the drastic reduction of CPU utilisation. The remaining resources can be used for other applications or participants, higher compression ratio or better video quality. Further information regarding the developed protocol design concepts as well as implementation details and performance results can be found in [2], [12] and [18]. In these publications relations to comparable approaches are discussed as well.

3 Design and Implementation of VTToolkit

In the previous section we presented our general conferencing system architecture and outlined several components. Next we will state design goals that were fundamental for the development of VTToolkit (Video Transmission Toolkit) - the multimedia processing library that is the core of our conferencing system. Afterwards, the basic architecture of VTToolkit as well as contained components, interfaces and object structures will be described. Then we will discuss used optimisation techniques and achieved results concerning the improved compression process. Based on that presentation we will compare our approach with related work.

3.1 Design Goals

Video frame resolutions of at least 352x288 (CIF) at 25 frames/s for every participant in a conference of three or more is regarded as high quality. Furthermore, these requirements have to be achieved with the lowest possible financial expenditure. This usually means without additional specialised hardware.

Investigations of different commercial [15], [14] as well as freely available platforms [13], [20] have shown that the mentioned requirements can not be fulfilled or only with high technical and financial effort. This result triggered the decision to design and develop an audio/video transmission toolkit (VTToolkit) as a foundation for advanced video conferencing applications. To bring the requirements of high quality combined with relatively low costs together, a performance oriented design from the beginning was necessary. A further objective was to address as many problem areas of a video transmission as possible in an integrated manner.

This includes:

- modular, freely configurable, object-oriented structure,
- support of heterogeneous environments concerning end systems and networks respectively,
- transparent support of different grabber and display hardware while concurrently exploiting offered video accelerator features,
- adaptive, optimised and standard-conform video compression codecs for different platforms → H.26x, MJPEG and MPEG for x86/MMX and Alpha,
- „real-time" aware transport system for a broad spectrum of networks and protocols → IP/RTP, H.320 and native transport over N-ISDN, (Fast, Gigabit)Ethernet, ATM and wireless LANs,
- configurable and integrated control component and
- enforcement of soft real-time with standard operating systems on end systems → Windows NT.

The development process quickly revealed numerous technical influences that arise due to the number of assembly opportunities of single components on end systems and networks. As a consequence, it can hardly be handled without a systematic design support. This lead to the parallel development of tools that account for the simplification of development and evaluation of VTToolkit's components as well as support for application development based on VTToolkit.

3.2 Basic Architecture of VTToolkit

Two main aspects were important for the design and development of VTToolkit. First, we concentrated on the process of video transmission and the involved components. On the other hand, we wanted to make the interfaces and structure of the objects as simple and uniform as possible to allow an abstraction of the specific operation and interactions. Therefore, the objects that make up a desired application are regarded as a streaming chain of objects as shown in figure 2.

All streaming objects but the Splitter Object possess exactly one input and output channel. Depending on the position of an object in a chain terminating objects (local sources and sinks) and intermediate objects are differentiated. Terminating objects may be input components like a Video Grabber. They are located at the beginning of a streaming chain. Hence, they do not have an accompanying logical input stream but a physical video signal input. A Streaming Object consumes an input stream, applies a processing according to its functionality and configuration (transformation function) and finally produces an output stream. The interactions of the single objects are regarded as a consumer-producer-model.

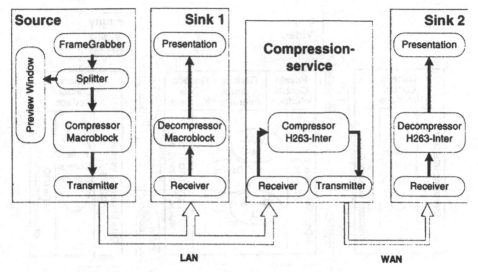

Fig. 2. Sample Configuration of a Streaming Chain

All VTToolkit objects are derived from the virtual class *CConsumerProducer*. It defines the standard interface that every object has to implement (figure 3). These methods can be grouped into three sections:

- Data Streaming Interface: This interface is used to exchange video data. Buffers can be transferred via a push or pull mechanism that can be different for the IN and OUT interface. By use of Buffer Objects components are interconnected and the specific transfer semantics is hidden. All required configuration is performed dynamically at run time.

- Generic Control Interface: Methods to control video streams like start, stop or pause are collected here. Furthermore, the configuration of IN and OUT interfaces can be performed.

- QoS Control Interface: To control and adjust runtime behaviour generic methods are defined here. They enable to get current performance data as well as the negotiation and restriction of resource usage.

Besides these interfaces every object might define a *Private Control Interface* to configure a specific parameter set like quality, frames per second or bandwidth on the fly.

Streaming Objects encapsulate one or more devices. These may be real hardware devices such as video grabber boards but also software „devices" like different compression codecs. The devices that are supported by an object and associated input/output formats may be inquired and configured via the *Generic Control Interface*.

A major optimisation approach is the avoidance of copy operations [7]. VTToolkit does not dictate whether producer, consumer or none of them prepares a certain buffer. Hence a buffer transfer technique was introduced which allows all possible

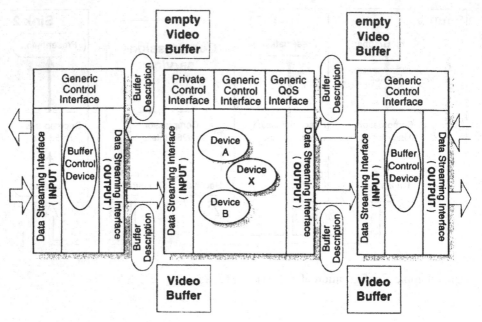

Fig. 3. Object Structure and Interactions

combinations and makes wrong usage almost impossible due to its in-band signalling. Another aspect in the design of the buffer transfer strategy was to find a compromise between optimisation and simple implementation. This was accomplished by delivering video buffers *by reference* amongst interfaces. All other information characterising the stream and the specific buffer are handed over *by value*. This avoids copy operations of large amounts of data and, on the other hand, significantly simplifies management of control information.

The described simple but well defined base architecture allowed to accomplish many design goals:

- All objects are simple to model since they basically represent a *Streaming Object* and all exchanged information covers a single video frame as a common processing unit.
- Due to the common interfaces all objects may be combined in almost any way, since their sequence is not determined in advance.
- Base objects can be automatically arranged and configured without adapting the specific control objects for every newly designed component.
- Reuse of single objects as well as of the common object structure is possible and already practised. This eases the development of new objects and additional devices.
- The design mandates the integration of a *QoS Control Interface*. Hence differentiated performance monitoring is supported.

- A number of optimisation techniques have been applied. However, by using adaptation layers like the *Buffer Objects*, a simple and clear implementation is facilitated.

4 Applied Optimisation Methods and Performance Results

This section discusses design decisions, measurements, performance tuning and applied optimisations. Since the resulting efficiency of the complete application is of primary concern layer integration as well as necessary compromises will be outlined. Furthermore, our performance oriented system design approach is reflected while concentrating on the video capture, compression and presentation components of VTToolkit.

4.1 Framegrabber and Videoview

Usually, software compression is considered quite expensive in terms of required CPU-capacity. On closer examination it becomes evident, that the processes of grabbing the frames and displaying them, that are involved in every live transmission of video, are quite time consuming in itself. Thus, closer attention has to be paid to the optimisation of these components as well.

In order to achieve maximum performance, optimisations have to consider a multitude of different hardware and device driver features. This is usually not done in research projects. Our approach uses a fallback strategy, where each component provides a list of possible modes, ordered by performance for the specific hardware and device driver. These parameters are mainly colour space and colour depth, with an additional parameter for the frame grabber operation mode. If the required conditions are not met (e.g. no MMX support), a less optimised version will be used.

The display component is based on Microsoft's DirectDraw API, avoiding all unnecessary copy operations and putting the decompressed image directly into the frame buffer of the graphic adapter. Another version integrates even the decompressor directly with the output object so that it can decompress directly into video memory. Thus, side-stepping the strict differentiation into objects, an even higher optimisation can be achieved. This is only one example for the use of integrated layer processing (ILP) according to [8]. Since one copy operation over the whole image can be avoided, it is possible to speed up processing significantly.

Nevertheless, the interface of the presentation component is able to communicate with all other objects in the normal fashion, so that it can be also used in configurations where other decompression objects are needed. In the same manner, a fall back to conventional GDI-output is possible, whenever a display driver is encountered that does not support the DirectDraw-API.

4.2 Colour Space Conversions

Codecs usually prefer some planar YUV colour space formats for input. However, some capture boards do not support these formats. Furthermore, the presentation requires specific formats, hence packed YUV or RGB data have to be produced. As a consequence colour space conversions are often required. However, as copy operations can not be fully avoided this processing can often be integrated. Thus, the added overhead can be almost completely hidden. This is especially the case if optimised MMX conversions are used.

4.3 Video Compression Codec Design

Our evaluation is based mainly on a H.263 codec we've implemented and tuned especially for the investigation of live compression and high quality video streams. Prior to this, first experiments with a H.261 codec were made, which are described in [9]. To improve flexibility our compression object does not only support uncompressed video data as an input but also possible intermediate formats of partly compressed video data. Within this environment our compression process makes the following assumptions with corresponding design constraints:

- The target video resolution should be CIF (352x288) with approximately 25 fps and relatively high image quality.
- Highly specialised codecs for the conferencing case, that means half-length portrait and very moderate motion. This allows to work with only motion detection schemes, since we found that in this case at 25 fps the motion is rarely more than 1 pixel. So there is no real need for a motion estimation, as would be for fast moving objects.
- We considered primarily LAN and Broadband WAN connections, so the size of the resulting data streams is not of main concern. This assumption is also common in other systems like vic [13] and NetMeeting when gaining for good video quality.
- With respect to network transportation, the size of the handled data units and possibilities for fast and simple error correction schemes must be evaluated.

Based on these constraints processing of all different stages from capturing/compression and decompression/displaying was highly integrated. Ideas from concepts such as integrated layer processing ILP [8] and ALF [1] were applied, where appropriate. Due to reduced copy operations, it was possible to improve speed significantly. The coding makes maximum use of available hardware features, for example the use of MMX processor extensions for motion detection, DCT and bit coding. By using these optimisations, a tremendous speedup was achieved.

For fast compression, ideas derived from several publicly available implementations were adopted. In this process complexity at specific processing stages compared with the improved compression ratio were evaluated. Another aspect was the introduction of fix point operations for the DCT and an optimised algorithm for the adaptation of smaller quantisation step sizes. As discussed in [9], [10], these optimisations led us to codecs that are significantly faster than comparable versions from commercial vendors.

4.4 Compression Codec Performance

This section discusses the applied compression codec optimisations and the resulting performance improvements. Figure 4 shows the block scheme of a H.263 inter codec [11]. The separate steps require different processing resources. The three most expensive parts are the DCT/IDCT, the half pixel search and the VLC (Variable Length Coding). We started our development based on the Telnor implementation. The first step we did was the optimisation of the internal structure. Furthermore, we investigated which features of the coding consume too much time in relation to their influence on the compression ratio. Based on this investigation we changed the behaviour of some components of the codec. One optimisation was the reduction of the half pixel search to the edges of the macroblock, which means that only a quarter of the search step must be performed. In most cases the influence on the compression ratio was less than 5%.

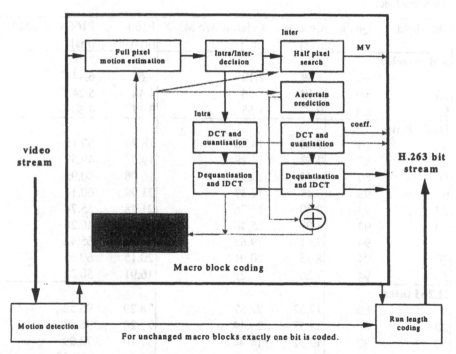

Fig. 4. H.263 inter coding scheme

After these basic steps, we replaced some parts of the codec with MMX code for modern Intel Pentium CPUs. We started with the DCT and IDCT process, which had required more than 70% of the coding time. After this replacement, the DCT only consumes 30% of the whole time. This leads to a massive shift in the distribution of the share that every part of the codec is using. We were amazed to find out, that the

VLC coding was the most expensive part now. That's why we replaced this part too by a highly optimised MMX/assembler based code. This leads to an impressive reduction of the overall coding time. After this step VLC and DCT are consuming a similar part of the computing power (~35%).

Table 1 shows the overall results of our optimisation. The table is divided in three blocks of different codec types. The macroblock encoder (MB encoder) performs only a motion detection algorithm as defined in H.263. No further steps, like DCT and VLC, are included here. We use this specific codec in order to support distributed compression schemas [10]. The next type is the so-called Intra Codec. This codec is similar to the H.261 codec and doesn't use motion compensation algorithms and the calculation of difference pictures (P- and B-frames). The last type, the Inter Codec, provides the full functionality of the H.263 standard except the coding of P-frames. For the measurements two different systems were used. One with Intel Celeron/400

Table 1. Compression codec performance measured on INTEL Celeron/400 and Pentium 166 MMX machines

Threshold	Quality [%]	Celeron [ms]	Celeron w/o MMX [ms]	P166 [ms]	P166 w/o MMX [ms]
MB encoder					
0	n/a	1.39	2.12	5.63	6.44
20	n/a	1.18	1.76	4.48	5.24
50	n/a	0.97	1.55	3.77	4.30
H.263 Intra					
0	80	7.00	14.38	18.98	53.18
20	80	7.82	14.89	19.27	49.53
50	80	6.75	12.97	16.08	40.95
0	93	8.49	17.37	21.99	60.11
20	93	9.30	17.76	21.78	55.74
50	93	8.07	15.48	18.17	46.26
0	94	7.61	19.63	20.03	65.42
20	94	8.35	20.01	20.15	60.65
50	94	7.26	17.29	16.91	50.22
H.263 Inter					
0	80	17.52	22.57	78.29	96.52
20	80	18.57	23.44	77.28	96.83
50	80	12.32	14.94	57.10	72.09
0	93	22.30	26.40	80.20	100.67
20	93	21.49	24.89	76.76	95.58
50	93	13.03	15.94	53.11	68.72
0	94	24.42	28.81	83.96	105.37
20	94	21.67	24.45	75.56	93.25
50	94	13.34	16.58	53.45	69.37

MHz and one with an older Intel Pentium-MMX with 200 MHz. Both systems had 64 Mbytes RAM. For every measurement the same video sequence was used. This sequence has a resolution of 352x288 and 25 frames per second. The content of the video was a typical video conferencing scenario with some slow and fast motion parts. The table shows the results with and without MMX/assembler optimisations.

The shown performance gains are based on an extensive use of optimisations and codec integration techniques. Due to the extensive use of optimised coding techniques, integrated layer processing and MMX coding performance gains up to 227% were achieved. But the possible improvements differ from case to case. For instance, the macro block encoder (MB encoder) shows very diverse results depending on the used platform. The reason for that is the different memory bandwidth of the used systems. The so-called MB encoder is derived from the H.263 coding process and consists only of a motion detection algorithm.

Further compression will not be performed by this codec. This means, that after the (MMX based) coder detects a motion in a macro block, the block will be copied to the output buffer. Therefore, a large amount of data must be transferred. Data movement produces the biggest part of workload of this codec. The Celeron processor has a much better second level architecture in comparison to the standard Pentium processor. Hence the Celeron benefits more from the MMX based coding of the motion detection algorithm. Another point is the difference between the results of the H.263 intra codecs with a motion threshold of zero or twenty. Of course, the coder running with a motion threshold of zero produces more data and needs more time for DCT and run length coding. Based on this it should need more processing time than less. But in our implementation the whole motion detection process will be skipped if a motion threshold of zero is configured. This leads to a reduction of coding time, which is higher than the additional cost of the DCT and run length processing.

As the table shows, the performance gains of the optimised coding are higher on the P166 using the H.263 intra codec. The reason for that is the different cache and bus architecture. The optimised code reduces the non-linear memory access significantly. The replacement and integration of table lookups do this by integrated layer processing. Therefore, we used MMX code and integrated some separate processing steps of the H.263 standard.

The shown performance gains for the H.263 inter codec are too low compared to the intra version. The optimisation process is not finished yet for this codec. The Inter codec version compresses the video stream much better than the Intra coder. Therefore, only different information between two consecutive pictures will be coded. In this process a DCT and IDCT processing is needed. Unfortunately, the used MMX based DCT isn't accurate enough. In the case of the simple Intra coding, where whole DCT results will be transferred, this does not matter. In contrast, using the same DCT in an Inter codec leads to a bigger output data stream (10 to 20% more). So we decided not to use a MMX based DCT and IDCT for the Inter coder up to now. Currently, we are working on a better implementation of the DCT, and we hope that this leads to similar performance gains as achieved with the Intra codec.

4.5 Related Work

A corresponding approach to VTToolkit is the CINEMA project [3] with the focus on the configuration of distributed multimedia applications. The enforcement of synchronisation and quality of service specifications in general is addressed here. In contrast, this paper concentrates on the configuration of video transmission systems with performance aspects as the focal point.

Another related system is Rendez-Vous, a successor of IVS, developed by the Rodeo group at INRIA. It includes RTP protocol support over multicast or unicast IP, H.261 video standard, high quality PCMU, ADPCM, VADPCM (with hi-fi support) and low bandwidth GSM and LPC audio coding. Additionally the issue of platform independence is of primary concern. A very interesting relation is the integrated scheduler for multilayer video and audio flow management and processing in Rendez-Vous. In opposition to our scheduling approach, that tries to guarantee that enough resources are available, in Rendez-Vous the scheduler handles intelligent dropping in case of overload. New transport protocols as well as distributed coding are not explicitly addressed. The approach focuses on new compression algorithms, while our VTToolkit tackles system level problems that are also relevant in a real world environment today.

Mash from the Berkeley university deals mainly with Scalable, Reliable Multicast (SRM), based on a collaboration tool, a mediaboard and a set of archival tools for standard RTP mbone sessions are developed. The support of heterogeneous network and end user capabilities is another topic, here gateways are deployed in an intelligent network scenario. This resembles the gateways used in VTToolkit to support heterogeneous environments with specific requirements. High end video transmission as well as use of new transport protocols is not a dominant topic within this project.

The focus on highest possible performance throughout all components maintaining low overall system costs is unique to our approach. Nevertheless, some quite innovative contributions as the use of new transport protocols in a near-product system, coexisting with the traditional IP and the distribution of compression are made.

5 Conclusions and Future Work

In this paper we introduced a framework for advanced conferencing solutions. Based on VTToolkit - an object oriented toolkit - it eases the handling of various aspects like audio/video capture, compression techniques, transmission and presentation. All significant components are designed to be flexible and complete enough to implement applications like a transport gateway or a remote compression server by simply connecting them. This is possible by hiding the complexity of various multimedia interfaces and compression algorithms as well as due to the exploitation of a common interface as outlined in section 3.

Demonstrated with a sample video transmission application, it was shown that the comprehensive observation of all involved components of a complex system enables a rich set of optimisations. These aspects include general techniques like application layer framing and integrated layer processing. Furthermore, the use of specialised multimedia system interfaces and distributed compression, the exploitation of processor extensions like MMX for improving the compression efficiency and the application of efficient transport systems over advanced network technologies were considered. Due to the applied optimisations and the integrated resource and quality of service management, it is now possible to perform high quality/multi-party conferences with moderate costs on modern platforms, since additional compression hardware is not required.

Video transmission is characterised by the trade-off between quality, compression effort and available network bandwidth. Together with the number of VTToolkit components and their configurations, like the used colour model, resolution, compression process or transport system, it is necessary to have a suitable design support.

Development of VTToolkit as well as other components of the framework will be consequently continued. This includes further improvements and optimisations as well as extensions concerning additional audio and video codecs and support of standardised conferencing protocols via gateways to enable interoperability. Moreover, it is intended to continue application development. This includes the practical use as a wide area video conferencing system in some teleteaching projects.

References

1. Ahlgren, B., Gunningberg, P., Moldeklev. K.: „Performance with a Minimal-Copy Data Path supporting ILP and ALF." 1995.
2. Benz, M., Engel, F.: „Hardware Supported Protocol Processing for Gigabit Networks", Workshop on System Design Automation, Dresden, 1998, pp. 23-30
3. Barth, I., Helbig, T., Rothermel K.: „Implementierung multimedialer Systemdienste in CINEMA" GI/ITG-Fachtagung "Kommunikation in Verteilten Systemen", February 22-24, 1995, Chemnitz, Germany
4. Braun, I., Hess, R., Schill, A.: Teleworking support for small and Medium-sized enterprises; IFIP World Computer Congress '98, Aug.-Sept. 1998
5. Braun, T., Diot, C.: „Protocol Implementation Using Integrated Layer Processing", ACM SIGCOMM '95, 1995
6. Clark, D. D., Tennenhouse, D. L.: „Architectural Considerations for a New Generation of Protocols", ACM SIGCOMM '90, 1990
7. Druschel, P., Abbot, M. B., Pagels, M. A., Peterson, L.: „Network Subsystem Design: A Case for an Integrated Data Path", IEEE network, 1993
8. Braun, T., Diot, C.: „Protocol Implementation using Integrated Layer Processing." SIGCOMM 1995.
9. Geske, D., Hess, R.: Fast and predictable video compression in software - design and implementation of an H.261 codec, Interactive Multimedia Service and Equipment - Syben 98 Workshop, Zürich, 1998
10. Hess, R., Geske, D., Kümmel, S., Thürmer, H.: „Distributed Compression of Live Video - an Application for Active Networks", AcoS 98 Workshop, Lissabon, 1998
11. ITU-T Recommendation H.263. Line Transmission of non-telephone signals. Video codec for low bitrate communication, March 96

12. Kümmel, S., Hutschenreuther, T.: „Protocol Support for Optimized, Context-Sensitive Request/Response Communication over Connection Oriented Networks", IFIP Int. Conf. On Open Distributed Processing, Toronto, Mai 1997, pp. 137-150

13. McCanne, St., Jacobson, V.: vic: A Flexible Framework for Packet Video. ACM Multimedia '95, San Francisco, CA, November 1995

14. http://www.picturetel.com/products.htm

15. http://www.intel.com/proshare/conferencing/products/

16. Audio-Video Transport Working Group: „RTP: A Transport Protocol for Real-Time Applications", 1996

17. Network Working Group: „Resource ReSerVation Protocol (RSVP)", 1997

18. Schill, A., Hutschenreuther, T.: „Architectural Support for QoS Management and Abstraction: SALMON", Computer Communications Journal, Vol. 20, No. 6, July 1997, pp. 411-419

19. Schill, A., Franze, K., Neumann, O.: An Infrastructure for Advanced Education: Technology and Experiences; 7th World Conference on Continuing Engineering Education, Turin, Mai 1998

20. Turletti, T.: The INRIA Videoconferencing System (IVS). ConneXions - The Interoperability Report Journal, Vol. 8, No 10, pp. 20-24, October 1994.

A Novel Replica Placement Strategy for Video Servers

Jamel Gafsi and Ernst W. Biersack

(gafsi,erbi)@eurecom.fr

Institut EURECOM, B.P. 193, 06904 Sophia Antipolis Cedex, FRANCE

Abstract. *Mirroring-based reliability as compared to parity-based reliability significantly simplifies the design and the implementation of video servers, since in case of failure mirroring does not require any synchronization of reads or decoding to reconstruct the lost video data. While mirroring doubles the amount of storage volume required, the steep decrease of the cost of magnetic disk storage makes it more and more attractive as a reliability mechanism. We present in this paper a novel data layout strategy for replicated data on a video server. In contrast to classical replica placement schemes that store original and replicated data separately, our approach stores replicated data adjacent to original data and thus does not require additional seek overhead when operating with disk failure. We show that our approach considerably improves the server performance compared to classical replica placement schemes such as the interleaved declustering scheme and the scheme used by the Microsoft Tiger video server. Our performance metric is the maximum number of users that a video server can simultaneously support (server throughput).*

Keywords: Video Servers, Data Replication, Performance Analysis

1 Introduction

In order to store a large number of voluminous video files, a video server requires numerous storage components, typically magnetic disk drives. As the number of server disks grows, the server mean time to failure degrades and the server becomes more vulnerable to data loss. Hence the need of fault tolerance in a video server. Two techniques are mainly applied in the context of fault tolerant video servers: **mirroring** and **parity**. Parity adds small storage overhead for parity data, while mirroring requires twice as much storage volume as in the non-fault tolerant case. Mirroring as compared to parity significantly simplifies the design and the implementation of video servers since it does not require any synchronization of reads or additional processing time to decode lost information, which must be performed for parity. For this reason, various video server designers [1–3] have adopted mirroring to achieve fault-tolerance. This paper only focuses on the mirroring-based technique.

The video server is assumed to use *round-based data retrieval*, where each stream is served once every time interval called the **service round**. For the data retrieval from disk, we use the SCAN scheduling algorithm that optimizes the time spend to seek the different video blocks needed. A video to store on the server is partitioned into video blocks that are stored on *all* disks of the server in a round robin fashion and that each video is distributed over all disks of the server.

The literature distinguishes two main strategies for the storage/retrieval of original blocks on/from server; the Fine Grained Striping (**FGS**) strategy and Coarse Grained Striping (**CGS**) strategy. FGS retrieves for one stream *multiple* blocks from many disks during *one* service round. A typical example of FGS is RAID3 as defined by Katz et al. [4]. Other researchers proposed some derivations of FGS like the streaming RAID of Tobagi et al. [5], the staggered-group scheme of Muntz et al. [6], and the configuration planner scheme of Ghandeharizadeh et al. [7], and our mean grained striping scheme [8]. FGS generally suffers from large buffer requirements. CGS, however, retrieves for one stream *one* block from a single disk during *each* service round. RAID5 is a typical CGS scheme. Oezden et al. [9, 10] have shown that CGS provides higher throughput than FGS (RAID5 vs. RAID3) for the same amount of resources (see also Vin et al. [11], Beadle et al. [12], and our contribution [8]). Accordingly, in order to achieve highest throughput, we adopt CGS to store and retrieve *original* blocks.

What remains to solve is the way original blocks of a single disk are replicated on the server. Obviously, original blocks of one disk are not replicated on the same disk. Mirroring schemes differ on whether a single disk contains original and/or replicated data. The **mirrored declustering** scheme sees two (many) *identical* disk arrays, where original content is replicated onto a distinct set of disks. When the server works in normal operation mode (disk failure free mode), only the half of the server disks are active, the other half remains idle, which results in load imbalances within the server.

Unlike mirrored declustering, **chained declustering** [13, 14] partitions each disk into two parts, the first part contains original blocks and the second part contains replicated blocks (copies): Original blocks of disk i are replicated on disk $(i + 1) \bmod D$, where D is the total number of disks of the server. **Interleaved declustering** is an extension of chained declustering, where original blocks of disk i are not entirely replicated on another disk $(i + 1) \bmod D$, but distributed over *multiple* disks of the server. Mourad [1] proposed the doubly striped scheme that is based on interleaved declustering, where original blocks of a disk are evenly distributed over *all* remaining disks of the sever. We can consider chained declustering as a special case of interleaved declustering having a distribution granularity of replicated blocks that equals 1.

We will restrict our discussion to interleaved declustering schemes, since these schemes distribute the total server load evenly among all components during normal operation mode. Note that interleaved declustering only indicates that the replica of the original blocks belonging to one disk are stored on *one, some,* or *all* remaining disks, but does not indicate *how* to replicate a *single* original block.

This paper is organized as follows. Section 2 classifies and studies several interleaved declustering schemes. We present our novel replica placement strategy in section 3. In section 4, we show that our approach outperforms the other existing schemes in terms of the server throughput. The conclusions are presented in section 5.

2 Interleaved Declustering Schemes

We present in Table 1 different interleaved declustering schemes. We adopt two classification metrics. The first metric examines *how a single block is replicated*. The second metric concerns *the number of disks* that store the replica of the original content of a single disk.

We consider for the first metric the following three alternatives:

1. The copy of the original block is entirely stored on a single disk (One).
2. The copy of the original block is divided into a set of sub-blocks, which are distributed among some disks building an independent group (Some).
3. The copy of the original block is divided into exactly $(D - 1)$ sub-blocks, which are distributed over all remaining $(D - 1)$ server disks (All).

We distinguish three alternatives for the second metric:

1. The original blocks that are stored on one disk are replicated on a *single* disk (One).
2. The original blocks of one disk are replicated on a set of disks that build an independent group (Some).
3. The original blocks of one disk are replicated on the remaining $(D-1)$ server disks (All).

The symbol "XXX" in Table 1 indicates combinations that are not useful for our discussion. The name of each scheme contains two parts. The first part indicates how an original block is replicated (the first metric) and the second part gives the number of disks, on which the content of one disk is distributed (the second metric). For instance, the scheme One/Some means that each original block is *entirely* replicated (One) and that the original content of one disk is distributed among a set of disks (Some).

	Single disk (One)	Set of disks (Some)	$(D-1)$ disks (All)
Entire block (One)	One/One	One/Some	One/All
Set of sub-blocks (Some)	XXX	Some/Some	XXX
$(D-1)$ sub-blocks (All)	XXX	XXX	All/All

Table 1. Classification of interleaved schemes

Let s assume a video server containing 6 disks (disks 0 to 5) and a video to store consisting of 30 original blocks. Each disk is partitioned into two equal-size parts, the first part stores original blocks and the second part stores replicated blocks (copies) (see Figures 1 and 2).

Figure 1(a) shows the One/One organization. For instance, original blocks of disk 0 are replicated on disk 1 (dashed blocks). During disk failures, the load of a failed disk is entirely shifted to another *single* disk, which results in load imbalances within the server. On the other hand, the One/One organization has the advantage of surviving up-to $\frac{D}{2}$ disk failures in the best case.

Figure 1(b) shows the One/All organization. The replication of original blocks of disk 0 are stored on the other disks $1, 2, 3, 4$, and 5 (dashed blocks). This organization allows, in the best case, to evenly distribute the load of one failed disk among all remaining disks. Its fault tolerance, however, is limited to a *single* disk failure.

We show in Figure 1(c) an example of the One/Some organization that divides the server into 2 independent groups, where each group contains a set $D_c = 3$ of disks. Original blocks of one disk are *entirely* replicated over the remaining disks of the group, e.g. original blocks of disk 0 are replicated on disks 1 and 2.

In order to ensure deterministic admission control, each disk of the server must reserve a proportion of its available I/O bandwidth, which is needed to retrieve replicated data during disk failure mode. The amount of I/O bandwidth that is reserved on each disk must respect the worst case scenario. Obviously, the One/One organization needs to reserve on each disk one half of the available I/O bandwidth for disk failure mode. For both, the One/Some and the One/All organizations, the original blocks of one disk are spread among multiple (some for One/Some and $(D-1)$ for One/All) disks. However, all blocks that would have been retrieved from the failed disk for a set of streams can, in the worst case, have their replica stored on the same disk. This worst case scenario therefore requires the reservation of one half of the I/O bandwidth of each disk. Consequently, all of the three schemes One/One, One/All, and One/Some must reserve the half of each disk's available I/O bandwidth in order to ensure *deterministic* service when operating in disk failure mode.

The Microsoft Tiger [2, 3] introduced a replication scheme, where an original block is not entirely replicated on a single disk. Indeed, the replica of an original block consists of a set of *sub-blocks*, each being stored on a different disk. Original blocks of one disk are replicated across the remaining disks of the group, to which this disk belongs. We have called this organization Some/Some in Table 1. Figure 2(a) illustrates an example of this organization, where dashed blocks show how original blocks of disk 0 (3) are replicated on disks 1 and 2 (4 and 5) inside group 1 (2). As the One/Some organization, the Some/Some organization allows to survive one disk failure inside each group.

The last organization of Table 1 is All/All, for which we show an example in Figure 2(b). Dashed original blocks of disk 0 are replicated as indicated. The main advantage of All/All is its *perfect* load balancing. In fact, the load of a

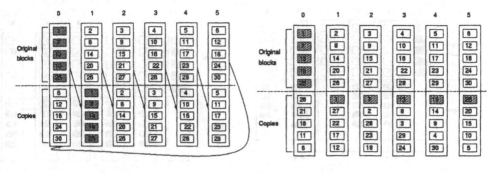

(a) One/One Organization.   (b) One/All Organization.

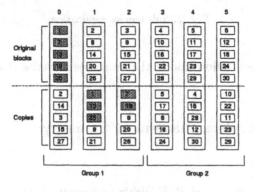

(c) One/Some Organization.

Fig. 1. Entire block replication.

failed disk is *always* evenly distributed among all remaining disks of the server. However, the All/All organization, as the One/All organization, only allows to survive a single disk failure, which might be not sufficient for large video servers.

Contrarily to the entire block replication organizations (Figure 1), the two sub-block replication organizations (Figure 2) avoid to reserve the half of each disk's I/O bandwidth to ensure deterministic service during disk failure mode. However, the number of seek operations will double for these two schemes when operating in disk failure mode compared to normal operation mode. Exact values of the amount of I/O bandwidth to be reserved are given in section 4.1.

The main drawback of all replication schemes considered in this section is their additional *seek overhead* when operating with disk failure as we will see in section 4.1. In fact, these schemes require additional seek times to retrieve replicated data that are stored separately from original data. Unfortunately, high seek overhead decreases disk utilization and therefore server performance. We present in the following our replication approach that resolves this problem by *eliminating* the additional seek overhead. In fact, we will see that our approach

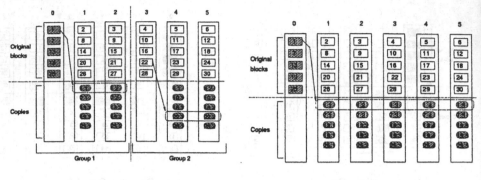

(a) Some/Some Organization.   (b) All/All Organization.

Fig. 2. Sub-blocks replication.

requires for the disk failure mode the same seek overhead as for the normal operation mode.

3 A Novel Replica Placement Strategy

3.1 Motivation

If we look at the evolution of SCSI disk's performance, we observe that (i) data transfer rates double every 3 years, whereas (ii) disk access time decreases by one third every 10 years [15]. Figure 3(a) shows data transfer rates of different Seagate disks' generations (SCSI-I, SCSI-II, Ultra SCSI, and finally Ultra2 SCSI) [16]. Figure 3(b) depicts the evolution of *average* access time for Seagate disks. We see that Figures 3(a) and 3(b) well confirm the observations (i) and (ii) respectively.

A disk drive typically contains a set of **surfaces** or platters that rotate in lockstep on a central spindle [1]. Each surface has an associated disk head responsible for reading data. Unfortunately, the disk drive has a single read data channel and therefore only one head is active at a time. A surface is set up to store data in a series of concentric circles, called **tracks**. Tracks belonging to different surfaces and having the same distance to the spindle build together a **cylinder**. As an example of today's disks, the seagate Barracuda ST118273W disk drive contains 20 surfaces; 7, 500 cylinders; and 150, 000 tracks.

The time a disk spends for performing seek operations is wasted since it can not be used to retrieve data. One seek operation mainly consists of a **rotational latency** and a **seek time**. Rotational latency is the time the disk arm spends inside one cylinder to reposition itself on the beginning of the block to be read.

---

[1] Regarding mechanical components of disk drives and their characteristics, we are based in this paper on [16] and also on the work done in [17].

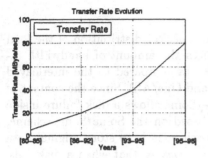
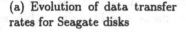

(a) Evolution of data transfer
rates for Seagate disks

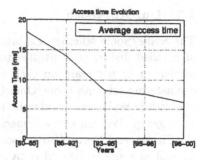

(b) Evolution of average access
time for Seagate disks

Fig. 3. Performance evolution of Seagate SCSI disks

The maximum value of the rotational latency t_{rot} is directly given by the rotation speed of the spindle. This rotation speed is actually at about $7,200$ rpm, which results in $t_{rot} = 10$ ms. Seek time t_{seek} as studied in [17, 18] is mainly composed of four phases: (i) a *speedup* phase, which is the acceleration phase of the arm, (ii) a *coast* phase (only for long seeks), where the arm moves at its maximum velocity, (iii) a *slowdown* phase, which is the phase to rest close to the desired track, and finally (iv) a *settle* phase, where the disk controller adjusts the head to access the desired location. Note the duration t_{stl} of the the settle phase is independent of the distance traveled and is about $t_{stl} = 3$ ms. However, the durations of the speedup phase (t_{speed}), the coast phase (t_{coast}), and the slowdown phase ($t_{slowdown}$) mainly depend on the distance traveled. The seek time t_{seek} takes then the following form:

$$t_{seek} = t_{speed} + t_{coast} + t_{slowdown} + t_{stl}$$

Let us assume that the disk arm moves from the outer track (cylinder) to the inner track (cylinder) to retrieve data during one service round and in the opposite direction (from the inner track to the outer track) during the next service round (CSCAN). If a single disk can support up-to 20 streams, at most 20 blocks must be retrieved from disk during one service round. If we assume that the different 20 blocks expected to be retrieved are *uniformly* spread over the cylinders of the disk, we then deal only with *short* seeks and the coast phase is neglected (distance between two blocks to read is about 300 cylinders). Wilkes et al. have shown that seek time is a function of the distance traveled by the disk arm and have proposed for short seeks the formula $t_{seek} = 3.45 + 0.597 \cdot \sqrt{d}$, where d is the number of cylinders the disk arm must travel. Assuming that $d \approx 300$ cylinders, the seek time is then about $t_{seek} \approx 13.79$ ms. Note that short seeks spend the most of their time in the speedup phase.

3.2 Our Approach

The Some/Some scheme (see table 1) ensures a perfect distribution of the load of a failed disk over multiple disks and reduces the amount of bandwidth reserved for each stream on each surviving disk as compared to the interleaved declustering schemes (One/One, One/Some, and One/All). Since the content of one disk is replicated inside one group, Some/Some allows a disk failure inside each group. We call our approach, which is based on the Some/Some scheme, the **Improved Some/Some** scheme. The basic idea is to store original data as well as some replicated data in *a continuous* way so that when a disk fails, no additional seeks are performed to read the replica. In light of this fact, our approach does not divide a disk in two *separate* parts, one for original blocks and the other for replicated blocks. Figure 4 shows an example of the Improved Some/Some scheme.

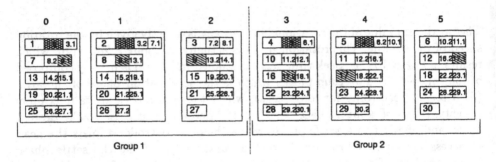

Fig. 4. Layout example of the Improved Some/Some scheme.

Let us consider only the disks' content of group 1 (disks 0, 1, and 2). Let us now consider original block 9 that is stored on disk 2 (dashed block). The replication is performed as follows. We divide the original block into $3 - 1 = 2$ [2] sub-blocks 9.1 and 9.2 that are stored *immediately* contiguous to the original blocks 7 and 8 respectively. Note that original blocks 7 and 8 represent the previous original blocks to block 9 inside the group. If we take original block 13, its previous original blocks inside the group are blocks 8 and 9. Now assume that disk 2 fails. Block 9 is reconstructed as follows. During the service round i where block 7 is retrieved, block 7 and sub-block 9.1 are *simultaneously* retrieved (neither additional seek time nor additional rotational latency, but additional read time). During the next service round $i + 1$, block 8 and sub-block 9.2 are simultaneously retrieved. Sub-blocks 9.1 and 9.2 are retrieved from server in advance and kept in buffer to be consumed during service round $i + 2$. Generally, sub-blocks that are read in advance are buffered for several service rounds before being consumed. The number of buffering rounds mainly depends on how large the server is (total number of server disks). If we assume that disk 0 is the failed

---

[2] 3 is the group size and therefore the number of su-blocks is 2.

disk, the reconstruction of block 19 is performed during the service rounds where blocks 14 (sub-block 19.1) and 15 (sub-block 19.2) are retrieved. The sub-blocks are kept in the buffer at most during 5 service rounds before they are consumed. The example shows that in order to simultaneously read one original block and one sub-block for one stream, data to be retrieved have a size of at most two original blocks. In order to ensure continuous read, one original block as well as the corresponding replicated blocks must be contained on the same track. Fortunately, today's disk drives satisfy this condition. In fact, the track size is continuously increasing. The actual mean track size for seagate new generation disk drives is about 160 KByte, which is about 1.3 Mbit. Hence the possibility to store inside *one* track the original block and the set of replicated sub-blocks as shown in Figure 4. Our approach therefore does not increase seek overhead, but *doubles*, in the worst case, the read time. Note that the very first blocks require special treatment: our approach *entirely* replicates the two first blocks of a video within each group, which is represented in Figure 4 with the dark-dashed blocks (block 1 on disk 1, block 2 on disk 0 for group 1 and block 5 on disk 3, block 4 on disk 4 for group 2). Let us take the following example to explain the reason of doing this. If disk 0 has already failed before a new stream is admitted to consume the video presented in the figure, the stream is delayed for one service round. During the next service round, the two first blocks 1 and 2 are simultaneously retrieved from disk 1. Based on the performance evolution of SCSI disks (see Figure 3), our approach will improve server performance in terms of the number of streams that the server can simultaneously admit (see section 4).

4 Performance Comparison

4.1 Admission Control Criterion

The admission control policy decides whether a new incoming stream can be admitted or not. The maximum number of streams Q that can be simultaneously admitted from server can be calculated in advance and is called server throughput. The server throughput depends on disk characteristics as well as on the striping/reliability scheme applied. In this paper, the difference between the schemes considered consists of the way original data is replicated. We consider the admission control criterion of Eq. 1. We first calculate disk throughput and then derive server throughput. Let Q_d denote the throughput achieved for a single disk. If we do not consider fault tolerance, the disk throughput is given in Eq. 1, where b is the block size, r_d is the data transfer rate of the disk, t_{rot} is the worst case rotational latency, t_{seek} is the worst case seek time, and τ is the service round duration [3].

---

[3] We take a constant value of τ, typically $\tau = \frac{b}{r_p}$, where b is the size of an original block and r_p is the playback rate of a video

$$Q_d \cdot \left(\frac{b}{r_d} + t_{rot} + t_{seek} \right) \leq \tau$$

$$Q_d = \frac{\tau}{\frac{b}{r_d} + t_{rot} + t_{seek}} \tag{1}$$

Introducing fault tolerance (mirroring-based), the disk throughput changes and becomes dependent on which mirroring scheme is applied. Three schemes are considered for discussion: our approach (Improved Some/Some), the One/Some scheme, and the Microsoft Some/Some scheme. Let Q_d^{OS}, Q_d^{SS}, and Q_d^{ISS} the disk throughput for One/Some, Some/Some, and our Improved Some/Some, respectively. Note that the disk throughput is the same during both, normal operation and disk failure mode.

For the One/Some scheme, half of the disk I/O bandwidth should be reserved in the worst case to reconstruct failed original blocks and thus Q_d^{OS} is calculated following Eq. 2.

$$Q_d^{OS} = \frac{Q_d}{2} = \frac{\left(\frac{\tau}{\frac{b}{r_d} + t_{rot} + t_{seek}} \right)}{2} \tag{2}$$

For the Some/Some scheme, in order to reconstruct a failed original block, the retrieval of sub-blocks requires small read overhead (small sub-blocks to read on each disk), but a *complete* latency overhead for each additional sub-block to read from disk. The admission control criterion presented in Eq. 1 is therefore modified as Eq. 3 shows. The parameter b_{sub} denotes the size of a sub-block.

$$Q_d^{SS} \cdot \left(\left(\frac{b}{r_d} + t_{rot} + t_{seek} \right) + \left(\frac{b_{sub}}{r_d} + t_{rot} + t_{seek} \right) \right) \leq \tau$$

$$Q_d^{SS} = \frac{\tau}{\frac{b + b_{sub}}{r_d} + 2 \cdot (t_{rot} + t_{seek})} \tag{3}$$

If we take our Improved Some/Some scheme and consider the case where a disk fails inside one group, we get the following admission control criterion (Eq. 4), where b_{over} denotes the amount of data (overhead) that should be simultaneously read with each original block. In the worst case $b_{over} = b$.

$$Q_d^{ISS} \cdot \left(\frac{b + b_{over}}{r_d} + t_{rot} + t_{seek} \right) \leq \tau$$

$$Q_d^{ISS} = \frac{\tau}{\frac{2 \cdot b}{r_d} + t_{rot} + t_{seek}} \tag{4}$$

Once the disk throughput Q_d is calculated, the server throughput Q can be easily derived as $Q = D \cdot Q_d$ for each of the schemes, where D denotes again the total number of disks on the server.

4.2 Throughput Results

We present in the following the results of the server throughput Q^{OS}, Q^{SS}, and Q^{ISS} respectively for the schemes One/Some, Some/Some, and our Improved Some/Some. In Figure 5, we keep constant the values of the seek time and the rotational latency and vary data transfer rate r_d of the disk. Figures 5(a) and 5(b) show that Improved Some/Some outperforms the Microsoft Some/Some scheme that, itself outperforms One/Some for all values of r_d (20, 80 MByte/sec). Figure 5 also shows that the gap between our Improved Some/Some and the two other schemes (One/Some and Some/Some) *considerably* increases with the increase of the data transfer rate r_d. Table 2 illustrates the benefit of our approach, where the ratios $\frac{Q^{ISS}}{Q^{OS}}$ and $\frac{Q^{ISS}}{Q^{SS}}$ are illustrated depending on r_d.

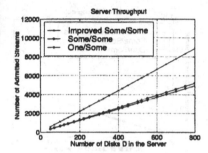

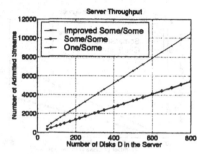

(a) Throughput for r_d = 20 MByte/sec.

(b) Throughput for r_d = 80 MByte/sec.

Fig. 5. Server throughput for One/Some, Some/Some, and Improved Some/Some with $t_{seek} = 13.79$ ms, $t_{rot} = 10$ ms, $b = 0.5$ Mbit, and $r_p = 1.5$ Mbit/sec.

	$\frac{Q^{ISS}}{Q^{OS}}$	$\frac{Q^{ISS}}{Q^{SS}}$
$r_d = 20$ MByte/sec	1.79	1.69
$r_d = 40$ MByte/sec	1.88	1.83
$r_d = 80$ MByte/sec	1.93	1.91

Table 2. Throughput ratios.

We focus now on the impact of the evolution of the seek time t_{seek} and the rotational latency t_{rot} on the throughput for the three schemes considered. We keep constant the data transfer rate that takes a relatively small value of r_d (40 Mbyte/sec). Figure 6 plots server throughput for the corresponding parameter values. The Figure shows that our Improved Some/Some scheme achieves highest server throughput for *all* seek time/rotational latency combination values adopted. Obviously, the decrease in t_{seek} and t_{rot} increases throughput for all schemes considered. We notice that the gap between our Improved Some/Some and the Microsoft Some/Some *slightly* decreases when t_{seek} and t_{rot} decrease as Table 3 depicts, where disk transfer rate has also the value r_d (40 Mbyte/sec). Note that th value $t_{rot} = 6$ ms corresponds to a spindle speed of 10000 prm, and the value $t_{rot} = 4$ ms corresponds to the speed of 15000 prm, which is a too optimistic value. We observe that even for very low values of t_{seek} and t_{rot}, our Improved Some/Some scheme outperforms the Microsoft Some/Some in terms of server throughput ($\frac{Q^{ISS}}{Q^{SS}} = 1.59$).

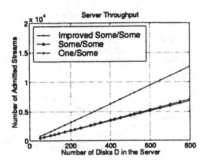

(a) Throughput for $t_{seek} =$ 10 ms, $t_{rot} = 8$ ms.

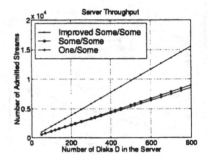

(b) Throughput for $t_{seek} =$ 8 ms, $t_{rot} = 6$ ms.

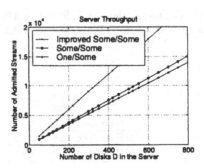

(c) Throughput for $t_{seek} =$ 4 ms, $t_{rot} = 4$ ms.

Fig. 6. Server throughput for different access time values with $r_d = 40$ MByte/sec, $b = 0.5$ Mbit, and $r_p = 1.5$ Mbit/sec.

	$\frac{Q^{ISS}}{Q^{SS}}$
$t_{seek} = 13,79$ and $t_{rot} = 10$	1.83
$t_{seek} = 10$ and $t_{rot} = 8$	1.78
$t_{seek} = 8$ and $t_{rot} = 6$	1.73
$t_{seek} = 4$ and $t_{rot} = 4$	1.59

Table 3. Throughput ratio between our Improved Some/Some and the Microsoft Some/Some.

4.3 Reducing Read Overhead for our Approach

The worst case read overhead of our Improved Some/Some scheme is the time to read redundant data of the size of a *complete* original block. We present in the following a method that reduces this worst case amount of data read down-to the half of the size of one original block. This method simply consists of storing different sub-blocks not only on one side of one original block, but to distribute them on the left as well as on the right side of the original block. Figure 7 shows an example, where each original block is stored in the middle of two replicated sub-blocks. Let us assume that disk 2 fails and that block 9 must be regenerated. While reading block 7, disk 0 continuous its read process and reads sub-block 9.1. On disk 1, the situation is slightly different. In fact, before reading block 8, sub-block 9.2 is read. In this particular example, no useless data is read, in contrast to the example in Figure 4.

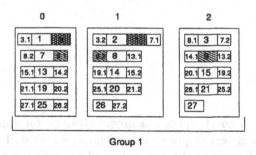

Fig. 7. Reducing read overhead for our approach.

Optimizing the placement of sub-blocks as presented in Figure 7 reduces the worst case read overhead to $\frac{b}{2}$ for disk failure mode. Accordingly, the admission control formula follows Eq. 5. Let us call this new method the **Optimized Some/Some** scheme and its disk throughput Q_d^{OSS}.

$$Q_d^{OSS} = \frac{\tau}{\frac{\frac{3}{2} \cdot b}{r_d} + t_{rot} + t_{seek}} \tag{5}$$

Figure 8 plots the server throughput results of the One/Some, Some/Some, Improved Some/Some, and Optimized Some/Some schemes for different values of data transfer rate r_d. We observe that Optimized Some/Some *slightly* improves server throughput as compared to Improved Some/Some. The ratio decreases as disk transfer rate increases. In fact we notice a 10% improvement in server throughput for $r_d = 20$ MByte/sec, 5% improvement for $r_d = 40$ MByte/sec, and only 2% improvement for $r_d = 80$ MByte/sec.

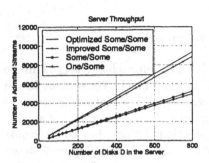
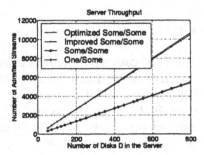

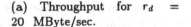

(a) Throughput for $r_d =$ 20 MByte/sec.

(b) Throughput for $r_d =$ 80 MByte/sec.

Fig. 8. Server throughput for One/Some, Some/Some, Improved Some/Some, and Optimized Some/Some with $t_{seek} = 13.79$ ms, $t_{rot} = 10$ ms, $b = 0.5$ Mbit, and $r_p = 1.5$ Mbit/sec..

5 Conclusion

We have proposed a novel replica placement strategy for video servers. In contrast to the classical replica placement schemes, where each single disk of the server is dedicated to original video data and redundancy data, our approach splices replicated data and original data and thus does not require any additional seek time and rotational latency when operating in the presence of disk failures. The results show that our replica placement strategy outperforms, in terms of the server throughput, classical interleaved declustering schemes like the one proposed for the Microsoft Tiger video server. Further, we have seen that our strategy, for pessimistic as well as for optimistic values of disk transfer rates and disk rotational and seek times, always achieves highest throughput as compared to the classical data replication schemes. Finally, we have enhanced our approach to reduce read overhead and noticed a slight increase in the server throughput.

References

1. A. Mourad, "Doubly-striped disk mirroring: Reliable storage for video servers," *Multimedia, Tools and Applications*, vol. 2, pp. 253–272, May 1996.
2. W. Bolosky *et al.*, "The tiger video fileserver," in *6th Workshop on Network and Operating System Support for Digital Audio and Video*, (Zushi, Japan), Apr. 1996.
3. W. Bolosky, R. F. Fritzgerald, and J. R. Douceur, "Distributed schedule management in the tiger video server," in *Proc. Symp. on Operating System Principles*, pp. 212–223, Oct. 1997.
4. D. A. Patterson, G. Gibson, and R. H. Katz, "A Case for Redundant Arrays of Inexpensive Disks (RAID)," in *Proceedings of the 1988 ACM Conference on Management of Data (SIGMOD)*, (Chicago, IL), pp. 109–116, June 1988.
5. F. A. Tobagi, J. Pang, R. Baird, and M. Gang, "Streaming raid(tm) – a disk array management system for video files," in *Proceedings of the 1st ACM International Conference on Multimedia*, (Anaheim, CA), August 1993.
6. S. Berson, L. Golubchik, and R. R. Muntz, "Fault tolerant design of multimedia servers," in *Proceedings of SIGMOD'95*, (San Jose, CA), pp. 364–375, May 1995.
7. S. Ghandeharizadeh and H. K. Seon, "Striping in multi-disk video servers," in *Proceedings in the SPIE International Symposium on Photonics Technologies and Systems for Voice, Video, and Data Communications*, 1995.
8. J. Gafsi and E. W. Biersack, "Data striping and reliablity aspects in distributed video servers," *In Cluster Computing: Networks, Software Tools, and Applications*, February 1999.
9. B. Ozden *et al.*, "Fault-tolerant architectures for continuous media servers," in *SIGMOD International Conference on Management of Data 96*, pp. 79–90, June 1996.
10. B. Ozden *et al.*, "Disk striping in video server environments," in *Proc. of the IEEE Conf. on Multimedia Systems*, (Hiroshima, Japan), pp. 580–589, jun 1996.
11. R. Tewari, D. M. Dias, W. Kish, and H. Vin, "Design and performance trade-offs in clustered video servers," in *Proceedings IEEE International Conference on Multimedia Computing and Systems (ICMCS'96)*, (Hiroshima), pp. 144–150, June 1996.
12. S. A. Barnett, G. J. Anido, and P. Beadle, "Predictive call admission control for a disk array based video server," in *Proceedings in Multimedia Computing and Networking*, (San Jose, California, USA), pp. 240, 251, February 1997.
13. H. I. Hsiao and D. J. DeWitt, "Chained declustering: A new availability strategy for multiprocessor database machines.," in *In Proceedings of the Int. Conference of Data Engeneering (ICDE), 1990*, pp. 456–465, 1990.
14. L. Golubchik, J. C. Lui, and R. R. Muntz, "Chained declustering: Load balancing and robustness to skew and failures," in *In Proceedings of the Second International Workshop on Research Issues in Data Engineering: Transaction and Query Processing, Tempe, Arizona*, (Tempe, Arizona), pp. 88–95, February 1992.
15. J. L. Hennessy and D. A. Patterson, *Computer Arcitecture A Quantative Approach.* Morgan Kaufmann Publishers, Inc., 1990.
16. Seagate Disc Home, http://www.seagate.com/disc/disctop.shtml.
17. C. Ruemmler and J. Wilkes, "An introduction to disk drive modeling," *IEEE Computer*, vol. 27, pp. 17–28, Mar. 1994.
18. B. L. Worthington, G. Ganger, Y. N. Patt, and J. Wilkes, "On-line extraction of scis drive characteristics," in *Proc. 1995 ACM SIGMETRICS*, (Ottawa, Canada), pp. 146–156, May 1995.

Network Bandwidth Allocation and Admission Control for a Continuous Media File Server

Dwight Makaroff, Gerald Neufeld, and Norman Hutchinson
{makaroff,neufeld,hutchinson}@cs.ubc.ca

Department of Computer Science, University of British Columbia
Vancouver, B.C. V6T 1Z4 Canada

Abstract

Resource reservation is required to guarantee delivery of continuous media data from a server across a network for continuous playback by a client. This paper addresses the characterization of the network bandwidth requirements of Variable Bit Rate data streams and the corresponding admission control mechanism at the server. We show that a characterization which sends data early, making intelligent use of client buffer space, reduces the amount of network bandwidth reserved per stream without creating any start-up latency. The results of performance experiments in a Continuous Media File Server find that operation with requests arriving over time can deliver up to 90% of the network bandwidth. The experiments also show that a system designer can configure a server so that the network and disk bandwidth can scale together.

Keywords: multimedia, file servers, variable bit rate, admission control, network transmission

1 Introduction

Continuous media file servers require that several system resources be reserved in order to guarantee timely delivery of the data to end-user clients. These resources include disk, network, and processor bandwidth. In a heterogeneous system accommodating variable bit-rate data streams, the amount of each resource differs for each stream and varies over time. A key component of determining the amount of a resource to reserve is characterizing each stream's bandwidth. Admission control is necessary to ensure adequate server resources for the duration of the playback requested by the user.

In this paper, we examine network bandwidth reservation from both the server's and the client's point of view. The provision of network bandwidth within the network between the server and the client is beyond the scope of this paper and has been addressed extensively in other work [6]. The server is only aware of problems with delivery through feedback from the client.

Two aspects of the network resource management issue are important: the bandwidth usage profile of each individual stream and the combined load on the network interface provided by the requests of all the clients of one server. The remainder of this paper is organized as follows. We begin with a description of the system model, then describe the comparative network allocation algorithms, followed by the network admission control algorithm. The description of the experimental model then provides a framework for the results. This is followed by a comparison of our approach with related work and finally, some conclusions and possible directions for future work.

2 System Model

This study takes place in the context of a Continuous Media File Server (CMFS). This system model is shown in Figure 1. The server is scalable in that multiple disks can be attached to each server node. Multiple server nodes can be configured with a single administrator node. The cumulative data traffic from the set of disks on a single server node provides the bandwidth that this paper characterizes.

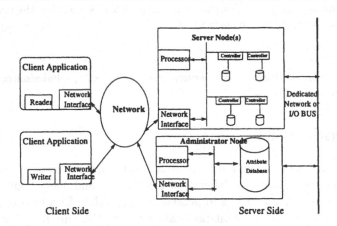

Figure 1: Organization of System

Client applications make requests for continuous media objects from the administrator node, which selects a copy of the object residing on one of the server nodes. A real-time data connection is then set up between the server node and the client to deliver the data at a rate that prevents the client application from starvation. If some small percentage of packets get corrupted or lost, the presentation can continue without loss of satisfaction from the user's point of view. Retransmissions can cause unacceptable latency [1].

Guaranteeing adequate bandwidth requires network resource reservation. This may be done in the form of a VBR connection in an ATM network, with statistical transmission guarantees. Cells may be lost due to transient overload. Such "capacity losses" (or "congestion losses") may invalidate the client's assumption on the expected error or loss rate, and may interfere with the client's ability to provide continuous playback [1][11]. We have selected instead to use CBR connections which can have bandwidth renegotiated, providing a small amount of overhead to the operation of the system.

If the network bandwidth cannot be maintained throughout stream delivery, some change to the delivery parameters is necessary. Unfortunately, the server does not know what adjustments would be appropriate for the client, nor if the client would be able to interpret the reduced amount of data that would be sent under the adjusted data rate. The client application requests a new delivery rate, which may have fewer frames per second, or involve skipping some sequences of the object. In this paper, we assume that the bandwidth of an individual server-client connection can always be maintained.

3   Network Block Schedule Creation

The network resource usage of a particular stream may be characterized in many ways. The tightest upper bound is the empirical envelope [6], which has been a basis for much of the previous work in this area. It results in a conservative, piece-wise linear function, specified by a set of parameters, but requires $O(n^2)$ time to compute (where n is the number of frames in the stream). Approximations have been developed based on leaky bucket schemes, but the results have still utilized the entire stream to calculate the bandwidth profile off-line. In the system model of the CMFS, the schedule must be created at request delivery time, because each play request could select different portions of the object (slow motion, skipping sequences) as in [9].

It is possible to give a single value for bandwidth characterization (such as the average bandwidth), and let the network infrastructure deal with transient overloads in the network. Such an allocation algorithm faces two main problems: client starvation and server buffer space. Sending at the average rate for the entire duration of stream delivery does not ensure that enough data will be present in the client buffer to handle peaks in the bandwidth which occur early in the stream. It is possible to prefetch data, but this introduces start-up latency and requires a large client buffer. Parameterized variants of average bandwidth allocation with intelligent discarding[1] of data at the server have shown reasonably good results. Both client buffer size and start-up latency have been parameters in previous research [13] where reductions in either buffer space or latency can

---

[1] requiring server knowledge of encoding formats.

be achieved. An approach which utilizes the VBR profile is essential to reduce both of these values simultaneously. A study of the effect of packet loss over the Internet for MPEG streams [1] shows that enhanced error concealment and/or error resilience techniques in the stream can reduce the apparent loss of quality in a manner transparent to a server such as the CMFS.

In keeping with the philosophy of admission control and resource usage characterization in the UBC CMFS, we have chosen to divide the time period during which data is transmitted into *network slots*, and provide a detailed schedule of the bandwidth needed in terms of a *network block schedule*. A network slot is an even multiple of the disk slot time. This schedule allows the system to transmit data at a constant rate within a network slot, known in other literature as Piecewise Constant Rate Transmission and Transport [2][8]. The size of a network slot is significantly larger than a disk slot for two main reasons: overhead of renegotiation and smoothing capability. A renegotiation takes a non-trivial amount of time and should be effective for more than a disk slot time. As well, the ability to smooth out the data delivery by sending data earlier in the network slot than is absolutely required increases performance. This utilizes the available client buffer space. Other research has experimented with the size of network slots in the range of 10 seconds to 1 minute [4] [14]. Zhang and Knightly [14] suggest that renegotiations at 20 second intervals provide good performance. We have used 20 seconds as the size of the network slot for the initial experiments.

Our initial algorithm (hereafter called *Original*) considers only the number of bytes that are required to be sent in each network slot. The cumulative average number of bytes per disk slot is calculated for each disk slot in the network slot. The maximum value encountered in the current network slot is rounded up to the next highest number of disk blocks (64 KBytes). This method has the advantage of absorbing peaks in the disk block schedule by assuming that the server can send at the specified rate for the entire network slot. Peaks which occur late in the network slot have marginally less influence in the cumulative average and will be absorbed easily as shown in Figure 2. Here, the first three large peaks in disk bandwidth at slots 68, 94, and 136 do not increase the reservation. If a peak in disk bandwidth occurs early in a network slot, then the maximum cumulative average is near this peak (disk slots 201 and 241).

Our server-based flow control policy [10] takes advantage of the client buffer by sending data to the client as early as possible, without overflow. Since the value used in the network block schedule is the maximum cumulative average, it is likely that some data will be present in the client buffer at the beginning of the next network slot.

The second algorithm improves on the first by explicitly accounting for sending data early. In nearly all cases, there is sufficient excess bandwidth to fill the client buffer. This reduces the amount of bandwidth that must be reserved for each subsequent slot, smoothing the network block schedule, and thus, we call it the *Smoothed* algorithm. A peak in disk bandwidth that occurs very early in a network slot could be merged with the previous network slot. Figure 3 shows the smoothed network block schedule for the same stream taking into account

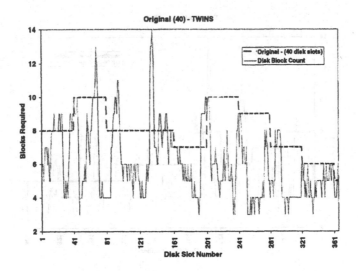

Figure 2: Network Block Schedule - Original

network send-ahead. Increased client buffer space enables smoothing to be more effective at reducing both the peaks and the overall bandwidth necessary [2].

A significant complication in the design is the assumption that the disk system has achieved sufficient read-ahead such that the buffers are available in memory for sending. The disk admission control algorithm utilized in the CMFS only guarantees that disk blocks will be available for sending at the end of the slot which they are required to be sent [9]. The disk subsystem guarantees a minimum bandwidth in every disk slot, for disk admission control. For a disk which is under heavy load, it is possible that the disk peaks which we have been trying to smooth at the network level will not be read off the disk when needed. If this is the case, the network bandwidth value must be increased above the cumulative average in order to transmit this peak amount when required.

The disk admission algorithm [9] guarantees that in steady state, the guaranteed bandwidth from the disk is always sufficient to service the accepted streams. In fact, the achieved disk bandwidth is greater than this value, because disk performance is variable and the average performance is somewhat above the guarantee. Thus, over time, all buffer space will be utilized by this aggressive read-ahead. The level of bandwidth for the accepted set of streams will always be lower than the capacity of the disk.

The issue of buffer space is slightly more complicated. In steady state, there are no buffers in which to read any blocks for the new stream, except those being returned to the system after being transmitted across the network. Buffers may be "stolen" from existing streams if the data is not needed until later than the deadline for the new stream. In the operation of the server, staggered request

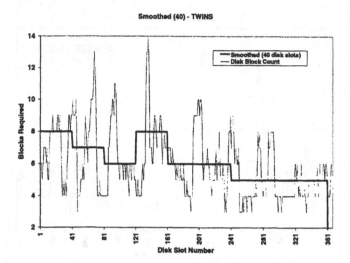

Figure 3: Network Block Schedule - Modified

arrivals and buffer stealing often results in a significant amount of contiguous reading when the new stream is accepted, increasing the bandwidth and the read-ahead achieved.

For example, consider video streams of approximately 4 Mbps, a typical value for average (TV) quality. If there are 5 currently accepted video streams and 64 MBytes of server buffer space, each stream would have approximately 12 MBytes of buffer space (or 24 seconds of video). If a new stream is accepted, there would be 10.6 MBytes per stream in steady state (or 20 seconds of video). This amount of data could accumulate from the disk in about 3 seconds, so that steady state is achieved rather quickly. The only time that the server would not have read ahead at least 20 seconds is during the first few slots of reading. With staggered arrival patterns, the server is reading from only one stream immediately after acceptance, and so the disk is substantially ahead after the first disk slot. The steady state will be reached soon enough that none of the borderline cases of buffer space and bandwidth will be encountered. Smoothing the bandwidth usage of each stream is a reasonable course of action, which reduces the resource reservation and potentially permits more simultaneous streams.

4   Network Admission Control Algorithm

Once we have achieved a suitable network bandwidth characterization for each stream, the stream requests are submitted to a network admission control algorithm that determines if there is enough outgoing network bandwidth to support these requests. The network admission control algorithm used in the CMFS is relatively simple. The maximum number of bytes that the network interface can transmit per second is easily converted to the number of blocks per disk slot,[2] which we hereafter refer to as *maxXmit*. The algorithm is shown in Figure 4 and can be summarized as follows: for each network slot, the bandwidth values for each stream are added, and as long as the sum is less than *maxXmit*, the scenario is accepted.

Requests which arrive in the middle of a network slot are adjusted so that the network slot ends for each stream simultaneously. Thus, such a stream has less opportunity to fill the client buffer in that first network slot. In the sample streams this made very little difference in the overall bandwidth required for the network block schedule, although the initial shape did differ somewhat. It did not change the overall distribution of bandwidth.

```
NetworkAdmissionTest( newStream, networkSlotCount )
begin
    for netwSlot = 0 to networkSlotCount do
        sum = 0

        for i = firstConn to lastConn do
            sum = sum + NetBlocks[netwSlot]
            if (sum > maxXmit) then return (REJECT)
        end
        return (ACCEPT)
    end
end
```

Figure 4: Network Admission Control Algorithm

The network admission control algorithm is the same algorithm that was called the "Instantaneous Maximum" disk admission control algorithm in our previous work [7]. This algorithm was rejected in favour of the *vbrSim* algorithm that took advantage of aggressive read-ahead in the future at the guaranteed rate or aggressive read-ahead in the past at the achieved rate. The *vbrSim* algorithm could be considered for network admission control. The smoothing effect enabled by sending data early could further eliminate transient network bandwidth peaks. One major benefit of *vbrSim* is the ability to use the server buffer space to store the data which is read-ahead. This buffer space is shared by all the streams and thus, at any given time, one connection can use several Megabytes, while another may use only a small amount of buffer space. For scenarios with cumulative bandwidth approaching capacity, significant server buffer space is required to enable acceptance.

---

[2]For disk blocks of 64 KBytes and disk slots of 500 msec, 1 Mbps is approximately 1 Block/slot.

If the same relative amount of buffer space was available at each client, then network send-ahead could be effective. The server model only requires two disk slot's worth of buffer space, and so, very little send-ahead is possible. Even this amount of buffer space is large compared with the minimum required by a decoder. For example, according to the MPEG-2 specifications, space for as few as three or four frames is required.

5   Experimental Design

In order to examine the admission performance of our network admission control algorithm, we loaded a CMFS with several representative VBR video streams on several disks. Each disk contained 11 streams. Then we presented a number of stream request scenarios for streams which were located on the same disk to determine which of the scenarios could be accepted by the vbrSim disk admission control algorithm. The initial selection of the streams for each scenario was done choosing a permutation of streams in such a manner as to have the same number of requests for each stream for each size of scenario. Thus, there were 33 scenarios that contained 7 streams and each of the 11 streams was selected $33*7/11 = 21$ times, and 33 scenarios that contained 6 streams. There were also 33 scenarios of 5 streams each and 44 of 4 streams each. When arrival times of the streams were staggered, the streams were requested in order of decreasing playback time to ensure that all streams in a scenario were active at some point in the delivery time. The scenarios for each disk were then combined with similar scenarios from other disks and the network admission control algorithm was used to determine whether or not the entire collection of streams could be accepted by a multi-disk server node. The admission control algorithm was not evaluated in a running CMFS, due to limitations in the measurement techniques employed.

A summary of the stream characteristics utilized in these experiments is given in Table 1. Each disk has a similar mix of streams that range from 40 seconds to 10 minutes with similar averages in variability, stream length, and average bandwidth. The variability measure reported is the coefficient of variation (Standard Deviation/Mean) of the number of blocks/slot.

6   Results

In this section, we compare the results of the Original algorithm with the Smoothed algorithm. The first observation that can be made is that the average bandwidth *reservation* is significantly greater than the average bandwidth *utilization*. When averaged over all scenarios, the Smoothed algorithm reserves significantly less bandwidth than the Original algorithm (113.3 Mbps versus 122.8 Mbps), both of which exceed the bandwidth utilization of 96.5 Mbps. Thus, it is reasonable to expect that the Smoothed algorithm will provide better admission performance results.

	Disk 1	Disk 2	Disk 3	Disk 4
Largest B/W	5.89 Mbps	6.03 Mbps	6.69 Mbps	7.28 Mbps
Smallest B/W	2.16 Mbps	3.33 Mbps	2.9 Mbps	1.71 Mbps
Average B/W	4.16 Mbps	4.89 Mbps	4.61 Mbps	4.61 Mbps
Std. Dev. B/W	1.15 Mbps	0.93 Mbps	1.07 Mbps	1.64 Mbps
Largest Variability	.43	.35	.354	.354
Smallest Variability	.184	.154	.185	.119
Average Variability	.266	.233	.251	.262
Longest Duration	574 secs	462 secs	625 secs	615 secs
Shortest Duration	95 secs	59 secs	52 secs	40 secs
Average Duration	260 secs	253 secs	311 secs	243 secs
Std. Dev of Duration	160 secs	139 secs	188 secs	181 secs

Table 1: Stream Characteristics

We grouped the scenarios with respect to the relative amount of disk bandwidth they request, by adding the average bandwidths of each stream and dividing by the bandwidth achieved on the particular run. The achieved bandwidth is affected by the placement of the blocks on the disk and the amount of contiguous reading that is possible.

In the first experiment, 193 scenarios were presented to a single-node CMFS configured with 4 disks. Each disk had a similar request pattern that issued requests for delivery of all the streams simultaneously. Table 2 gives a summary of admission performance with respect to number of scenarios in each request range that could be accepted by both the network admission algorithm and the disk admission algorithm on each disk, which were fewer than 193.

Pct Band	Number of Scenarios	Disk Accepted	Original Accepted	Smoothed Accepted
95-100	0	0	0	0
90-94	8	0	0	0
85-89	6	0	0	0
80-84	9	3	0	0
75-79	26	7	0	3
70-74	21	15	2	14
65-69	34	33	18	33
60-64	25	25	23	25
55-59	10	10	10	10
50-54	2	2	2	2
Total	141	95	56	87

Table 2: Admission Performance: Simultaneous Arrivals (% of Disk)

The four disks were able to achieve between 110 and 120 Mbps. The scenario with the largest cumulative bandwidth that the Smoothed algorithm could accept was 93 Mbps, as compared with 87.4 Mbps for the Original algorithm. In this set of scenarios, the requested bandwidth varied from approximately 55% to 95% of the achievable disk bandwidth. The original algorithm accepts only a small percentage (2/15) of the scenarios within the 70-74% request range and approximately half the requests in the band immediately below. With the Smoothed algorithm, about half the requests in the 75-79% request range are accepted, and nearly all in the 70-74% range. The Smoothed algorithm increases network utilization by approximately 10 to 15%.

One major benefit of *vbrSim* is the ability to take advantage of read-ahead achieved when the disk bandwidth exceeded the minimum guarantee. This is enhanced when only some of the streams are actively reading off the disk, reducing the relative number of seeks, producing a significant change in admission results. The achieved bandwidth of the disk increases by approximately 10%, with only 9 of the 193 scenarios rejected by the disk system and the network block schedules are slightly different.

Pct Band	Number of Scenarios	Disk Accepted	Original Accepted	Smoothed Accepted
95-100	29	22	0	0
90-94	9	7	0	0
85-89	12	12	0	0
80-84	14	14	0	0
75-79	5	5	0	3
70-74	22	22	1	7
65-69	23	23	5	21
60-64	34	34	10	34
55-59	25	25	24	25
50-54	17	17	17	17
45-49	2	2	2	2
Total	193	184	59	103

Table 3: Admission Performance: Staggered Arrivals (% of Disk)

Table 3 shows admission decisions under staggered arrival. The Original algorithm performed significantly worse in terms of percentage of bandwidth requests that are accepted. As mentioned before, many of the scenarios move to a lower percentage request band, due to the increase in achieved bandwidth from the disk. This shows that the increase in disk bandwidth achieved due to stagger was greater than the increase in the amount of accepted network bandwidth. For the Smoothed algorithm, relative acceptance rates are unchanged. The ability to accept streams at the network level and at the disk level have kept up with the increase in achieved bandwidth off the disks.

Another experiment examined the percentage of the network bandwidth that can be accepted. The results of admission for the simultaneous arrivals and the staggered arrivals case are shown in Tables 4 and 5. We see that smoothing is an effective way to enhance the admission performance. A maximum of 80% of the network bandwidth can be accepted by the Original algorithm on simultaneous arrivals, although most of the scenarios in that range are accepted. The smoothing operation allows almost all scenarios below 80% to be accepted, along with a small number with greater bandwidth requests.

Pct Band	Number of Scenarios	Original Accepted	Smoothed Accepted
95-100	0	0	0
90-94	5	0	0
85-89	4	0	2
80-84	18	1	17
75-79	32	19	32
70-74	19	18	19
65-69	11	11	11
60-64	2	2	2
Total	91	51	85

Table 4: Admission Performance: Simultaneous Arrivals (% of Network)

Pct Band	Number of Scenarios	Original Accepted	Smoothed Accepted
95-100	5	0	0
90-94	19	0	2
85-89	15	2	14
80-84	27	3	27
75-79	29	18	29
70-74	22	22	22
65-69	11	11	11
60-64	2	2	2
Total	131	59	106

Table 5: Admission Performance: Staggered Arrivals (% of Network)

In Table 5, we see that the maximum bandwidth range requested and accepted by the disk subsystem approaches 100 Mbps. None of these high bandwidth scenarios are accepted by either network admission algorithm. A few scenarios between 80% and 90% can be accepted with the Original algorithm. The Smoothed algorithm accepts nearly all requests below 90% of the network bandwidth, due to the fact that a smaller number of streams are reading and transmitting the first network slot at the same time. With staggered arrivals, all streams but the most recently accepted stream are sending at smoothed rates, meaning lower peaks for the entire scenario.

The results of these experiments enable an additional aspect of the CMFS design to be evaluated: scalability. It is desirable that the disk and network bandwidth scale together. In the configuration tested, 4 disks (with $minRead = 23$) provided 96 Mbps of guaranteed bandwidth with a network interface of 100 Mbps. At this level of analysis, it would seem a perfect match, but the tests with simultaneous arrivals did not support this conjecture. A system configured with guaranteed cumulative disk bandwidth approximately equal to nominal network bandwidth was unable to accept enough streams at the disk in order to use the network resource fully. There were no scenarios accepted by the disk that requested more than 94% of the network bandwidth. In Table 4, there are only 4 scenarios in the 85-89% request range, that were accepted by the disk system. In Table 5, there were 15 such scenarios. This increase is only due to the staggered arrivals as the same streams were requested in the same order.

With staggered arrivals, the network admission control became the performance limitation, as more of the scenarios were accepted by the disk. There were no scenarios that requested less than 100 Mbps that were rejected by the disk. This arrival pattern would be the common case in the operation of a CMFS. Thus, equating disk bandwidth with network bandwidth is an appropriate design point which maximizes resource usage for moderate bandwidth video streams of short duration if the requests arrive staggered in time.

7   Related Work

The problem of characterizing the network resource requirements of Variable Bit Rate audio/video transmission has been studied extensively. Zhang and Knightly [14] provide a brief taxonomy of the approaches from conservative peak-rate allocation to probabilistic allocation using VBR channels of networks such as ATM.

The empirical envelope is the tightest upper bound on the network utilization for VBR streams, as proven in Knightly et al. [6], but it is computationally expensive. This characterization has inspired other approximations [3] which are less accurate, less expensive to compute, but still provide useful predictions of network traffic.

Traffic shaping has been introduced to reduce the peaks and variability of network utilization for inherently bursty traffic. Graf [3] examines live and stored video and provides traffic descriptors and a traffic shaper based on multiple leaky-buckets. Traffic can be smoothed in an optimal fashion [12], but requires a-priori calculation of the entire stream. If only certain portions of the streams are retrieved (i.e. I-frames only for a fast-motion low B/W MPEG stream delivery), the bandwidth profile of the stream is greatly modified.

Four different methods of smoothing bandwidth are compared by Feng and Rexford [2], with particular cost-performance tradeoffs. The algorithms they used attempt to minimize the number of bandwidth changes, and the variability in network bandwidth, as well as the computation required to construct the schedule. They do not integrate this with particular admission strategies other than peak-rate allocation. This bandwidth smoothing can be utilized in a system that uses either variable bit rate network channels or constant bit rate channels. Recent work in the literature has shifted the focus away from true VBR on the network towards variations of Constant Bit-Rate Transmission [8],[14]. Since the resource requirements vary over time, renegotiation of the bandwidth [4] is needed in most cases to police the network. This method is used by Kamiyama and Li [5] in a Video-On-Demand system. McManus and Ross [8] analyze a system of delivery that prefetches enough of the data stream to allow end-to-end constant bit rate transmission of the remainder without starvation or overflow at the client, but at the expense of substantial latency in start-up. Indications are that minimum buffer utilization can be realized with a latency of between 30 seconds and 1 minute [13]. For short playback times (less than 5 minutes) that may be appropriate for news-on-demand, such a delay would be unacceptable.

8 Conclusions and Further Work

In this paper, we have presented a network bandwidth characterization scheme for Variable Bit Rate continuous media objects which provides a detailed network block schedule indicating the bandwidth needed for each time slot. This schedule can be utilized to police the bandwidth allocated for each network channel via sender-based rate control, or network-based renegotiation.

We observed that the Original algorithm was susceptible to disk bandwidth peaks at the beginning of network slots. The Smoothed algorithm was introduced, taking advantage of client buffer space and excess network bandwidth that must be reserved, for a reduced overall reservation.

The network admission algorithm provides a deterministic guarantee of data transmission, ensuring that no network slot has a cumulative bandwidth peak over the network interface bandwidth. Scenarios with simultaneous arrivals were limited by the disk subsystem. The disk admission control method [7] used in the CMFS, when combined with staggered arrivals, showed that the same disk configuration shifted the bottleneck to the network side. The network admission control algorithm and the smoothed network bandwidth stream characterization combined to provide an environment where scenarios that request up to 90% of the network interface can be supported.

These experiments utilized a single value for the size of the network slot and a single granularity for the block size. Extensions to this work could include comparing admission results with different values for these two parameters.

References

[1] Jill M. Boyce and Robert D. Gaglianello. Packet Loss Effects on MPEG Video Sent Over the Public Internet. In *ACM Multimedia*, Bristol, England, September 1998.

[2] Wu-Chi Feng and Jennifer Rexford. A Comparison of Bandwidth Smoothing Techniques for the Transmission of Prerecorded Compressed Video. In *IEEE Infocomm*, pages 58–66, Los Angeles, CA, June 1997.

[3] Marcel Graf. VBR Video over ATM: Reducing Network Requirements through Endsystem Traffic Shaping. In *IEEE Infocomm*, pages 48–57, Los Angeles, CA, June 1997.

[4] M. Grossglauser, S. Keshav, and D. Tse. RCBR: A Simple and Efficient Service for Multiple Time-Scale Traffic. In *ACM SIGCOMM*, pages 219–230, Boston, MA, August 1995.

[5] N. Kamiyama and V. Li. Renegotiated CBR Transmission in Interactive Video-on-Demand Systems. In *IEEE Multimedia*, pages 12–19, Ottawa, Canada, June 1997.

[6] E. W. Knightly, D. E. Wrege, J. Liebeherr, and H. Zhang. Fundamental Limits and Tradeoffs of Providing Deterministic Guarantees to VBR Video Traffic. In *ACM SIGMETRICS '95*. ACM, 1995.

[7] D. Makaroff, G. Neufeld, and N. Hutchinson. An Evaluation of VBR Admission Algorithms for Continuous Media File Servers. In *ACM Multimedia*, pages 143–154, Seattle, WA, November 1997.

[8] J. M. McManus and K. W. Ross. Video on Demand over ATM: Constant-Rate Transmission and Transport. In *IEEE InfoComm*, pages 1357–1362, San Francisco, CA, October 1996.

[9] G. Neufeld, D. Makaroff, and N. Hutchinson. Design of a Variable Bit Rate Continuous Media File Server for an ATM Network. In *IST/SPIE Multimedia Computing and Networking*, pages 370–380, San Jose, CA, January 1996.

[10] G. Neufeld, D. Makaroff, and N. Hutchinson. Server-Based Flow Control in a Continuous Media File System. In *6th International Workshop on Network and Operating Systems Support for Digital Audio and Video*, pages 29–35, Zushi, Japan, 1996.

[11] Ranga Ramanujan, Atiq Ahamad, and Ken Thurber. Traffic Control Mechanism to Support Video Multicast Over IP Networks. In *IEEE Multimedia*, pages 85–94, Ottawa, Canada, June 1997.

[12] J. D. Salehi, Z. Zhang, J. F. Kurose, and D. Towsley. Supporting Stored Video: Reducing Rate Variability and End-to-End Resource Requirements through Optimal Smoothing. In *ACM SIGMETRICS*, May 1996.

[13] Subrahata Sen, Jayanta Dey, James Kurose, John Stankovic, and Don Towsley. CBR Transmission of VBR Stored Video. In *SPIE Symposium on Voice Video and Data Communications: Multimedia Networks: Security, Displays, Terminals, Gateways*, Dallas, TX, November 1997.

[14] H. Zhang and E. W. Knightly. A New Approach to Support Delay-Sensitive VBR Video in Packet-Switched Networks. In *5th International Workshop on Network and Operating Systems Support for Digital Audio and Video*, pages 381–397, Durham NH, April 1995.

Design and Evaluation of Ring-Based Video Servers

Christophe Guittenit and Abdelaziz M'zoughi

Institut de Recherche en Informatique de Toulouse (IRIT),
Université Paul Sabatier,
31400 Toulouse, France
{guitteni,mzoughi}@irit.fr

Abstract. Video-on-demand servers (VODS) must be based on cost-effective architectures. Therefore architectures based on clusters of PCs will probably be the most suitable to build VODS; they provide the same performance of a large server through the aggregation of many smaller, inexpensive nodes. In this paper, we show that the interconnection network used to build the cluster can be very expensive if it is based on a set of switches. To obtain the most cost-effective architecture, we argue that VODS must be Dual Counter-Rotating Ring-based (DCRR-based). DCRR are very inexpensive and fulfill all the basic criteria needed for VODS architectures except scalability. To address the scalability issue, we propose to enhance the design of DCRR by partitioning it logically in combination with three new policies ("fast stream migration", "stream splitting" and "distributed XORing"). This design brings very cost-effective and scalable VODS able to play up to 13500 MPEG-2 concurrent streams using 252 nodes.

1 Introduction

Video-on-demand is a mainly commercial application; the more inexpensive it is, the more it will succeed. Building cost-effective video-on-demand servers (VODS) is a complex task; these storage systems are those implying many constraints. A VODS must serve many continuous data streams simultaneously, store a large amount of data, be fault-tolerant and offer a latency lower than one second for some applications (e.g. News-on-Demand).

In order to reduce costs, *clustered video server* [11] is the solution generally used to interconnect a large number of hard drives in an expandable way. It provides the equivalent performance of a large server through the combination of many smaller, relatively inexpensive nodes. The disks are grouped within storage nodes (SN). Accessing the delivery network outside the VODS is carried out via interface nodes (IN). All the nodes are distributed over the *interconnection network* (figure 1).

So as to get the most inexpensive server, PCs are generally used to make the nodes. It is the case in this paper. The nodes are sufficiently autonomous so

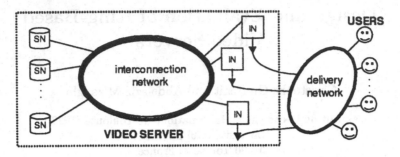

Fig. 1. Clustered video server.

that the extension of VODS is as simple as just adding SNs or INs. There are no centralized resources: a resource failure cannot interrupt all the users' services.

The interconnection network is the most important resource of clustered architectures. To build expandable architectures, it is necessary that the interconnection network links many nodes easily, that is, its bandwidth as well as the number of ports must be sufficient. To build fault-tolerant architectures, the interconnection network must allow hot swap in order to add or remove nodes without interrupting the VOD services. Similarly, it must have enough redundant paths to avoid service interruptions in case of a link or a node failure.

To obtain the most cost-effective architecture, we suggest in this paper to build Dual Counter-Rotating Ring-based (DCRR-based) VODS, principally because this interconnection network is very inexpensive. At first sight, DCRR does not include enough links to build scalable VODS. To address the scalability issue, we propose to enhance the design by partitioning logically the DCRR. However partitioning causes load imbalance in the VODS, so we propose and evaluate two new policies to replicate videos dynamically when the load imbalance occurs. These policies (*fast stream migration* and *stream splitting*) accelerate the replication process by up to 10 times. We propose also a third policy (*distributed XORing*) that reduces the bandwidth used to recover lost data when an SN fails in a parity RAID. These three policies allow the use of DCRR to build scalable VODS able to play up to 13500 MPEG-2 concurrent streams (using a cluster of 252 nodes).

In section 2, we evaluate quantitatively the cost of some interconnection networks and show that DCRR-based architectures are very cost-effective. Section 3 presents the DCRR, and exposes and compares several types of DCRR-based interfaces. It gives the advantages of a DCRR and also its drawbacks: its insufficient scalability and the overload caused by a rupture of links. In section 4, we introduce the concept of DCRR logical partitioning and show the trade-off concerning this method: the partitioning lessens load on the DCRR but increases the workload imbalance. In section 5, we present the system model we use to simulate the behavior of the DCRR-based VODS. We explore, in section 6, the partitioning trade-off with experimental evaluations. We show in section 7 how to balance the load in a partitioned DCRR with fast video replications by using *fast stream migration* and *stream splitting*. In section 8, we measure the overload caused by a failure in the system and we propose *distributed XORing* to reduce this overload. We end with a summary and some additional remarks in section 9.

2 Cost evaluation of switch-based networks and ring-based networks

Interconnection networks built using standard devices are less expensive than dedicated networks like those in MPPs. Thus we focus in this section on the following standard interfaces: ATM-based LAN networks [2, 5], Fibre Channel (FC) [12], Serial Storage Architecture (SSA) [1] and Sebring Ring [13]. Most of these interfaces can be found in two classes of topologies: switch-based and ring-based (see figure 2).

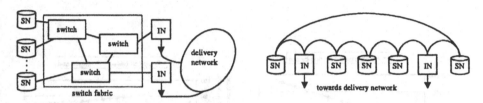

Fig. 2. A switch-based interconnection network and a ring-based interconnection network. Switch-based networks include generally more links than ring-based networks for the same number of nodes.

ATM-based LAN networks can be switched (as with ATM WAN or MAN) or ring-based (MetaRing, ATMR, CRMA-II,...). FC and SSA interfaces are usually used to replace the SCSI-bus in I/O architectures. However these interfaces are greatly versatile and can be applied to interconnection networks: switch-based or ring-based. Fibre Channel becomes a FC-Arbitred Loop (FC-AL) in its ring-based variant. The Sebring Ring is only ring-based.

In this section, we evaluate quantitatively the cost difference between switch-based ATM interfaces and ring-based ATM interfaces. We estimate the cost of the interconnection network and of the VODS as a function of the number of SNs interconnected. We choose to evaluate ATM-based interfaces because given that the delivery network is also ATM based, the INs can be removed by connecting the interconnection outputs directly to ATM switches in the delivery network. Note that the IN removal has an impact on the SN management. SNs must then be able to achieve operations dedicated to INs. For example, QoS protocol between the proxy and users, service aggregation or migration...

To evaluate the switch-based ATM interconnection network, we need to build $n \times n$ switch fabrics (where n is the number of SNs). A $n \times n$ switch fabric is split, into smaller switches, using a Clos network. Note that this network has a much larger bandwidth than a ring-based network and thus, we do not compare two interconnection networks equivalent in performance. Our objective is to show that the interconnection network can be the most expensive resource in a VODS.

Figure 3 shows the results of our cost evaluation, and table 1 shows the parameters used to determine the cost of the architectures.

Table 1. Cost parameters (source: www.buymicro.com).

Device	Price in $US
ATM switch 8 ports (OC-3 155 Mbit/s)	6660
ATM switch 14 ports (OC-3 155 Mbit/s)	9500
ATM switch 52 ports (OC-3 155 Mbit/s)	89000
PCI/ATM interface card	500
SN (PC Intel PentiumII 400, 128 MB, 4 IDE 10 GB disks)	3000

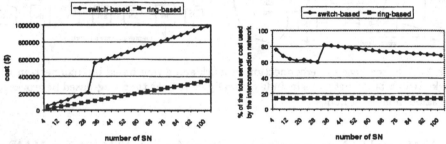

Fig. 3. Cost comparison of a Clos network-based interconnection network and a ring-based interconnection network.

Our evaluation confirms that Clos network-based architectures are much more expensive than ring-based architectures. It is due to the high cost of the switches that corresponds to close to 80% of the total VODS cost. The interconnection network multiplies the VODS cost by more than 2.

Figure 3 also shows that the ring cost evolves proportionally to the number of SNs, contrary to switch-based networks for which the cost evolves in fits and starts. These jerks are due to the fact that more powerful (and so more expensive) switches must be used when the traffic grows. These results can be extended to FC or SSA.

This evaluation shows that the use of an inexpensive interconnection network reduces drastically the price of a VODS: ring-based networks can lead to truly cost-effective VODS. The ring network is inexpensive because it includes few links. Our objective in the following sections, is to find policies that can deal with the lack of links to build scalable VODS.

3 Dual Counter-Rotating Rings (DCRR)

For fault-tolerance reasons, the ring is actually made up of two unidirectional sub-rings; data transmission on a sub-ring is carried out in the opposite direction to transmission on the other ring: it forms a dual counter-rotating ring. This topology is equivalent to a single ring composed of bi-directional links. The DCRR is thus tolerant of a node or a link failure. When a node or a link fails, both sub-rings are broken in the same place. The four extremities of the two broken rings are joined to form a new ring (figure 4).

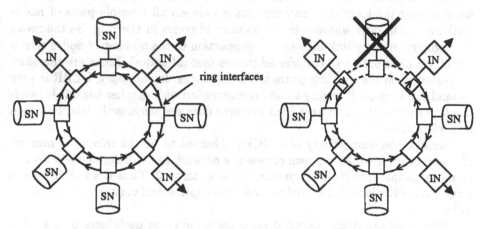

Fig. 4. A Dual Counter-Rotating Ring working in normal mode and in degraded mode.

Table 2 shows interfaces that could be used to form a DCRR-based interconnection network. A very important feature concerning this type of network is *spatial reuse*, i.e., the ability to carry out concurrent transmissions on independent links. FC-AL is the only interface that does not allow spatial reuse. In this case, the ring is viewed as a unique resource and a node must get control of the ring before transmitting its packet (like on a bus). We will see in section 4, that we need spatial reuse to partition the DCRR and then to build scalable VODS.

Table 2. Comparison of ring-based interfaces. Every interface allows hot swap.

Interfaces	ATMR	SSA	FC-AL	Sebring Ring
Max. throughput on one sub-ring (MB/S)	312 (OC-48)	20	125	532
Spatial reuse	Yes	Yes	No	Yes

SSA allows spatial reuse but the maximum throughput on its links is not sufficient to build large servers. We will see in section 6 that high throughputs on links reduce the load imbalance in large VODS.

Both ATMR and Sebring Ring have enough bandwidth and concurrency to be good interconnection networks for VODS. At IRIT (Institut de Recherche en Informatique de Toulouse, Computer Science Research Institute of Toulouse), we are developing a prototype[1] of a VODS based on the Sebring Ring. We chose the Sebring Ring because of its large bandwidth and because this network is a PCI-PCI bridge. It makes the construction of PC-based architecture easier and it allows PCI cards to be directly connected on the ring (i.e. the cards do not have to be in a PC). This feature is very useful to alleviate the potential bottleneck that exists in each IN.

The main drawback of a DCRR is that its diameter (defined as the maximum among the lengths of shortest paths between all possible pairs of nodes [14]) evolves in $o(n)$ where n is the number of nodes in the ring. So the mean throughput on every link increases in proportion to the number of nodes (given a uniform distribution of transfer addresses over all nodes). Links are therefore saturated when the ring integrates too many nodes. In summary, DCRR is very expandable because it is easy to add resources but on the other hand it is insufficiently scalable because an added resource does not necessarily lead to a gain in performance.

Besides, the connectivity of a DCRR (defined to be the minimum number of nodes or links whose removal causes the network to be disconnected into two or more components [14]) is too small: it is equal to 2. Thus if a link breaks, a large part of the load is diverted to avoid the rupture and clutters the remaining links.

These two drawbacks could dramatically limit the usefulness of a DCRR for large VODS. In the next sections, we will show how to lessen these two drawbacks.

4 Logical partitioning of a DCRR-based VODS

Disk array partitioning has been proposed in [3, 6, 7] to improve fault-tolerance and to make the management of the RAIDs easier. Partitioning increases the performance of storage systems that include a large number of disk drives [3]. In this section we show that partitioning is also very interesting to build large clustered VODS based a on low-performance interconnection network like DCRR. Figure 5 illustrates our proposition to adapt the partitioning to a DCRR-based VODS.

Each SN is the PC described in table 1 and INs contain PCI/ATM 155 Mb/s interface cards. Data are striped in order to distribute the workload uniformly over all SNs. Redundant information is distributed on every SN and allows data to be recovered in case of a SN failure: a RAID is created. So different SNs can serve a user during the same service. On the other hand, during a service, data is always transmitted by the same IN. Indeed, transfers on the delivery

---

[1] This project is financed by the Conseil Régional de Midi Pyrénées (Regional Council of Midi-Pyrénées) and the CNRS (Centre National de la Recherche Scientifique, National Center of Scientific Research).

network, between the INs and the users, are carried out using connected mode protocols with resource reservation at the connection time (e.g. ATM protocol or IPv6 when using RSVP). Changing the IN during a service would require a new connection and therefore new resource reservation. Thus, a packet emitted by an SN can cover many links (at maximum: half of the total number of links in one sub-ring) before reaching the right IN.

In order to reduce the accumulation of throughput on links near INs (as shown in RAID A, figure 5), *spatial reuse* is used to partition the ring into several independent RAIDs and thus to form a logically partitioned DCRR. SNs are grouped around one or several INs. Every RAID contains whole videos.

The *partitioning degree* is defined as the number of independent RAIDs that the VODS includes.

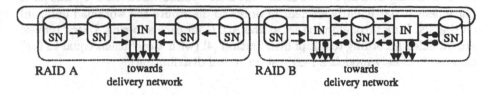

RAID A   towards   RAID B   towards
        delivery network   delivery network

Fig. 5. Partitioning of a DCRR into two independent RAIDs.

DCRR-partitioning improves scalability and fault-tolerance. Scalability is no longer limited by the maximum throughput on a link and fault-tolerance is improved because the server can tolerate multiple failures provided that they occur in different RAIDs. Moreover, fewer users are concerned by an SN failure (since RAIDs are smaller, multiple and independent).

Note that logical partitioning is very different from physical partitioning. In the latter case, the DCRR is physically divided into several DCRRs; thus every data transmission needed for load balancing or data rebuilding after a failure (see sections 7 and 8) is done using the delivery network and could overload it. Moreover, a physically partitioned ring offers less versatility in case of an addition or a removal of nodes. It cannot be logically reorganized to take advantage of the new resources. We therefore think that partitioning must be logical.

However, partitioning has a drawback: the load on SN could be unbalanced since videos are not striped on all SNs but only on SNs belonging to the same RAID: a RAID can be overloaded if it stores very popular videos. Thus, there is a tradeoff based on the partitioning degree. If the partitioning degree is high (i.e. the RAIDs are small), the load on RAIDs would probably be unbalanced. If the partitioning degree is low (i.e. the RAIDs are large), throughput on the links near INs would be cumbersome and fault tolerance would be poor. To study and take advantage of this tradeoff, we have run simulations on a model of the VODS designed in this section.

5 System model

User streams are served in rounds. During a round, every SN generates a given number of *packets* that are sent to the users through INs. A packet must contain enough data to allow the user to "wait" for the end of the round. To simplify the model, we assume that the size of a packet is equal to the striping unit (SU) of the RAID. In our system, SNs are PCs able to produce 30 MB/s. Every stream has a throughput equal to 500 KB/s. So every SN serves 60 MPEG-2 streams simultaneously. SU size is equal to 50 KB; therefore every round lasts 1/10th of second. In order to use all the disks bandwidth, some buffers are used on SNs to transfer data from disk to interconnection network (so a block, read from a disk in one access, is larger than an SU). Videos last 3600 seconds and every SN can store 25 videos.

Concerning data layout, we chose Flat-Left-Symmetric RAID 5 [9]; this parity placement is particularly efficient in order to distribute the workload uniformly on the SN. Link throughput is set at 250 MB/s for each sub-ring. Users' requests are modeled according to a Zipf distribution [4] (with a coefficient equal to 0.271 because it closely models a movie store rental distribution).

6 Evaluation of DCRR partitioning

A DCRR is logically partitioned into independent RAIDs in order to reduce the throughput of the most loaded link. To evaluate the effect of partitioning the DCRR, we carried out simulations on VODS with partitioning degrees of 1, 2, 4, 8, 16 and 32. We measure the number of streams that the architecture is able to play as a function of the number of SNs interconnected.

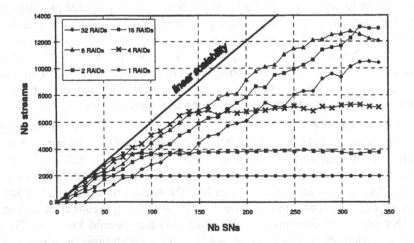

Fig. 6. Scalability of logically partitioned DCRR-based VODS.

Figure 6 shows that partitioning improves greatly the scalability of a DCRR-based VODS. With a small partitioning degree (1, 2 et 4), links of the DCRR are overloaded because of the throughput accumulation that occurs on links near INs. On the other hand, the improvement of the scalability is not linear with the partitioning degree (for high partitioning degrees). As pointed out in section 4, some RAIDs contain more popular videos than others; these RAIDs are then too small to deal with the load. The workload becomes unbalanced and that severely diminishes the scalability of the VODS. Figure 6 clearly shows that a partitioning degree of 32 yields architectures less scalable than architectures which use a partitioning degree of 8 or 16.

With high partitioning degrees, the device that limits the scalability of a VODS is not the DCRR anymore: it is the RAID. In order to effectively improve scalability using high partitioning degrees, it is necessary to decrease the sensitivity of the partitioned ring to the user demand imbalance. As popularity of videos is very difficult to foresee, it is necessary to balance the load dynamically by moving data in order to avoid popular videos being concentrated in some RAIDs. Thus, we use a well-known technique used in server array load-balancing: replications.

7   Load balancing

So as to dynamically balance the load, we replicate some videos onto several RAIDs. The use of replications to balance the load in VODS has already been studied [6–8]. However in this paper, we put the emphasis on policies designed for accelerating the replication process. Since replications are done once the load imbalance occurs, they must be completed as soon as possible in order to reduce the load imbalance time. Replications are done from overload RAIDs (*original RAIDs*) to less loaded RAIDs (*target RAIDs*). If we want the replications to be fast enough, the bandwidth part dedicated to replications must be increased in the target RAID and more critically, in the original RAID (already overloaded). In this section, we propose and evaluate two new policies that accelerate the replication process without consuming extra bandwidth of the original RAID.

First, we propose that video should be dynamically replicated by taking advantage of the streams used for service to users (*service streams*). We call this policy, *stream splitting*. The packets belonging to a service stream are duplicated: the original service stream goes towards the IN (service to the user) and the *duplicated load balancing stream* goes towards the SNs of the target RAID. Thus a large part of each replication is done using disk accesses already used to serve customers. In order to deal with data that are not replicated by the duplicated load balancing streams, *dedicated load balancing streams* are created. Note that there are much less dedicated streams than duplicated streams.

Second, in order to move services from the original RAID to the target RAID, it is not necessary to wait for the video to be entirely replicated to the target RAID. Assuming that the throughput of a dedicated load balancing stream is equal to or greater than the throughput of the service stream, a service could be moved *as soon as the user needs an* SU^2 *that is already on the target RAID*. This policy, *fast stream migration*, accelerates the load balancing.

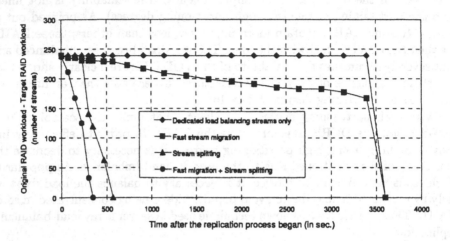

Fig. 7. Evaluation of the time used to balance completely the load on two unbalanced RAIDs.

Figure 7 shows the time necessary to balance the load on two unbalanced RAIDs. The original RAID plays 989 service streams and the target RAID, 750. 1% of the original RAID bandwidth is used for dedicated load balancing streams. 2% of the storage capacity is reserved in the target RAID to store the replicated videos.

Our evaluations show that *stream splitting* is very efficient. It divides by 6 the time necessary to balance the load in comparison with a policy using dedicated load-balancing streams only. These good results are due to the fact that, since the videos concerned by the replications are inevitably popular, a large amount of service streams can be split to accelerate the replication process.

Fast stream migration is actually efficient only when it is used in combination with stream splitting. The load balancing time is reduced by 39% in comparison with a policy using stream splitting only. And above all, a partial load balancing occurs much faster: the load imbalance is reduced by 30% in 2.5 minutes (2% in 2.5 minutes with a policy using only stream splitting).

Dynamic replications using stream splitting combined with fast stream migration reduces dramatically the load imbalance time. Thus, it increases significantly the scalability of the DCRR by allowing the design of highly partitioned DCRR.

---

[2] replicated by a dedicated load balancing stream

8 Fault tolerance

We have seen in section 3 that a simple DCRR has another drawback: its bad fault-tolerance. Partitioning increases the fault-tolerance of the server, but is it fault-tolerant enough for VOD applications that need guarantees for real-time services?

Several components of the VODS could fail: links, IN and SN. The three types of failure cause a rupture of the DCRR; some packets are diverted to avoid the rupture.

When an SN fails (the most problematical failure), it is necessary to recover the lost data located on the failed disk by reading the remains of the stripe as well as the corresponding parity. Video-on-demand has a property that is often exploited by VODS designers: videos are read in a sequential manner. So the data needed to recover the lost SUs have already been read or are going to be read. On every SN, by preserving the SU read last in some buffers and by prefetching the SU that will soon be read, the lost SU can be recovered without needing new disk accesses (except for reading of parity) [10]. When an SU belonging to a failed SN must be recovered, every still working SN in the RAID sends an SU to the SN that has been chosen to calculate the XOR of every valid SU in the stripe. If the RAID holds n_d SN, then $n_d - 1$ packets are sent on the DCRR for every SU that must be recovered.

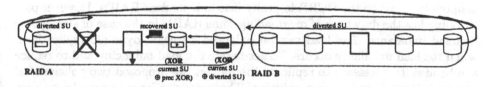

Fig. 8. Diverted load and recovery of an SU after an SN failure using the *distributed XORing*.

To decrease this load that arises on the DCRR after an SN failure, we propose *distributed XORing* (see figure 8). This policy is managed as follows: an SU of the stripe passes from SN to SN, every SN concerned makes an XOR with the SU stored in its buffers; once all the SNs of the stripe have done their XOR, the lost SU is recovered and sent to the user through an IN.

In order to evaluate the diverted load and the efficiency of distributed XORing, we did simulations on a server including 4 RAIDs each one containing 8 SNs and 2 INs.

Figure 9 show that distributed XORing reduces by 22% the throughput of the most loaded link. But even with this optimization, the diverted load is still cumbersome. The throughput of the most loaded link increases by 112%. However this increase in load can easily be absorbed by a DCRR which has a high partitioning degree (that is, in which links are loaded lightly before failure).

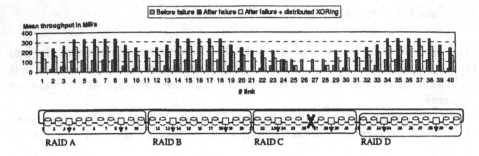

Fig. 9. Mean throughput on DCRR links before and after an SN failure (between links #26 and #27).

9 Conclusion

In this paper, we propose an architecture based on a dual counter-rotating ring (DCRR). This topology is used by many interfaces: MetaRing, ATMR, Fibre Channel Arbitred Loop, Serial Storage Architecture, Sebring Ring, etc... Our cost evaluation shows that the DCRR-based architectures are very cost-effective and that the price of the system increases linearly with the number of resources it uses.

On the other hand its scalability is insufficient because of the throughput accumulation problem that arises on links. In order to remedy this aspect, we propose to partition the DCRR logically into independent RAIDs. However partitioning has the drawback of overloading some RAIDs in the case of strong user demand imbalance. The load can be balanced by using dynamic replications, when the load imbalance occurs. To accelerate the load balancing and to reduce the bandwidth necessary to replicate the movies, we proposed two policies: *fast stream migration* and *stream splitting*. These policies have proven to be very efficient in our evaluations. When they are combined, they divide by 10 the load balancing time and above all, the load balancing process has an impact from the first seconds after the replications began.

The load increase on DCRR caused by a storage node failure is large: 175%. We proposed the distributed XORing, a recovery policy adapted to the DCRR. This policy reduces the load increase to 112%. The load increase is still cumbersome but a partitioned DCRR is able to absorb this additional load provided that RAIDs are not too large: a partitioned enough DCRR is fault-tolerant enough for VOD applications. However, we are studying the use of mirroring-based redundancy schemes to alleviate this additional load on DCRR.

We made supplementary simulations on a realistic VODS based on the Sebring Ring; it contains 252 nodes (14 RAIDs each containing 16 storage nodes and 2 interface nodes - the number of nodes on the Sebring Ring is limited to 256). This VODS can play 13500 MPEG-2 streams. The internal throughputs measured are sustainable by the DCRR, even during a node failure.

Based on our study, we believe that the DCRR-based cluster of PCs is one of the most efficient architecture to achieve truly cost-effective VODS.

References

1. ANSI X3T10.1/0989D revision 1.0, "Information Technology - Serial Storage Architecture - Transport Layer 1 (SSA-TL1)", American National Standard Institute, April 1996.
2. Bach C. and Grebe A., "Comparison and performance evaluation of CRMA-II and ATMR", Eleventh Annual Conference on European Fibre Optic Communications and Networks, EFOC&N '93, June 1993.
3. Shenoy P.J. and Vin H.M., "Efficient Striping Techniques for Multimedia File Servers", in Proceedings of the Seventh IEEE International Workshop on Network and Operating System Support for Digital Audio and Video (NOSSDAV'97), pp. 25-36, May 1997.
4. Chervenak A.L., "Tertiary Storage: An Evaluation of New Applications", Phd Thesis, University of California at Berkeley, 1994.
5. Cidon I. and Ofek Y., "MetaRing - A Full-Duplex Ring with Fairness and Spatial Reuse", IEEE Transactions on Communications, Vol.41, No.1, Jan 1993.
6. Dan A., Kienzle M.G. and Sitaram D., "A Dynamic Policy of Segment Replication for Load-Balancing in Video-On-Demand Servers", Multimedia Systems, Vol. 3, No. 3, pp. 93-103, July 1995.
7. Tetzlaff W. and Flynn R., "Block Allocation in Video Servers for Availability and Throughput", Multimedia Computing and Networking, 1996.
8. Dan A. and Sitaram A., "An Online Video Placement Policy based on Bandwidth to Space Ratio (BSR)", Proceedings of the ACM SIGMOD International Conference on Management of Data, pp. 376-385, San Jose, May 1995.
9. Lee E.K. and Katz R.H., "Performance Consequences of Parity Placement in Disk Arrays", Proceedings of the Fourth International Conference on Architectural Support for Programming Languages and Operating Systems, ASPLOS-IV, pp. 190-199, April 1991.
10. Shenoy P.J. and Vin H.M., "Failure Recovery Algorithms for Multi-Disk Multimedia Servers", Univ. of Texas at Austin, Department of Computer Sciences, Technical Report 96-06, April 1996.
11. Tewari R., Mukherjee R. and Dias D.M., "Real-Time Issues for Clustered Multimedia Servers", IBM Research Report, RC20020, June 1995.
12. Sachs M.W. and Varma A., "Fibre Channel and Related Standards", IEEE Communications Magazine, Vol. 34, No. 8, pp. 40-50, August 1996.
13. "SRC3266DE Preliminary Data Sheet, Sebring Ring Connection for PCI 32", http://www.sebringring.com, 1999.
14. Varma A. and Raghavendra C.S., "Interconnection Networks for Multiprocessors and Multicomputers - Theory and Practice", IEEE Computer Society Press, 1994.

Pricing for Differentiated Internet Services

Zhong Fan and E. Stewart Lee

Centre for Communications Systems Research
University of Cambridge
10 Downing Street, Cambridge CB2 3DS, UK
z.fan@ccsr.cam.ac.uk

Abstract. As the Internet evolves into a global commercial infrastructure, there is a growing need to support more enhanced services than the traditional best-effort service. Differentiated services are a suitable solution to quality of service provisioning in the Internet while the number of users keeps growing. However, recent advances in differentiated service models have not spawned much work in pricing yet. This is due to the complexity of economic elements involved and the fact that the shape of the differentiated services Internet is still not fully defined. In this position paper, we briefly review the state of the art in this relatively new area and compare some of the main existing pricing approaches. The impact of pricing schemes on differentiated services is emphasized. Based on a discussion of unsolved issues, possible research directions are identified.

1 Introduction

The current Internet is based on the so-called best-effort model where all packets are treated equally and the network tries its best to achieve reliable data delivery. Although this simple model is very easy to implement, it has a number of undesirable consequences when the Internet is evolving towards a multi-service network with heterogeneous traffic and diverse quality of service (QOS) requirements. The flat rate charging structure attached to the best-effort model has undoubtedly contributed to the growing problem of congestion on the Internet. Since the network resources are completely shared by all users, the Internet tends to suffer from the well-known economic problem of "tragedy of the commons". The greedy users will try to grab as much resources as possible, leading to an unstable system and eventually congestion collapses. The Internet has been successful till now because most end systems use TCP congestion control mechanisms and back off during congestion. However, as the number of TCP-unfriendly users increases, such dependence on the end systems' cooperation is becoming unrealistic. The lack of explicit bandwidth policing and delay guarantees in the current Internet also prevents Internet service providers (ISP) from creating flexible packages to meet the different needs of their customers [1].

As the Internet evolves into a global commercial infrastructure, there is a growing need to support more enhanced services than the traditional best-effort service. To address this issue, recently there are new efforts in the IETF to develop a new class of service models called Differentiated Services or Diffserv models. The key difference between previously proposed Integrated Services (Intserv) models and Diffserv is that while Intserv provides end-to-end QOS on a per flow basis, Diffserv is intended to provide long term service differentiation among the traffic aggregates to different users. In particular, Diffserv pushes the complexity to the network edge, and requires very simple priority scheduling/dropping mechanisms inside the core. While the number of Internet users keeps growing, the Diffserv solution is more suitable because it scales well with increasing number of network users and it does not alter the current Internet paradigm much. Unfortunately, recent advances in differentiated service models have not spawned much work in pricing yet. This is due to the complexity of economic elements involved and the fact that the shape of the differentiated services Internet is still not fully defined.

In this position paper, we briefly review the state of the art in this relatively new area and compare some of the main existing pricing approaches. The impact of pricing schemes on differentiated services is emphasized. Based on a discussion of unsolved issues, future research directions are also identified.

2 Differentiated Services

Two examples of differentiated service models are the Assured Service proposed by Clark and Wroclawski [2] and the Premium Service proposed by Nichols *et al.* [3]. The assured service is based on the profile tagging scheme which uses drop priority to differentiate traffic. Each user is assigned with a service profile that describes the "expected capacity" from the ISP. At the ingress points of an ISP user traffic is monitored by a profile meter. Packets that are out of the profile ("out" packets) can still go through but they are tagged as such by the profile meter. When congestion occurs inside the network, the routers drop the "out" packets first. It is envisaged that, when traffic conforms to the agreed profile ("in" packets), a user can have predictable levels of service. One desirable feature of this assured service model is that both "in" and "out" packets share the same queue, so there is no fundamental change to the current Internet architecture and it is easier to implement than Intserv models (e.g., RSVP). To control congestion, Clark suggests the use of a packet dropping scheme called RED with In and Out (RIO), which is a variant of the Random Early Detection (RED).

The premium service is based on explicit resource reservations and a simple priority queuing scheme. The scheme creates a "premium" class that is provisioned according to the worst case requirements (peak rate) and serviced by a high priority queue. The premium traffic is strictly policed at the entry points to an ISP: out-of-profile packets are dropped or delayed until they are within the service profile. Inside the network, packets with the premium bit set in the packet header are transmitted prior to other packets.

There are a number of issues with the above Diffserv models, among which pricing is our main concern. We will address this topic in the next section.

3 Pricing Schemes for the Differentiated Services Internet

Current pricing practices are very different on the Internet and the PSTN (Public Switched Telecommunication Network) due to the difference between the underlying service paradigms. Calls over the former generally incur no usage price while calls over the latter incur a usage price which is time, distance and destination dependent. Currently most of the Internet service providers use flat rate pricing (users pay only an access capacity dependent connection fee), which is an important factor in stimulating traffic growth and the development of new applications. Its major advantage is simplicity leading to lower network operating costs. A disadvantage is its inherent unfairness in which a light user has to pay as much as a heavy user. Furthermore, in this scheme there is no restraint to prevent users from reserving excessive bandwidth, exacerbating the state of network congestion. Apparently, flat rate pricing is no longer tenable in an environment of rapid commercial growth, rising demand for high bandwidth and rising need for service differentiation. To provide effective and fair resource allocation and congestion control, some form of congestion related pricing is desirable.

3.1 Developments in Internet Pricing

In this section, we consider several leading popular pricing schemes proposed so far, discussing their strengths and weaknesses. Here we do not attempt a comprehensive survey of network pricing proposals, but would rather present the most important trend in the area. The impact of the adopted pricing scheme on the feasibility of fulfilling QOS requirements is also examined.

In so-called *congestion pricing*, network usage can be controlled by the introduction of usage sensitive charging with rates determined by the level of congestion. A well known congestion pricing scheme is the "smart market" [4]. The intention is that the usage price should be zero when the network is uncongested, but when there is congestion the price should reflect the incremental social cost determined by the marginal delay cost to other users. At the equilibrium price the user's willingness to pay for additional data packets equals the marginal increase in delay cost generated by those packets. This is implemented by requiring all users to include a bid in each packet. In case of congestion, the users offering the lowest bids are discarded first and accepted packets are charged at a rate determined by the highest bid among the rejected packets. The smart market has been proved to be an optimal charging model, but the complexity of the scheme and the implied accounting system have ruled out its actual implementation. Nevertheless, it can be used as a theoretical benchmark against which other pricing schemes can be measured [5].

Recently Odlyzko's "Paris Metro pricing" scheme [6] has received a lot of attention. It mimics the system that was used some years ago on the Paris Metro. The idea is to partition the network into several logical networks, with different usage charges applied on each sub-network. There would be no formal guarantees of service quality, but on average networks charging higher prices will be less congested. Users will sort themselves according to their preferences for congestion and the prices charged on the sub-network. The main attraction of this scheme is its simplicity: there is no need to engineer QOS, and price would be the primary tool of traffic management. The implementation of this scheme is thus much easier and less expensive than the "smart market". However, initial results in [5] indicate that networks will not use it if they want to maximize profits and they face competition from other networks.

Here at Cambridge, Kelly proposes the "proportionally fair pricing (PFP)" scheme [7], in which each user declares a price per unit time that he is willing to pay for his flow. In that sense, capacity is shared among the flows of all users in proportion to the prices paid by the users. It has been shown in [7] that in a weighted proportionally fair system where the weights are the prices the users pay per unit time, when each user chooses the price that maximizes the utility she gets from the network, the system converges to a state where the total utility of the network is maximized. In other words, in an ideal environment, the PFP proposal is able to decentralize the global optimal allocation of congestible resources. Another result in [7] is that rate control (such as TCP) based on additive increase and multiplicative decrease, achieves proportional fairness. We will discuss PFP in more detail in the next section.

3.2 Issues, Research Directions and Plan

The aim of our research is to search for a simple pricing model capable of providing service differentiation and proper incentive to utilize network resources efficiently. This could be achieved by some form of congestion pricing, in which the quality perceived by price will vary as a function of the congestion of the network.

Recently, the optimal charging paradigm (such as the one proposed in [4]) has been criticized by Shenker *et al.* [8]. A major problem is that congestion costs are inherently inaccessible to the network and so cannot reliably form the basis for pricing. It is argued in [8] that the research agenda on pricing in the Internet should shift away from the optimality paradigm and focus on more structural aspects of the underlying service model. It has been suggested that it is sufficient to offer a number of differentially priced service classes with charges increasing with the expected level of QOS. Users regulate their charge by choosing to (or not to) use a higher quality of service class in times of congestion. The simplest service model fulfilling this objective has just two differently charged service classes. In this regard, we feel that pricing in a differentiated service framework could be a right direction to pursue.

Despite its importance, charging in the context of differentiated services has been ignored so far with a few exceptions. Both assured and premium service models have a number of problems. First, choosing a proper profile for a user is not a trivial task. Consequently, charging is complicated as the profile relates to the prices the user would like to pay. When a network does not have uniform bandwidth provisioning, profiles are likely to be destination-specific and hence it is difficult to choose a fixed profile for a user [1]. To implement any charging scheme in a profile-based Diffserv model, the following aspects need further exploration: (a) traffic specifications of the service profile, (b) spatial granularity of the service, and (c) probability of assurance, i.e., how likely is an in-profile packet to be delivered to the destination. Second, best-effort traffic in the premium service scheme may get starved completely [1]. Finally, in current proposals, the connection (and the interaction) between the price being paid for a given service and the congestion situation in the network bottlenecks is not clear.

Among the three pricing schemes discussed earlier, Kelly's proportionally fair pricing seems to be most suitable for differentiated services. It can provide differential quality of service without some of the limitations mentioned above. Its philosophy is that "users who are willing to pay more should get more". As the network makes no explicit promises to the user there is no need for over-provisioning in the core of the network. Moreover, there would be no need for admission control (which is required in the premium service model), reservations or priority queues [9]. Therefore, at present our particular focus is on investigating the feasibility of the PFP scheme in the context of the IETF's Diffserv architecture, e.g., how to implement it in a way that is compatible with currently employed Internet protocols or their Diffserv variants?

Conceptually there are two ways to achieve proportionally fair sharing. One approach is to give control to end systems (users), as suggested by Kelly et al. [7]. In this scheme, the TCP algorithm is modified to incorporate congestion prices by means of protocols like explicit congestion notification (ECN) [10]. Upon receiving feedback signals, $f(t)$, which are related to shadow prices (in terms of marks), the users are free to react as they choose, but will incur charges when resources are congested. An end system can adjust its rate using a willingness-to-pay (WTP) parameter w:

$$x(t+1) = x(t) + \kappa(w - f(t)) \tag{1}$$

where κ affects the rate of convergence of the algorithm. As different users can react quite differently, whether the optimal result of proportional fairness can be achieved in a real heterogeneous network is still an open question. Fortunately, as suggested in [10], a distributed game could be constructed as an experimental framework to test different control algorithms and strategies.

In the other scheme, it is the network who determines each user's share of the resources based on users' WTPs. In the case of a set of users sharing a single resource with capacity C, if the WTP of user i is w_i, then the share ratio for user i is $s_i = \frac{w_i}{\sum_i w_i}$. The exact amount of bandwidth allocated to the user is then $C * s_i$. This *relative* sharing representation has a number of advantages. In particular, it scales well with multiple bottlenecks with different bandwidth provisioning and hence can be regarded as a flexible profile. It also ensures that bandwidth allocation is always fair with respect to user shares or prices users are willing to pay. The major disadvantage of this scheme is the increased complexity of the routers in the network core since a WFQ-like (weighted fair queuing) scheduling algorithm is required to enforce proportional bandwidth sharing. Note that this scheme is similar to the USD (User-Share Differentiation) proposal of Wang [1].

Currently we intend to do a comparative evaluation of the above two schemes through simulations for a number of different scenarios. Possible topics include: the performance in the presence of non-responsive users, the impact of control granularity and traffic aggregation on scalability, how to form service level agreements between inter-domain boundaries, the implementation details (e.g., marking strategy, how to distribute user prices/shares across the network). Another interesting problem is to incorporate the pricing information into appropriate path cost functions to do multi-path routing and load balancing, which is our future work.

4 Conclusion

In this paper we have briefly reviewed the state of the art in differentiated services and pricing schemes for the Internet. We believe as the Diffserv architectures evolve, the important factor of charging should be explicitly incorporated into the framework and new economic models are needed to ensure an efficient and fair allocation of network resources. The work presented in this paper is still at its early stage. The ideas are being developed and enhanced as part of an ongoing project aimed at devising effective pricing schemes for differentiated services in the Internet.

References

1. Wang, Z.: Towards scalable bandwidth allocation in the Internet. Technical report, Bell Labs, Lucent Technologies (1998)
2. Clark, D., Wroclawski, J.: An approach to service allocation in the Internet. Internet Draft (1997)
3. Nichols, K., Jacobson, V., Zhang, L.: A two bit differentiated services architecture for the Internet. Internet Draft (1997)
4. MacKie-Mason, J., Varian, H.: Pricing the Internet. In: Kahin, B., Keller, J. (eds.): Public Access to the Internet. Prentice Hall, New Jersey (1995)
5. Gibbens, R., Mason, R., Steinberg, R.: An economic analysis of Paris Metro pricing. Technical report, University of Southampton (1998)
6. Odlyzko, A.: A modest proposal for preventing Internet congestion. Technical report, AT&T Labs – Research (1997)
7. Kelly, F., Maulloo, A., Tan, D.: Rate control for communication networks: shadow prices, proportional fairness and stability. Journal of the Operational Research Society. 49 (1998) 237–252
8. Shenker, S., Clark, D., Estrin, D., Herzog, S.: Pricing in computer networks: reshaping the research agenda. Computer Communication Review. 26 (1996) 19–43
9. Crowcroft, J., Oechslin, P.: Differentiated end-to-end Internet services using a weighted proportional fair sharing TCP. ACM Computer Communication Review. 28 (1998) 53–69
10. Key, P., McAuley, D.: Differential QoS and pricing in networks: where flow control meets game theory. IEE Proceedings Software. 146 (1999)

Acknowledgements: This work is supported by Hitachi Europe Ltd.

An Agent-Based Adaptive QoS Management Framework and Its Applications

Masakatsu Kosuga, Tatsuya Yamazaki, Nagao Ogino, and Jun Matsuda

ATR Adaptive Communications Research Laboratories, 2-2 Hikaridai, Seika-cho,
Soraku-gun, Kyoto 619-0288 Japan
{kosuga, yamazaki, ogino, matsuda}@acr.atr.co.jp
http://www.acr.atr.co.jp

Abstract. For distributed multimedia applications, adaptive QoS management mechanisms need to be developed to guarantee various end-to-end QoS requirements. In this paper, we propose an adaptive QoS management framework based on multi-agent systems. Adaptation to system resource changes and various user requirements is achieved by the direct or indirect collaboration of agents in the respective phases. In the flow establishment and re-negotiation phase, application agents determine the optimal resource allocation through QoS negotiations to maximize the total users' utility. In the media-transfer phase, stream agents collaborate to adjust each stream QoS reactively. In addition, personal agents help novice users specify stream QoS without any *a priori* knowledge on QoS. Some multimedia applications have been designed based on the proposed framework.

1. Introduction

With the recent advent of new network services like the Internet as well as multimedia applications, and the rapid spread of wireless telephony, we can expect future communication environments to become extremely complicated and changeable, where various types of networks, terminals, applications, and methods to use services will be intermixed. Therefore, the development of adaptive and self-organizing communication systems will become increasingly important for the achievement of greater capabilities of services and a higher network efficiency.

Our group is working in the research area of "adaptive communication application and its platform." Our objective is to deploy a framework and the key technologies for coordinating communications, adapting to heterogeneous and dynamically changing open system environments, e.g., networks, end-systems, and user environments. This type of application is highly applicable to distributed multimedia applications with QoS (Quality of Service) adaptation.

It is especially important to satisfy end-to-end QoS requirements for continuous media such as video and audio in those applications. However, the available system resources are changeable because of various reasons, e.g., the throughput of the best-effort network decreases as network traffic increases, the error rate of wireless links fluctuates according to the electromagnetic wave propagation environment, and the available CPU performance is reduced when other applications are in operation.

Moreover, the QoS requirements of users change according to when, where, and with which terminals and networks the users use. Consequently, adaptable mechanisms for variations in the system performance and user requirements are indispensable for end-to-end QoS management.

In this paper, we propose an adaptive QoS management framework for distributed multimedia. The framework is based on multi-agent systems, which are applicable to many domains for which centralized systems cannot be applied because of their useful features such as parallelism, robustness, and scalability [1]. In the framework, agents directly or indirectly collaborate to adaptively manage the media QoS according to the available network and terminal resources as well as the user requirements. The particular point of the framework is that it provides two-tier QoS management. Namely, long-term and global QoS adaptation is executed in one tier, while short-term and local QoS adjustment is executed in the other tier. Some multimedia applications have been designed on the basis of the proposed framework and the effectiveness of the agents' interworking has been checked in experiments.

The rest of the paper is organized as follows. In Section 2, we propose the agent-based adaptive QoS management framework, and in Section 3, the agents' behaviors in this framework are described. Section 4 refers to some applications based on the framework. Finally, Section 5 concludes this paper.

2. An agent-based adaptive QoS management framework

Recently, Aurrecoechea et al. [2] proposed a generalized QoS framework based on a set of principles governing the behaviors of QoS architectures. This generalized QoS framework is composed of three types of QoS mechanisms: QoS provision mechanisms, QoS management mechanisms, and QoS control mechanisms. The QoS provision mechanisms perform static resource management in the flow establishment and QoS re-negotiation phase. On the other hand, the QoS management and control mechanisms deal with dynamic resource management in the media-transfer phase. QoS control is set apart from QoS management by an operational time-scale, and operates on a faster time-scale than QoS management.

Fig. 1 presents a QoS management flow on the basis of the generalized QoS framework. In the flow establishment and re-negotiation phase, the QoS mapping module translates user QoS requests into QoS candidates that are understandable to the system (terminals and networks). The QoS negotiation module selects the QoS for each media stream from among the QoS candidates via intra-terminal and inter-terminal negotiations. The QoS admission module tests whether the selected QoS will be guaranteed or not for the system, and reserves the necessary system resources by resource reservation protocols if possible. Otherwise, the QoS admission module issues a re-negotiation message for the QoS negotiation module.

Then, the modules in the media-transfer phase take over the QoS management. The selected QoS is transferred to QoS control and management mechanisms. A real-time flow control module in the QoS control mechanism tries to maintain the QoS through flow filtering, flow shaping, flow scheduling, and so forth. In the QoS management mechanism, the QoS monitoring module notifies the QoS management module upon perceiving fluctuations in the system resources. The QoS management module deals with any QoS adjustment within an admissible range, which is specified by the user, using the resource information. The real-time flow control module receives the adjusted QoS, and continues the QoS maintenance. When the QoS management module can no longer perform QoS adjustment because, for example, the resource changes are too severe to recover, it issues a re-negotiation request message to the QoS negotiation module.

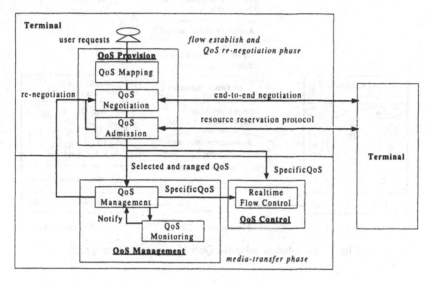

Fig. 1. QoS Management Flow

We propose an agent-based adaptive QoS management framework as a common platform for various communication-intensive applications (Fig. 2). The proposed framework, whose basis is the QoS management flow presented in Fig. 1, consists of three kinds of agents, a resource manager module, and a QoS interface.

When a user has little knowledge on how to set the application-level QoS, the user should provide abstract QoS requests for media streams. A Personal Agent (PA) then interprets the abstract user-level QoS using the user profile database, which reflects the user's preference, and transfers it to the QoS provision mechanisms as the application-level QoS. PA also updates the user profile database using a learning mechanism.

An Application Agent (AA) selects the best viable QoS for each media stream from among the application-level QoS candidates by intra-terminal and inter-terminal negotiations. Since QoS negotiations do not need to be executed in real-time, it is desirable for the AA to be deliberative in taking the optimality into consideration from a long-term viewpoint.

A Stream Agent (SA) adjusts the selected QoS within the admissible range specified by the user. Since QoS adjustment by the SA is carried out while multimedia applications are in operation, this adjustment must be done in real-time. Hence, the SA must be reactive.

A Resource Manager (RM) performs scheduling and reservation for the terminal resources, such as the CPU and memory. A QoS interface mediates between the terminals and network, and enables the terminals to reserve network resources.

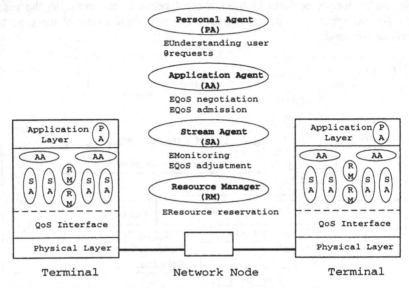

Fig. 2. Agent-based Adaptive QoS Management Framework

3. Agents' behaviors in the framework

In this section, we describe the agents' behaviors according to the QoS flow.

AA selects the best and most viable QoS from among the multiple QoS candidates by intra-terminal and inter-terminal negotiations. In the intra-terminal negotiations, AA negotiates the allocation of resources to maximize the total user utility. The procedure for the QoS negotiations is as follows. The AA who requests the intra-terminal negotiations sends a QoS negotiation request message to all of the AAs concerned on the terminal. This AA is called the master agent. If an AA receiving the request message is able to participate in the negotiations, it returns its multiple QoS candidates and utility parameters to the master agent. The master agent selects a QoS set for the streams so that the total utility U defined in (1) is maximized under the resource constraint conditions in (2).

$$U \quad \sum_{S} w(S) \log u(S, q), \qquad (1)$$

$$\sum_{S} r_m(S, q) \le R_m, \qquad (2)$$

where $u(S, q)$ is a utility parameter when a stream S has a QoS of q, $w(S)$ is the priority of stream S taking the priority of the concerned application into consideration, and $r_m(S, q)$ is the amount of the m-th resource required by the processing of stream S

with the QoS of q, and R_m indicates the maximum availability of the m-th resource. In (1) and (2), the summation is operated for all streams involved in the negotiations.

After the intra-terminal QoS negotiations, the AAs together execute the inter-terminal QoS negotiations to resolve any QoS conflicts between the stream sender and the stream receivers. The cooperative game theory is applied to the inter-terminal QoS negotiations. In this scheme, each AA exchanges the terminal's utility we have newly defined and finds a compromise between the sender and the receivers.

After a ranged QoS candidate is selected by the AAs, the SAs adjust the selected QoS parameters within the range provided by the AAs and determine a specific QoS for each media stream. The SAs use the streams' priority order to order the QoS adjustment, and the priority threshold parameter Th is shared among the SAs as a common datum.

The SA monitors the resources. If the SA recognizes a shortage or surplus of resources, it refers to Th. By comparing its own stream priority with Th, it decides whether to execute the QoS adjustment or not. If the SA undertakes the QoS adjustment, it updates the value of Th after the adjustment. The behavior of the SA for the QoS adjustment differs according to whether there is a resource shortage or a resource surplus. In the former, the SA decreases the QoS parameters stepwise by increasing the QoS parameters' priority order and the SA increases the value of Th after the QoS adjustment. In the latter case, the SA increases the QoS parameters stepwise by decreasing the QoS parameters' priority order and the SA decreases the value of Th after the QoS adjustment. The initial value of Th is set to the maximum, minimum, or average value of all stream priority parameters. Th updating is done by increasing (decreasing) a constant value or by setting the average value of the priority parameters the SAs excluded from the QoS adjustment process.

4. Multimedia applications based on the framework

To verify the effectiveness of the agents' interworking, we have implemented a Video-on-Demand system called MARM-VoD (Multi-Agent Resource Management Video-on-Demand) on the basis of the agent-based adaptive QoS management framework. Fig.3 depicts the interrelationship among the MARM-VoD modules. A VoD client is connected to two video servers, and two MARM-VoD applications run on the client to communicate with the respective servers. It is assumed that at most two video streams are coming from each server to the client, because the video server can handle real-time videos for monitoring as well as accumulated videos. In this system, the client determines the stream QoS according to the existing resource allocation and notifies the server of the QoS parameters. Then, the server transfers the video stream encoded by the QoS parameters to the client.

As future work, we are designing and implementing MARM-VC (Multi-Agent Resource Management Video-Conference) based on our framework. Since a terminal must receive and send media streams in videoconferencing applications, the inter-terminal negotiations will become more complicated than in the case of VoD applications. Moreover, we plan on designing and implementing a proxy application for the WWW based on the framework. It will handle QoS mapping and QoS negotiation for data (plain text, JPEG, and so forth) QoS.

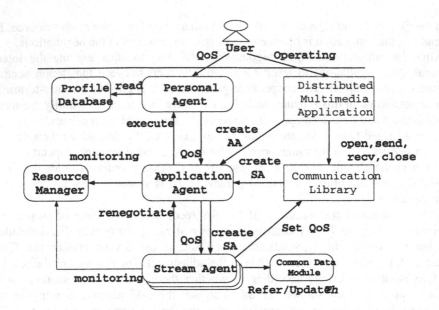

Fig. 3. Interrelationship among the MARM-VoD Mdules

5. Conclusion

For communication-intensive applications using distributed multimedia, we have proposed an agent-based adaptive QoS management framework. The target here is adaptability towards various users' QoS requirements and network and performance fluctuations. The adaptive QoS management task is accomplished by the direct and indirect collaboration of agents. The most remarkable and characteristic point is the mutually supplemental cooperation of AAs and SAs.

As an example of a multimedia application with the proposed framework, we developed a video-on-demand system called MARM-VoD. We have evaluated the performance of MARM-VoD in a laboratory testbed. We are now working on the implementation of MARM-VC and a proxy application for the WWW based on the framework in the testbed.

References

1. G. Weiss, "Adaptation and Learning in Multi-Agent Systems: Some Remarks and a Bibliography (in Lecture Notes in Artificial Intelligence 1042)," Springer-Verlag, Tokyo, 1996.
2. C. Aurrecoechea, A. T. Campbell and L. Hauw, "A Survey of QoS Architectures," ACM/Springer Verlag Multimedia Systems Journal, Special Issue on QoS Architecture, Vol.6, No.3, pp. 138-151, May 1998.

Improving the Quality of Recorded Mbone Sessions Using a Distributed Model

Lambros Lambrinos, Peter Kirstein and Vicky Hardman

Department of Computer Science - University College London
Gower Street, London WC1E 6BT, UK
{L.Lambrinos, P.Kirstein, V.Hardman}@cs.ucl.ac.uk

Abstract. Multicast conference recording currently uses a single, arbitrarily placed recorder. The recordings have less than perfect quality because they are often not loss-free, and the recording is as perceived by the recorder (delays from senders to the recorder have been imposed on the streams). An ultimate recording consists of the loss-free data set, extra information about loss patterns received by individual participants, and information about the delays between participants. The delay information is required so that we can re-create the conference from a number of perception points. In this paper, we initially consider off-line recording of material, which produces loss-free recording, can provide timing information, and is simple, but does suffer from deficiencies. Our second (and preferred) solution uses a distributed arrangement of a number of co-operating recording caches, and sender interaction for local repair.

1 Introduction

Multimedia facilities are available on a variety of platforms for desktop conferencing; users interact with each other in real-time using audio, video and shared workspace tools. The ability to archive multimedia data is required in many collaborative applications. The current systems designed to record multicast conferences assume a single recorder arbitrarily placed in the Mbone[1]. This is unsatisfactory; it does not provide loss-free recordings and the recording is stored as perceived by the recorder (delays from senders to the recorder are imposed on the stored streams). The imposed delays do not represent the perception of any of the conference participants. An ultimate recording consists of the loss-free data set, information about loss patterns seen by participants, and information about the delays between senders.

We consider initially off-line material recording, with data being temporarily stored by each sender during the conference. This solution produces a loss-free recording, and is uncomplicated but its deficiencies are that no information about delays between senders is easily available (this can only be provided using global clock synchronisation e.g. GPS[2]) and that storage is required on each sender's host.

Our second solution is a hybrid approach, which uses a number of distributed recording caches. A cache nearby to a source uses an agent on the participant's host for local repair; a second stage of repair is accomplished when recording caches co-operate to produce a complete data set on each cache. This solution results in loss-free recording, does not require large storage facilities on the local host, and provides either exact delay information (using GPS synchronised clocks), or (without GPS) recordings from a number of points near to receivers, which will afford an approximate view of delays. This model is more complicated than our first solution.

The paper is structured as follows: initially, we describe our experiences of Mbone recording tools. Then we identify the ultimate requirements and look into current relevant research in other areas. In the rest of the paper we propose some solutions that use distributed recorders in order to create better quality recordings.

2 Recording Mbone Streams - Our Experiences

The ability to record live sessions and to introduce recordings into conferences is important. It is crucial to obtain a good quality recording of the initial session, since most VoD players assume loss free material. A number of MBone recording tools already exist (MVoD[3], MARS[4], MMCR[5]). These have common characteristics (client-server-based architecture, remote real-time control, record programming, VCR features). The tools have advantages and disadvantages relative to one another [5].

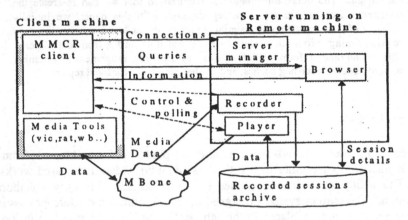

Fig. 1. The MMCR architecture

The Multicast Multimedia Conference Recorder (MMCR) is a typical Mbone recording system and has been used over the last 3 years as a basis for our research work. MMCR's architecture is shown in Figure 1 (for completeness we also show the playback and browsing modules). Indexing facilities are an integral part of the system, and an index is created for each recorded stream. The index includes receiver timestamps, which represent the recorder's perception of the original session.

A centralised recorder has the same problems as a session's participants; received data is susceptible to random loss and delay. Data loss is a problem that the Mbone community is trying to resolve, as it can significantly degrade the quality of received data in a session. For a recorder, the problem is exacerbated since there may be further data loss during playback, leading to greater quality degradation. An initial solution to this problem might be to use the same mechanisms used in the tools for repair [6,7,8], but Mbone loss measurements [9,10] have shown that loss can exceed levels at which repair mechanisms (e.g. audio redundancy) can fulfil their task.

Network delay between receivers varies with time, giving each one a different perception of the session. A faithful view of a conference can only be afforded by exactly capturing the timing intricacies between data packets at each perception point. Lack of delay information in multicast speech can lead to talkspurt misalignment in playout. This is important since humans in a two-party interaction change their conversational behavior patterns depending upon the delay between them [11].

3 Application Requirements and Relevant Research

Multimedia conferencing applications have differing requirements with respect to a recording system. In a multicast multimedia recording, the ultimate aim is not simply loss-free data and no delay, but rather a spectrum of different requirements:

- In small-group distance learning sessions, the recording may serve to inform students not able to attend the lecture of the content, or it may serve as revision material. Both situations require a loss-free recording, with revision material also requiring that a student can observe the conference as originally experienced.
- If the application is a recording of broadcast material, the requirement is one of a loss-free recording, but there is no requirement to re-create the original perception.
- Transmission and recordings of emergency and military services, which are beginning to find IP routing attractive because of its inherent robustness. These applications require loss-free recordings and the ability to re-create the session as experienced at specific perception points.
- Quality assessment of multi-way collaboration is difficult to accomplish at the time the session occurs. What is needed is the ability to assess the quality perceived by each participant; this requires conference observation at all perception points (delay and loss information for all participants, and loss free recordings).
- Remote job interviews, where recordings may be required to allow assessors to revisit interviews during the decision process. The application requires the ability to reconstruct the session as experienced by the interviewee.

Recording of Mbone streams requires considerable processing. Hence, data may be cached with the minimum of changes in order to preserve recorder processing power. Once recorded, it may be preferable to change the data format to that of a VoD system (e.g. QuickTime [12]) that is optimised for multiple simultaneous retrieval operations.

The Internet community is currently considering mechanisms for providing Quality of Service (QoS). Intserv [13] provides a small discrete set of qualities suitable for different traffic types. Diffserv [14] provides different classes of service, which may have identifiable characteristics, or may just be able to provide a better service for one flow compared to another. Diffserv looks to be far easier to deploy, as it aggregates traffic and provides class of service on the aggregated group, rather than on a 'per flow' basis. It is not certain which scheme will emerge as favourite (a combination may also be deployed), but the result will be a spectrum of quality/cost choices; often different choices will be made by different users in a multicast group.

In today's Internet / Mbone, no QoS mechanism has been widely deployed yet. QoS availability may lead to differential delays between flows e.g. for a price, audio from one participant may be transferred with low and tightly bound delays. The current rationale behind multimedia conference recorders (i.e. using a single recorder) does not translate well into a QoS enabled Mbone/Internet. This has important consequences for multicast recording, since the need is for loss free recording, preferably without the recorder having to pay for quality.

The current monolithic software tools approach to multimedia conferencing (audio, video and text programs) is giving way to a tool-kit [15] approach, where media components or engines can be controlled using a conference bus [16]. This is an important development for multicast recording, as it provides a means to incorporate an agent on the local host that can assist in recording.

4 Distributed Recorders

For the ultimate set of requirements, we require that a recorder can re-create a perception point in the conference, often at better quality than actually received by the original listener. This implies a loss-free recording with information about timing and loss in the material. One solution is to extend the single multicast conference recorder to have a component on each participant's workstation. We investigate this model and propose a hybrid solution with fewer requirements on the local host.

Single multicast session recorder with recording agents at sender workstations.
The material from the conference may be collected out-of-band of the main conference; individual contributions are recorded at the sender (by a recording agent), and delivered later or in frequent stages, using point-to-point TCP transfer. This approach has storage implications at individual hosts, but this may not be a problem in view of the power of such workstations. Loss information, as seen by a participant, can be collected by the recording agent, and sent with the recorded media data.

To obtain accurate delay information, the workstation clocks for each participant should be synchronised using GPS. Alternatively, if we assume that the transmit time is roughly symmetrical, then we can obtain approximate clock synchronisation [17]. Timing information is provided in the form of Network Time Protocol (NTP) timestamps and the difference between the sender's and the receiver's timestamps gives the delay over the network. To this, the playout delay in the tool is added to calculate the perceived delay between each receiver and the source. Adjustments to reconstruct a perception point are then made to the data prior to playing it back.

It is not required that the reconstructed sessions be available in a short time; the loss-free data can be deposited on the recorder in an off-line or delay tolerant way. The mechanism is relatively simple (just requires a TCP connection), but will need a way of staggering uploading in order to be scalable. If communication were achieved with the recorder during the session, then staggered transmission would also be needed. The disadvantage here is that the data is sent twice over the network.

This approach does not rely on complicated protocols and reconstruction of a perception point is relatively simple; calculated delays are used to pace the playback. However, this solution implies that single points of failure exist; if a host crashes, a participant's contribution (or part of) to the conference recording may be lost. The solution needs synchronised clocks but no network-wide clock synchronisation mechanism is currently widely deployed, and real-time delay-sensitive traffic is concerned only with a relative delay estimate between the receiver and an individual sender (no accurate absolute value of delay can be calculated, only the relative change). The other problems of this scheme are the processing requirements on the host, and the availability of storage space for real-time media data during the session.

Multiple Conference Recorders and light-weight Recording Agents

This solution consists of a combination of recording agents and distributed caches. The client communicates directly with the controlling recorder (R) which propagates recording control commands to the set of strategically placed recording caches (C). The recording caches will operate simultaneously at various sites (preferably local or with good network connectivity to participants). The recording agents (S) (located at each sender's workstation) must have the ability to repair data lost by a nearby cache and have access to a small storage space for storing a few seconds worth of data.

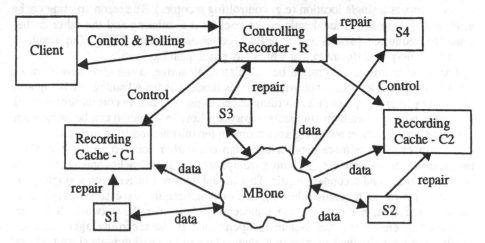

Fig. 2. Schematic of architecture of system with distributed recording

The various operations are to be performed in a number of stages: existence, recording control by the client, recording caches record all streams, allocating sources to a nearby cache, first repair from recording agents, second repair between caches.

Existence: All recorders need to be aware of the existence/location of other recorders via an existence announcement protocol [3].

Recording control by the client: The client contacts a local recorder, which then becomes the controlling recorder. Identification of suitable caches might be achieved using the method proposed in [18] or by user allocation. The mechanism proposed in [18] advertises received loss patterns, and nearby receivers seeing less loss respond. This could also be used to identify better recorders as the conference proceeds.

Recording caches record all streams: The identified caches (C1,C2, and R) start recording all the streams of the session.

Allocating sources to a nearby cache: Local repair will take place between every source (S1, S2, S3, S4) and its nearest cache. There is an allocation stage where a source is bound to a nearby cache. Allocation might be achieved by a Time-To-Live (TTL) based expanding ring search algorithm; the recording agent at the source sends allocation request messages with increasing TTL, until it receives a response from a cache. If the actual cache changes during the conference, then the new cache notifies the affected sources of the change.

First repair from recording agents: When a cache detects lost data from the source, it asks the source's recording agent to re-send it (via a unicast connection). The recording agent will only hold a limited amount of data (e.g. the last 5 seconds), as correct reception can be periodically acknowledged. In this way, we ensure that the complete data set for each source is recorded on at least one of the caches.

Second repair between recording caches: The second stage (right after recording has finished) involves the co-operation of the caches. The repair method depends on the current network conditions. Over a low loss network, we can perform repair at all the caches resulting in multiple loss-free copies of the recording. This could be achieved by using the Scalable Reliable Multicast (SRM) [17] mechanism. However, it is pointless to attempt multiple location repair over a lossy network; in this case repair occurs at a single location (e.g. controlling recorder). The reconstructing cache multicasts (using SRM) the details of the packets it is missing and the other caches send the requested packets via a unicast connection (using TCP). The resulting recordings may be different to that which any participant has seen.

Exact delay information might be calculated, as above, using GPS synchronised clocks. If this mechanism is not available, this model has the advantage of being able to use the perception point of a recording cache to provide nearly correct timing (with respect to what the nearby participants observed). Loss information can be acquired in the same way as above; recording agents obtain the information from media tools.

This mechanism is more complicated than our earlier solution, since it includes relatively complex protocols to bind participants to a local cache, and protocols to achieve primary and secondary repair. This model does not suffer from a single point of failure (as in our previous solution), and does not require excessive storage space on the local host, nor does it send unnecessary data over the network. It is very important to ensure that the system is operational if the recording agents are not available at the individual sources (e.g. due to lack of processing power). In this case the caches will record any data they receive and at the end of the recording co-operate to construct the best possible data set.

5 Conclusion

Collaborative multicast applications require higher quality recordings than that provided by current Mbone conference recorders. The recording must be loss-free (sometimes with information about loss patterns experienced by individuals) and the ability to reconstruct the exact timing relationships between participants is sometimes required in order to re-create the conference from an individual's perception point. We have presented two possible solutions to this problem. Our first solution promotes the existence of a recording agent running on the local host that records source material. The recording agent then communicates off-line with a central conference recorder. Although this solution does not require any complicated protocols it relies absolutely on GPS to calculate timing values, it requires (possibly large) storage facilities on the local host, and data is sent twice on the network.

Our second (and preferred) solution promotes the use of a number of co-operating conference recorders. Recording agents are used also in this solution, but here to provide repair of streams to a nearby recording cache. Recording caches then co-operate off-line to provide further repair leading to a loss-free recording at every recording cache. The main advantages of this solution are that it does not explicitly require GPS to provide delay information, data is only primarily sent once onto the network, and it only requires limited storage facilities on the local host. The main disadvantage of this solution is that it is much more complicated than our first solution, but not prohibitively so.

6 Acknowledgements

We acknowledge the support of the Defence Advanced Research Project Agency (DARPA) under contract D079/00 and of the European Commission with the MERCI (RE1007) and MECCANO (RE4007) projects.

References

1. H.Eriksson, "MBone:The multicast backbone", Communications ACM, vol.37, Aug. 1994
2. See: http://www.eecis.udel.edu/~ntp
3. W.Holfelder, "Interactive Remote Recording and Playback of Multicast Videoconferences", IDMS'97
4. A.Schuett et al, "A Soft State Protocol for Accessing Multimedia Archives", NOSSDAV 98, Cambridge, UK
5. L.Lambrinos, P.Kirstein, V.Hardman, "The Multicast Multimedia Conference Recorder", 7th Intl. Conference on Computer Communications and Networks, Oct 98, Louisiana, USA
6. J.Rosenberg, H.Schulzrinne, "An RTP payload format for Generic Forward Error Correction", draft-ietf-avt-fec-06.txt
7. Perkins et al, "RTP Payload for Redundant audio", RFC 2198
8. C.Perkins, O.Hodson, V.Hardman, "A Survey of Packet-loss recovery techniques for streaming audio", IEEE Network Magazine, September/October 1998
9. M.Handley, "An examination of MBone performance", USC/ISI report: ISI/RR-97-450
10. M.Yajnik et al, "Measurement and Modelling of the Temporal Dependence in Packet Loss, IEEE Infocom 1999
11. P.T.Brady, "Effects of Transmission delay on Conversational behavior on echo-free telephone circuits", The Bell system Technical Journal, January 1971
12. See: http://www.apple.com/quicktime
13. See: http://www.ietf.org/html.charters/intserv-charter.html
14. See: http://www.ietf.org/html.charters/diffserv-charter.html
15. S.McCanne et al, "Toward a Common Infrastructure for Multimedia-Networking Middleware", NOSSDAV 97, Missouri, USA
16. J.Ott, C.Perkins, D.Kutscher, "A message bus for conferencing systems", draft-ietf-mmusic-mbus-transport-00.txt
17. S.Floyd et al, "A Reliable Multicast Framework for Light-weight Sessions and Application Level Framing, Scalable Reliable Multicast (SRM)", ACM SIGCOMM 95
18. I.Kouvelas, V.Hardman, J.Crowcroft, "Network Adaptive Continuous-Media Applications Through Self Organised Transcoding", NOSSDAV 98, Cambridge, UK

Author Index

Lecture Notes in Computer Science

For information about Vols. 1–1622
please contact your bookseller or Springer-Verlag

Vol. 1667: J. Hlavička, E. Maehle, A. Pataricza (Eds.), Dependable Computing – EDCC-3. Proceedings, 1999. XVIII, 455 pages. 1999.

Vol. 1668: J.S. Vitter, C.D. Zaroliagis (Eds.), Algorithm Engineering. Proceedings, 1999. VIII, 361 pages. 1999.

Vol. 1670: N.A. Streitz, J. Siegel, V. Hartkopf, S. Konomi (Eds.), Cooperative Buildings. Proceedings, 1999. X, 229 pages. 1999.

Vol. 1671: D. Hochbaum, K. Jansen, J.D.P. Rolim, A. Sinclair (Eds.), Randomization, Approximation, and Combinatorial Optimization. Proceedings, 1999. IX, 289 pages. 1999.

Vol. 1672: M. Kutylowski, L. Pacholski, T. Wierzbicki (Eds.), Mathematical Foundations of Computer Science 1999. Proceedings, 1999. XII, 455 pages. 1999.

Vol. 1673: P. Lysaght, J. Irvine, R. Hartenstein (Eds.), Field Programmable Logic and Applications. Proceedings, 1999. XI, 541 pages. 1999.

Vol. 1674: D. Floreano, J.-D. Nicoud, F. Mondada (Eds.), Advances in Artificial Life. Proceedings, 1999. XVI, 737 pages. 1999. (Subseries LNAI).

Vol. 1675: J. Estublier (Ed.), System Configuration Management. Proceedings, 1999. VIII, 255 pages. 1999.

Vol. 1976: M. Mohania, A. M. Tjoa (Eds.), Data Warehousing and Knowledge Discovery. Proceedings, 1999. XII, 400 pages. 1999.

Vol. 1677: T. Bench-Capon, G. Soda, A. M. Tjoa (Eds.), Database and Expert Systems Applications. Proceedings, 1999. XVIII, 1105 pages. 1999.

Vol. 1678: M.H. Böhlen, C.S. Jensen, M.O. Scholl (Eds.), Spatio-Temporal Database Management. Proceedings, 1999. X, 243 pages. 1999.

Vol. 1679: C. Taylor, A. Colchester (Eds.), Medical Image Computing and Computer-Assisted Intervention – MICCAI'99. Proceedings, 1999. XXI, 1240 pages. 1999.

Vol. 1680: D. Dams, R. Gerth, S. Leue, M. Massink (Eds.), Theoretical and Practical Aspects of SPIN Model Checking. Proceedings, 1999. X, 277 pages. 1999.

Vol. 1682: M. Nielsen, P. Johansen, O.F. Olsen, J. Weickert (Eds.), Scale-Space Theories in Computer Vision. Proceedings, 1999. XII, 532 pages. 1999.

Vol. 1683: J. Flum, M. Rodríguez-Artalejo (Eds.), Computer Science Logic. Proceedings, 1999. XI, 580 pages. 1999.

Vol. 1684: G. Ciobanu, G. Păun (Eds.), Fundamentals of Computation Theory. Proceedings, 1999. XI, 570 pages. 1999.

Vol. 1685: P. Amestoy, P. Berger, M. Daydé, I. Duff, V. Frayssé, L. Giraud, D. Ruiz (Eds.), Euro-Par'99. Parallel Processing. Proceedings, 1999. XXXII, 1503 pages. 1999.

Vol. 1687: O. Nierstrasz, M. Lemoine (Eds.), Software Engineering – ESEC/FSE '99. Proceedings, 1999. XII, 529 pages. 1999.

Vol. 1688: P. Bouquet, L. Serafini, P. Brézillon, M. Benerecetti, F. Castellani (Eds.), Modeling and Using Context. Proceedings, 1999. XII, 528 pages. 1999. (Subseries LNAI).

Vol. 1689: F. Solina, A. Leonardis (Eds.), Computer Analysis of Images and Patterns. Proceedings, 1999. XIV, 650 pages. 1999.

Vol. 1690: Y. Bertot, G. Dowek, A. Hirschowitz, C. Paulin, L. Théry (Eds.), Theorem Proving in Higher Order Logics. Proceedings, 1999. VIII, 359 pages. 1999.

Vol. 1691: J. Eder, I. Rozman, T. Welzer (Eds.), Advances in Databases and Information Systems. Proceedings, 1999. XIII, 383 pages. 1999.

Vol. 1692: V. Matoušek, P. Mautner, J. Ocelíková, P. Sojka (Eds.), Text, Speech and Dialogue. Proceedings, 1999. XI, 396 pages. 1999. (Subseries LNAI).

Vol. 1693: P. Jayanti (Ed.), Distributed Computing. Proceedings, 1999. X, 357 pages. 1999.

Vol. 1694: A. Cortesi, G. Filé (Eds.), Static Analysis. Proceedings, 1999. VIII, 357 pages. 1999.

Vol. 1695: P. Barahona, J.J. Alferes (Eds.), Progress in Artificial Intelligence. Proceedings, 1999. XI, 385 pages. 1999. (Subseries LNAI).

Vol. 1696: S. Abiteboul, A.-M. Vercoustre (Eds.), Research and Advanced Technology for Digital Libraries. Proceedings, 1999. XII, 497 pages. 1999.

Vol. 1697: J. Dongarra, E. Luque, T. Margalef (Eds.), Recent Advances in Parallel Virtual Machine and Message Passing Interface. Proceedings, 1999. XVII, 551 pages. 1999.

Vol. 1698: M. Felici, K. Kanoun, A. Pasquini (Eds.), Computer Safety, Reliability and Security. Proceedings, 1999. XVIII, 482 pages. 1999.

Vol. 1699: S. Albayrak (Ed.), Intelligent Agents for Telecommunication Applications. Proceedings, 1999. IX, 191 pages. 1999. (Subseries LNAI).

Vol. 1700: R. Stadler, B. Stiller (Eds.), Active Technologies for Network and Service Management. Proceedings, 1999. XII, 299 pages. 1999.

Vol. 1701: W. Burgard, T. Christaller, A.B. Cremers (Eds.), KI-99: Advances in Artificial Intelligence. Proceedings, 1999. XI, 311 pages. 1999. (Subseries LNAI).

Vol. 1702: G. Nadathur (Ed.), Principles and Practice of Declarative Programming. Proceedings, 1999. X, 434 pages. 1999.

Vol. 1703: L. Pierre, T. Kropf (Eds.), Correct Hardware Design and Verification Methods. Proceedings, 1999. XI, 366 pages. 1999.

Vol. 1704: Jan M. Żytkow, J. Rauch (Eds.), Principles of Data Mining and Knowledge Discovery. Proceedings, 1999. XIV, 593 pages. 1999. (Subseries LNAI).

Vol. 1705: H. Ganzinger, D. McAllester, A. Voronkov (Eds.), Logic for Programming and Automated Reasoning. Proceedings, 1999. XII, 397 pages. 1999. (Subseries LNAI).

Vol. 1707: H.-W. Gellersen (Ed.), Handheld and Ubiquitous Computing. Proceedings, 1999. XII, 390 pages. 1999.

Vol. 1708: J.M. Wing, J. Woodcock, J. Davies (Eds.), FM'99 – Formal Methods. Proceedings Vol. I, 1999. XVIII, 937 pages. 1999.

Vol. 1709: J.M. Wing, J. Woodcock, J. Davies (Eds.), FM'99 – Formal Methods. Proceedings Vol. II, 1999. XVIII, 937 pages. 1999.

Vol. 1710: E.-R. Olderog, B. Steffen (Eds.), Correct System Design. XIV, 417 pages. 1999.

Vol. 1718: M. Diaz, P. Owezarski, P. Sénac (Eds.), Interactive Distributed Multimedia Systems and Telecommunication Services.. Proceedings, 1999. XI, 386 pages.